CW00358014

IES
P

Springer Series in Electrophysics
Volume 7

Edited by Walter Engl

Springer Series in Electrophysics

Editors: Günter Ecker   Walter Engl   Leopold B. Felsen

Insulating Films on Semiconductors

Proceedings of the Second International
Conference, INFOS 81,
Erlangen, Fed. Rep. of Germany,
April 27–29, 1981

Editors: M. Schulz and G. Pensl

With 240 Figures

Springer-Verlag Berlin Heidelberg New York 1981

Professor Dr. Max J. Schulz · Dr. Gerhard Pensl

Institut für Angewandte Physik, Universität Erlangen-Nürnberg, Glückstraße 9,
D-8520 Erlangen, Fed. Rep. of Germany

Series Editors:

Professor Dr. Günter Ecker

Ruhr-Universität Bochum, Theoretische Physik, Lehrstuhl I,
Universitätsstrasse 150, D-4630 Bochum-Querenburg, Fed. Rep. of Germany

Professor Dr. Walter Engl

Institut für Theoretische Elektrotechnik,
Rhein.-Westf. Technische Hochschule, Templergraben 55,
D-5100 Aachen, Fed. Rep. of Germany

Professor Leopold B. Felsen Ph.D.

Polytechnic Institute of New York, 333 Jay Street, Brooklyn, NY 11201, USA

Conference Chairman: M. Schulz (Univ. Erlangen, FRG)

Program Committee: J. Cave (Caswell, UK); G. Dorda (München, FRG); A. Goetzberger
(Freiburg, FRG); G. G. Roberts (Durham, UK); M. Schulz (Erlangen, FRG); J. Simonne (Toulouse, F);
J. Verwey (Eindhoven, NL)

International Advisory Committee: H. Flietner (Berlin, GDR); G. Lucovsky (Raleigh, USA);
W. Spicer (Stanford, USA); T. Sugano (Tokyo, Japan); D. Young (Yorktown Heights, USA)

Local Organising Committee: K.-P. Frohmader (Univ. Erlangen); P. Glasow (Siemens, Erlangen);
R. Helbig, R. Hezel, G. Pensl, M. Schulz (Univ. Erlangen)

The Conference was sponsored by: European Physical Society (EPS); International Union of Pure and
Applied Physics (IUPAP); Institute of Electrical and Electronics Engineers (IEEE Region 8);
German Physical Society DPG)

The Conference was supported by: Deutsche Forschungsgemeinschaft, Bonn; Bundesministerium für
Forschung und Technologie, Bonn, Bayerisches Ministerium für Unterricht und Kulturs, European
Research Office (Army, Air Force and Navy), London; IBM, Böblingen; Siemens, München;
Sparkasse, Erlangen, Vlavo, Hamburg

ISBN 3-540-11021-6 Springer-Verlag Berlin Heidelberg New York
ISBN 0-387-11021-6 Springer-Verlag New York Heidelberg Berlin

This work is subject to copyright. All rights are reserved, whether the whole or part of the material is
concerned, specifically those of translation, reprinting, reuse of illustrations, broadcasting, reproduction
by photocopying machine or similar means, and storage in data banks. Under § 54 of the German
Copyright Law where copies are made for other than private use, a fee is payable to "Verwertungs-
gesellschaft Wort", Munich.
© by Springer-Verlag Berlin Heidelberg 1981
Printed in Germany
The use of registered names, trademarks etc. in this publication does not imply, even in the absence
of a specific statement, that such names are exempt from the relevant protective laws and regulations
and therefore free for general use.
Offset printing: Beltz Offsetdruck, 6944 Hemsbach/Bergstr.
Bookbinding: J. Schäffer OHG, 6718 Grünstadt.
2153/3130-543210

Preface

The INFOS 81 Conference on Insulating Films on Semiconductors was held at the University of Erlangen-Nürnberg in Erlangen from 27 to 29 April 1981. This conference was a sequel to the first conference INFOS 79 held in Durham. INFOS 81 attracted 170 participants from universities, research institutes and industry. Attendants were registered from 15 nations. The biannual topical conference series will be continued by INFOS 83 to be held in Eindhoven, The Netherlands, in April 1983.

The conference proceedings include all the invited (9) and contributed (42) papers presented at the meeting. The topics range from the basic physical understanding of the properties of insulating films and their interface to semiconductors to the discussion of stability and dielectric strength as well as growing and deposition techniques which are relevant for technical applications. Strong emphasis was given to the semiconductor silicon and its native oxide; however, sessions on compound semiconductors and other insulating films also raised strong interest. The proceedings survey the present state of our understanding of the system of insulating films on semiconductors. As a new aspect of the topic, the properties of semiconductors deposited and laser processed on insulating films was included for the first time.

The organisational load of the conference was distributed on many collaborators. The service of the session chairmen who together with many participants reviewed all the publication manuscripts during the conference, was very much appreciated. Special thanks go to all the colleagues and students of the Institute of Applied Physics who assisted in the local organization. The conference administration and the retyping of many manuscripts in the proceedings was performed with great care and engagement by Mrs. Gabriele Loy. All the help is gratefully acknowledged.

Erlangen, April 1981   *M.Schulz · G.Pensl*

Contents

Part X   Films on Compound Semiconductors

Part I:

Si–SiO$_2$ Interface

Electronic Structure of the Si–SiO$_2$ Interface

Frank Herman
IBM Research Laboratory
San Jose, CA 95193, USA

After sketching various theoretical models that have been proposed for describing the Si/SiO$_2$ interface on an atomic scale, we will discuss an idealized model which is designed to simulate the average contact between a silicon
crystal and its oxide overlayer. According to this model, the Si/SiO$_2$ interface is represented by the boundary between two crystalline domains, the first
being the silicon substrate, and the second an idealized crystalline form of
SiO$_2$, diamond-like beta cristobalite. Nearly perfect registry between Si and
SiO$_2$ is obtained by placing the (100) face of the former next to the 45° rotated (100) face of the latter. Half the Si atoms at the interface are fourfold coordinated, connecting the Si and SiO$_2$ regions to one another. The remaining half are connected only to the silicon substrate and are thus twofold
coordinated. Localized electronic states in the thermal gap at the interface
are associated with the dangling bonds at the twofold coordinated Si atoms.
If H, O, or OH groups are attached to these Si atoms, eliminating the dangling bonds, the thermal gap interface states are also eliminated. Even though
this model ignores structural disorder, it does provide considerable insight
into the nature of the Si/SiO$_2$ interface. We will discuss some of the implications of this model, as well as the results of model-based theoretical calculations. In particular, we will consider the electronic structure of the
interface including the effects produced by the introduction of Si and O vacancies and interstitials and other types of localized defects.

1. Introduction

Because of the important role played by the Si/SiO$_2$ interface in modern semiconductor technology, considerable experimental and theoretical effort has
been devoted to the study of this interface [2-8]. Oxidation of a silicon surface produces an amorphous oxide layer whose atomic arrangement is not well

2

understood. Many authors have suggested that there is a non-stoichiometric transition layer, often described as SiO_x, separating the silicon substrate from the stoichiometric SiO_2 layer. (Here x lies between 1 and 2.) The detailed atomic structure of the transition layer, including its uniformity and width, its structural disorder, its affinity for impurity atoms, etc., continues to be a subject of debate. Of course, the nature of the transition layer is expected to depend on the manner in which the oxide layer is formed. By the same token, a better understanding of the transition layer could lead to further improvements in the preparation of Si/SiO_2 interfaces.

The technological importance of the Si/SiO_2 interface arises from the fortunate circumstance that suitably oxidized silicon substrates are protected from environmental contaminants and have rather low densities of localized interface states in the thermal gap. There continues to be considerable interest in the nature and identity of the residual defects that occur near well-prepared interfaces, giving rise to residual trapping sites and fixed charge. Many imaginative models of defects at the Si/SiO_2 interface and in SiO_2 have been proposed [2-11], but relatively few of these defects have been unambiguously identified, even by such precise measurements as ESR [11]. In view of the difficulties encountered in practice in attempting to interpret most experimental measurements in terms of specific defects, we thought it would be useful to develop a simple model for the Si/SiO_2 interface which could provide a conceptual framework for thinking about interface defects as well as a theoretical framework for carrying out numerical calculations.

In the first paper on this subject [12], we described the construction of idealized Si/SiO_2 interfaces and superlattices and demonstrated by first-principles band structure calculations that there are no localized interface states in the thermal gap if all the Si atoms at the interface are saturated. In a second paper [13] we showed how localized defects in SiO_2 could be investigated theoretically by introducing periodic arrays of noninteracting defects in a SiO_2 supercell normally containing 24 atoms. In a third paper [14] we introduced periodic arrays of noninteracting defects in idealized Si/SiO_2 superlattices, thereby simulating the presence of localized defects at or near the Si/SiO_2 interface. In the present paper we will first mention some recent attempts by others to develop a theoretical model of SiO_2 and the Si/SiO_2 interface, and then we will discuss our own model briefly and summarize some of our principal conclusions.

2. Theoretical Models of the Si/SiO$_2$ Interface

The extensive literature on crystalline and non-crystalline SiO$_2$ and on Si/SiO$_2$ interfaces [2-11] contains a variety of models of the oxide layer and the Si/SiO$_2$ interface. In this paper, we will focus on atomic scale rather than on phenomenological models. Nearly all the models that have been proposed for describing the atomic arrangements at the interface are speculative. Some of the proposals are quite specific. They assume particular types of linkages between Si and SiO$_2$, well-defined structural imperfections arising from modifications in the normal linkages, as well as chemical imperfections due to the presence of specific foreign species. The remaining proposals tend to be descriptive or impressionistic, invoking dislocations, disjunctions, varying degrees of stoichiometry and crystallinity, etc. The first set of proposals can be distinguished from the second by the ease with which ball-and-stick models can be constructed to illustrate the proposed arrangements.

There has also been considerable debate regarding the initial formation of the oxide layer, and how this layer might change as it grows [2-8]. Some authors believe that a better understanding of the initial stages of oxidation could provide valuable clues concerning the nature of the ultimate oxide layer; other authors disagree. In any event, it is very important to have heuristic atomic models of the formation and ultimate structure of the Si/SiO$_2$ interface, because such ideas could lead to improved processing techniques, new device concepts, and better understanding of still more complex semiconductor/insulator interfaces.

Up to now, most Si/SiO$_2$ interface modeling studies have been qualitative, being based on simple physical and chemical reasoning. Many of the arguments put forward in favor of specific atomic models seem quite plausible, but it is usually difficult to verify these interface models experimentally. Therefore, a constructive step would be to study some of these models theoretically and obtain quantitative estimates of crucial electronic properties. If one wishes to carry out detailed electronic structure calculations for model systems, it is obviously necessary to have specific atomic arrangements in mind. The essential difficulty is that we do not know the actual structure of the Si/SiO$_2$ interface on an atomic scale.

The overall problem can be dealt with in three parts: In the first and easiest part, one makes certain assumptions about the silicon substrate. For example, one chooses a specific crystalline orientation, and then assumes that the silicon is structurally perfect up to the interface. Relaxation and reconstruction effects can be deferred to a later stage or expressly ignored.

4

The second part concerns the description of the oxide film property, allowing if possible for its noncrystalline nature. In dealing with non-crystalline SiO_2, one can build on our improved knowledge of amorphous and glassy semi-conductors [9,15-19]. The third part, the most difficult of all, involves connecting the silicon substrate to the non-crystalline oxide layer in a physically and chemically plausible manner.

One popular model for noncrystalline SiO_2 and the Si/SiO_2 interface is the cluster Bethe lattice model [20,21], which takes topological aspects of structural disorder into account in an elegant manner. Because this model is rather simple in form and can be easily parameterized, it can be used to study electronic and lattice vibrational spectral properties associated with ideal and defective interfaces. But because these models are based on adjustable parameters that are assigned ad hoc values, they are not particularly trustworthy from a quantitative point of view [22]. Nevertheless, they do have considerable pedagogical value.

Another popular model of non-crystalline SiO_2 is the continuous random network model [15-19], according to which SiO_2 is composed of a randomly oriented collection of SiO_4 tetrahedra which are linked to one another to form a continuous network of Si-O-Si bonds. It is assumed that the SiO_4 tetrahedra tend to maintain their structural integrity (bond lengths and bond angles) in different environments, and that the necessary structural versatility is provided by the Si-O-Si bond, whose angle may assume a wide range of values at little cost in energy [4]. Models of this type have recently been used to study bulk SiO_2 [23] as well as the Si/SiO_2 interface [24]. In the study of bulk SiO_2, cyclic boundary conditions were imposed on a large disordered molecular SiO_2 cluster, producing a computationally tractable model of non-crystalline SiO_2. The object here was to investigate electron and hole localization induced by the structural disorder.

In the study of the Si/SiO_2 interface, it was demonstrated that a rather abrupt interface could be constructed within the framework of a continuous random network model, a rather important result. Unfortunately, the resulting atomic model is much too complex for detailed electronic structure studies. In both [23] and [24], the construction of continuous random network models is based on ideas regarding the nature of non-crystallinity which are highly subjective and which may not necessarily describe the experimental situation. It is possible that non-crystalline SiO_2 near a Si/SiO_2 interface does not actually form a completely continuous random network. That is to say, the network may contain discontinuous elements such as non-bridging O atoms or three-

5

fold coordinated Si atoms connected to H or OH groups, or voids composed of aggregates of such discontinuous elements [19].

There are many theoretical efforts currently underway [18,19,25] whose aim is to formulate general guiding principles for constructing atomic models of non-crystalline materials. Combined with intuitive notions [26] and computer modeling [27,28], these theoretical studies may eventually lead us to objective models which we can use with confidence for electronic structure studies of non-crystalline SiO_2. In the meanwhile, the atomic architecture of the Si/ SiO_2 interface remains an open question.

We already know from extensive experimental and theoretical studies [2,3, 29] that the gross features of the electronic structure of non-crystalline and various crystalline forms of SiO_2 are essentially the same, reflecting the common SiO_4 building block. Only the finer details of electronic structure are affected by the special manner in which the SiO_4 tetrahedra are linked together in a particular structural modification. In view of the overwhelming importance of the short-range order (linked SiO_4 tetrahedra) in determining the essential features of the electronic structure of SiO_2, and the likelihood that the long-range disorder affects only the finer details, we will explicitly ignore the structural disorder of SiO_2 and concentrate on models of the interface which represent the average contact between Si and SiO_2.

3. Idealized Si/SiO_2 Superlattices and Interfaces

Our overall program is as follows. First, we will disregard structural disorder in SiO_2, and represent SiO_2 by a suitably chosen periodic structure. We will then form ordered Si/SiO_2 interfaces by attaching crystalline SiO_2 to crystalline Si. Next, we will construct a Si/SiO_2 superlattice by stacking Si and SiO_2 slabs on top of one another. The slabs will be made sufficiently thick so that successive interfaces will not interact with one another. Because the Si/SiO_2 superlattice is a periodic structure in three dimensions, we can use the highly developed methods of band theory [30] to calculate the electronic structure. The results of such calculations include a description of localized interface states, including their energy levels and charge distributions. Finally, we can introduce structural and chemical imperfections, and study their effects on the electronic structure.

To carry out the above program, we begin by representing the oxide layer by an idealized crystalline form of SiO_2 having as simple a unit cell as pos-

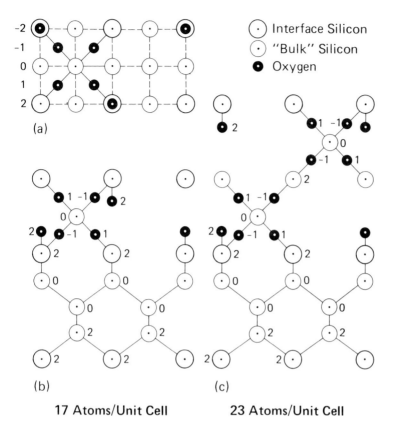

Fig. 1a-c. Si/SiO$_2$ superlattices and interfaces. Panels (a) and (b) depict the top and front views of a superlattice having 17 atoms per unit cell, and panel (c) depicts the front view of a superlattice having 23 atoms per unit cell. Both superlattices have five layers of silicon (2 atoms each) and two oxygen atoms saturating the twofold coordinated Si substrate atoms. In panel (b) there is one tier of SiO$_4$ tetrahedra (5 atoms), while in panel (c) there are three SiO$_4$ tiers (11 atoms). The numbers denote the planes in which the various atoms lie, the 2 representing the front-most plane, etc. Another view of these superlattices appears in [8]

sible. For this purpose we will use beta cristobalite [31], except that we will straighten out the Si-O-Si bonds in the actual structure, making them all linear rather than crooked (all bond angles 180° rather than about 140°). Assuming an Si-O bond length of 1.633 Å [32], the linear Si-O-Si bond length becomes twice this, and the unit cube edge of idealized diamond-like SiO$_2$ becomes 7.543 Å, which is 5 percent larger than the unit cube edge of actual beta cristobalite, 7.16 Å. This 5 percent difference reflects the contraction produced by the bent bonds in the actual structure. Except as otherwise noted, we will henceforth regard SiO_2 as having the idealized diamond-like beta cristobalite structure with linear Si-O-Si bonds.

It is an interesting exercise to find ways of attaching crystalline SiO_2 to crystalline Si so that as many bonds as possible are formed. The simplest arrangement of all is to place the (100) face of Si next to the 45° rotated face of (100) SiO_2. This produces almost perfect registry because the ratio of the lattice constants of SiO_2 and Si is very nearly equal to the square root of 2 (7.543/5.431=1.39). To put it more directly, we obtain nearly perfect registry in this fashion because the ratio of the straight Si-O-Si bond length in SiO_2 to the Si-O-Si bond length in silicon is very nearly equal to the square root of 2. (For actual calculations, we will increase the Si-O-Si bond length slightly so that the registry is exact.)

We thus obtain an extremely simple model of the Si/SiO_2 interface, namely, a fully coherent, ideally abrupt interface between silicon and crystalline SiO_2 [12-14]. This construction leads to an interface at which half the Si substrate atoms are fourfold coordinated, the remaining half being twofold coordinated, with two dangling bonds each. We can describe the interface as a checkerboard with Si atoms common to Si and SiO_2 occupying the red squares, and unsaturated Si atoms (two dangling bonds each) occupying the black squares. It is possible to use other Si and SiO_2 crystal faces and join them together in a similar manner, but these alternate constructions lead to more complex interface matching patterns, stretched Si-O-Si bonds, and larger repeat periods, making them less suitable for modeling. It is interesting, nevertheless, that even these more complex models exhibit the property that roughly half the Si substrate atoms are attached to the SiO_2, while the remaining half are unattached. If we are to adopt an idealized model, we might as well adopt the simplest one possible, and this is unquestionably the (100) 45° rotated geometry already described.

Because of the 45° rotation at the interface, the four bonds emanating from the common Si atoms deviate from ideal tetrahedral geometry. Earlier studies [13] suggest that the silicon lattice is considerably more rigid than the SiO_2 lattice, so that most of the strain would be relieved by lattice relaxation in the SiO_2 region. We have made some progress in taking lattice relaxation effects into account, but on the basis of our work to date [13] we view the distortion from ideal tetrahedral geometry at the common Si atoms as a relatively minor shortcoming of our model. The thermal (Si) band gap is determined in part by the common Si atoms and the fact that they are saturated. The deviation from local tetrahedral geometry appears to be only of secondary importance.

8

It is possible to avoid the lattice strain associated with the 45° rotation altogether by considering a more exotic interface model. Here each common Si atom is replaced by a Si=Si molecule, one end of which terminates the silicon substrate, and the other end the SiO_2 region. The 45° rotation is then absorbed by the Si=Si double bonds. In this model the Si-Si bond length is 2.352 Å in the silicon crystal and slightly less (2.252 Å) in Si=Si [32]. We have not adopted this more exotic model for two reasons. First, the Si=Si bond is a weak one [33], so there is some question as to whether it would be stable in the environment just described. Secondly, we have carried out comparison calculations using the more exotic as well as the original model, and we find that the Si double bonds lead to a significant reduction in the Si/SiO_2 thermal gap, which is contrary to experiment. So we will stick with the original model.

We do not regard the use of straight rather than crooked Si-O-Si bonds in SiO_2 as a serious defect in our model. We have already said that there is very little energy required to change the Si-O-Si bond angle from 140° to 180° [4]. Some bond distortion is inevitable at the Si/SiO_2 interface anyway. Moreover, the Si-O-Si bond angles are not expected to play a major role in determining the electronic structure [29].

By using a periodic model and introducing an idealized interface, we have focused attention on the average contact between Si and SiO_2, highlighting the fact that this average contact requires approximately half the Si substrate atoms to be attached to the SiO_2, and the remaining half to be unattached. The unattached (twofold-coordinated) Si atoms face onto cavities in the adjoining SiO_2 region, presumably making it easy for species such as O, H, or OH to find their way to the beckoning dangling bonds. Because of the periodicity, and the simple manner in which we have been able to attach SiO_2 to the silicon substrate, we can describe various forms of the transition layer, so-called nonstoichiometric SiO_x, on an atomic scale. According to our model, this transition layer does not actually exist if each twofold coordinated Si atom at the interface is saturated by two H atoms. That is to say, the passage from Si to SiO_2 is essentially abrupt. By saturating some or all of these twofold coordinated Si atoms with O or OH groups, we can introduce varying O concentrations in one atomic layer just beyond the silicon substrate, this now becoming the transition layer. Perhaps simple geometrical pictures such as these will prove more helpful in interpreting experimental measurements than complex models of a nonstoichiometric transition layer.

4. Electronic Structure Calculations for Intrinsic Interfaces

Having constructed an idealized interface model, we proceed to study its electronic structure by considering an Si/SiO_2 superlattice composed of alternating Si and SiO_2 slabs which are sufficiently thick to isolate adjacent interfaces from one another. Most of our studies are based on a 21 atom per unit cell superlattice consisting of 5 Si layers (2 atoms per layer) alternating with 3 tiers of SiO_4 tetrahedra (11 atoms). This superlattice with two additional O atoms forming Si=O double bonds with the twofold coordinated Si atoms is shown in [12].

Our studies are based on the first-principles extended muffin tin orbital [EMTO] method [34]. All of the calculations reported in [12-14] and summarized below are based on superlattice charge distributions constructed from neutral free atom self-consistent charge densities [35]. Although the superlattice charge distributions themselves are not exactly self-consistent, such calculations give a good account of Si, SiO_2, and many other types of semiconductors and insulators. A fully self-consistent EMTO computer program has recently been developed [36], and this program will be used in subsequent studies of Si/SiO_2 interfaces. We believe that the EMTO calculations done to date on Si/SiO_2 provide reasonable estimates of the electronic structure of ideal and defective Si/SiO_2. Refinements based on fully self-consistent EMTO calculations will be reported in due course.

The essential results for an intrinsic Si/SiO_2 superlattice can be summarized as follows: a) Localized interface states occur in the thermal gap of Si if the dangling bonds are left unsaturated. b) These localized states are removed from the thermal gap if the dangling bonds are saturated by pairs of H atoms, single O atoms, OH groups, etc. These conclusions apply to unrelaxed lattice geometries. We have made some attempts to relieve the lattice strain using Monte Carlo methods [13]. Because of the limited width of the SiO_2 region and the severe constraints imposed by the rigid intervening Si regions, the SiO_2 regions relax only slightly, leading to negligible changes in the energy level structure and to no change in conclusions a) and b) above.

5. Si and O Vacancies and Interstitials

Within the framework of a band structure approach it is natural to introduce
periodic arrays of noninteracting defects in the SiO_2 region near the interface. Before embarking on such calculations, we first performed exploratory
studies of this type using a 24-atom SiO_2 supercell to simulate bulk SiO_2,
as reported in an earlier paper [13]. Similar studies were carried out some
time ago using the extended Hückel method [37]. The essential idea is to introduce one defect such as a Si or O vacancy or interstitial per supercell,
and then calculate the energy band structure at two or more well-separated
points in the reduced zone. In this way one obtains not only the valence and
conduction bands of the host SiO_2, but also impurity bands corresponding to
the defects introduced. If the supercell is sufficiently large, and the defects sufficiently localized, the impurity bands will have negligible dispersion, and the centers of gravity of these impurity bands will describe the
energy levels of isolated defects.

An important advantage of the periodic defect array approach over molecular cluster methods is that the energy levels of the defects are located
relative to the valence and conduction band edges of the host crystal. We
will now briefly recapitulate our earlier SiO_2 defect investigations [13].

By trial and error we find that a unit cube of SiO_2 containing 24 atoms
places adjacent defects sufficiently far apart (7.68 Å) to produce impurity
bands with dispersions of the order of 0.1 eV, which is small enough for our
purposes. For some defects it is even possible to use a smaller unit cell
(SiO_2 with 6 atoms per unit cell) and still end up with acceptably small impurity band dispersions. To illustrate our approach, we consider the 24-atom
supercell, which is constructed by placing Si atoms at (0,0,0), (1/4,1/4,1/4),
etc., and O atoms at (1/8,1/8,1/8), etc., where distances are measured in
units of the cube edge. The intrinsic SiO_2 supercell contains 128 valence
electrons, so there are normally 64 filled bands. We test the dispersion
properties of defect-related bands by determining the band structure of the
supercell at the zone center and also at the zone corner. The essential results are as follows.

5.1 Interstitial Silicon Atoms

Let us introduce a neutral silicon atom at the interstitial position located
at the center of the SiO_2 cage in the SiO_2 supercell, (1/2,1/2,1/2). The 64
filled valence bands are perturbed only slighlty. There is a filled inter-

stitial band roughly midway in the SiO_2 band gap arising from the Si 3s level, containing 2 of the 4 valence electrons contributed to the SiO_2 supercell by the Si interstitial. There are also empty interstitial bands located well above the SiO_2 conduction band minimum (CBM). The lowest (perturbed) conduction band is also filled, accounting for the remaining 2 electrons associated with the Si interstitial.

Taking into account the nature of a band-theoretic description of a periodic array of non-interacting defects in an otherwise perfect crystal, the above results can be interpreted as follows: If we introduce a Si^{4+} interstitial ion core into SiO_2, plus the 4 electrons required to establish charge neutrality, 2 of these electrons will occupy a level in the midrange of the forbidden band, while the other 2 will be ionized, occupying low-lying conduction band states. This view is somewhat naive, reflecting the limitations of a model in which the defects are separated just far enough so that they do not interact to any appreciable degree. In practice, the final 2 electrons required for local charge neutrality would not be uniformly distributed throughout the crystal, but would be concentrated on the atoms defining the cage surrounding the Si interstitial atom.

It is interesting to find that a Si interstitial produces a level roughly midway in the SiO_2 forbidden band. This suggests that Si interstitials, if they actually exist in SiO_2, could create levels within striking distance of the thermal gap of the Si/SiO_2 interface. Our calculations are too crude at this stage and the possible positions of the Si interstitial too numerous in actual SiO_2 for us to make predictions more precise than this just now.

In a subsequent study [14], we investigated the effect of putting a neutral interstitial Si atom at the center of a SiO_2 cage lying next to a Si/SiO_2 interface. In this particular study we added the interstitial Si atom to a 17 atom Si/SiO_2 superlattice (cf. Fig.1 in [12]) composed of 5 Si layers (2 atoms each), one tier of SiO_4 tetrahedra (5 atoms), and 2 O(xygen) atoms doubly bonded to the twofold coordinated Si atoms at the interfaces. For this particular geometry, the interstitial Si atom is close enough to the doubly bonded O atoms to interact with them, leading to a lowering of the Si 3s and 3p atomic levels, relative to their positions in the SiO_2 supercell. The net result is that the Si 3s band lies well below the Si/SiO_2 thermal gap, while the Si 3p bands lie within this gap.

These studies represent the first stage of an investigation of interstitial Si atoms at an interface interacting with O atoms originally doubly bonded to the silicon substrate. The object is to study the formation of Si-O-Si-O-Si bridges in which the O atoms are singly bonded to the Si substrate atoms as

well as to the bridging Si atom (originally the interstitial). The bridging
Si atom could be doubly charged, representing one form of fixed positive
charge at the interface.

5.2 Interstitial Oxygen Atoms

We have also introduced a neutral O interstitial atom at the center of one
of the cages in the 24-atom SiO_2 supercell, position (1/2,1/2,1/2). This
leads to an occupied interstitial O 2s band about 11 eV below the SiO_2 valence band maximum (VBM), two occupied O 2p bands about 2 eV above the VBM,
and an empty O 2p band about 3 eV above the VBM. In practice, such interstitial O atoms would be expected to complete their valence by attaching themselves to a pair of H atoms, for example, to form water.

   If we place the neutral O interstitial atom at the center of the SiO_2 cage
in the 17-atom Si/SiO_2 superlattice described above, we find that the O 2s
and 2p impurity bands lie about 7 to 8 eV below their corresponding values in
the SiO_2 supercell. The reason again is the proximity of the interstitial O
to the nearby O atoms attached to the silicon substrate. This situation represents a step in the direction of creating a Si-O-O-O-Si bridge, where the
interstitial O lies at the middle of the span. Because the O interstitial has
acquired its full valence of 8 electrons, the highest valence band is empty,
simulating a shallow acceptor level.

   These studies suggest that the O interstitial is not likely to produce
localized levels within the range of the thermal (Si) gap, in contrast to
the Si interstitial. Roughly speaking, the highest occupied interstitial O
levels will tend to line up with the highest occupied O levels in SiO_2 (within a few eV), this energy range being well below the thermal gap.

5.3 Neutral Si Vacancies

Returning to the normal 24-atom SiO_2 supercell, let us now introduce a neutral
Si vacancy by removing a Si atom, say from position (0,0,0). This removes 4
valence electrons from the supercell, so the lowest 62 rather than tHe lowest
64 valence bands are occupied. Because the uppermost valence bands in the
normal supercell are built up from O 2p non-bonding orbitals, the removal of
a Si atom has negligible influence on the energies of these bands. As a consequence, the 2 vacant acceptor levels (bands 63 and 64) remain essentially
degenerate with the topmost valence band (band 62). If these acceptor levels
become occupied, a negatively charged Si vacancy would be created. The situa-

tion remains essentially the same if we remove a Si atom from the center of the SiO_4 tetrahedron in the 17-atom Si/SiO_2 superlattice.

However, if in this superlattice we remove the fourfold coordinated Si atom at the interface, i.e., the atom common to the Si and SiO_2 regions, the dominant effect is the creation of two dangling Si bonds and the concomitant formation of localized electronic states at the interface. The levels associated with the nonbridging O atoms are similar to those generated by the removal of a Si atom from the center of a SiO_4 tetrahedron far away from the interface (as in the 24-atom SiO_2 supercell).

5.4 Neutral Oxygen Vacancies

Returning once more to the SiO_2 supercell and removing a neutral O atom, we are left with $128 - 6 = 122$ valence electrons in the unit cell. These electrons fill the lowest 60 valence bands as well as one O vacancy band located about 2 eV below the CBM. On the other hand, if we remove an O atom from the Si-O-Si bond which connects directly to the Si/SiO_2 interface, we create dangling Si bonds and localized interface states in the thermal gap. In all the studies reported here, we have not attempted to relax the lattice after the introduction of a vacancy or interstitial. Some of the most interesting defects of all involve vacancies and interstitials associated with relaxed lattices [9-11] and vacancies leading to broken bonds which are subsequently saturated by the attachment of H or OH groups [9-11].

6. Interstitial O_2, SiO, and H_2O Molecules

We have also placed neutral O_2, SiO, and H_2O molecules inside the SiO_2 cages in the 17-atom Si/SiO_2 superlattice [14]. Each molecule was inserted so that its center of gravity coincides with the cage center, and its longest dimension is perpendicular to the interface. For O_2, which normally has 12 valence electrons, we find that there are 7 occupied O_2-related bands (2 electrons of opposite spin per band). Our calculations thus indicate that the O_2 molecule has acquired 2 additional electrons, becoming doubly negatively charged. The highest valence band is now empty, corresponding to a shallow acceptor level. We are presently investigating the possibility that O_2 molecules close to an interface can form bridges between adjacent pairs of twofold coordinated Si substrate atoms. To form Si-0=0-Si bridges, the terminal Si atoms would have to contribute one dangling bond each to the spanning 0=0 molecule.

For SiO, which normally has 10 valence electrons, we find that there are six occupied SiO-related bands, so the SiO molecule acquires 2 additional electrons, emptying the highest valence band and producing a shallow acceptor level. For H_2O, which normally has 8 valence electrons, there are 4 occupied H_2O-related bands, so the molecule remains neutral. Of course, these results apply to the particular interstitial position chosen, but the conclusions are instructive nevertheless.

7. Substitutional Al and Substitutional Al + Interstitial H

Finally, let us consider the substitution of a neutral Al atom for a neutral Si atom in the SiO_2 supercell. We choose Al because it is a common impurity in SiO_2. The valence and conduction bands of SiO_2 are not affected significantly by this replacement. The removal of one electron from the supercell leads to a half-empty highest valence band (the 64 highest band). Within the present framework, this represents a partially filled acceptor level located slightly above the VBM of SiO_2. We can fill this acceptor level by supplying an additional electron to the supercell. For purposes of illustration, let us introduce a neutral interstitial H atom close to the substitutional Al atom in the SiO_2 supercell, at position (-1/8,-1/8,-1/8). The H atom donates its electron to the Al defect, completing the occupation of all the valence bands, and producing an empty H level roughly midway within the SiO_2 forbidden band. The exact position of the empty level depends on the relative positions of the Al and H atoms, but the midrange of the SiO_2 band gap is the indicated range. Thus we see that such a complex could produce localized states within the thermal gap of Si/SiO_2, which also lies in the midrange of the SiO_2 band gap. Similar effects would be expected if we introduced an interstitial monovalent atom such as Na instead of H.

8. Speculations on the U-Shaped Continuum of Interface States

Various authors [38] have called attention to the fact that there is often a sharp peak in the density of Si/SiO_2 interface states at about 0.3 eV above the Si VBM, and that this peak can be removed by exposing the interface to hydrogen. Once this peak is removed, the distribution of interface states in the thermal gap of Si appears to be U-shaped, rising toward the valence and conduction band edges and falling to minimal values in between. A plausible explanation [38] for the disappearance of the sharp peak is that the dangling Si bonds at the interface responsible for this peak are removed by the formation of Si-H bonds.

The origin of the U-shaped continuum is not understood. There are obviously many possible explanations. Using our own studies as a guide, and in particular the fact that substitutional Al + interstitial M can lead to interface states in or near the thermal gap (M=monovalent metal atom or H), we will go one step further and ask: Are there different Si positions at or near the interface at which substitutional Al + interstitial M could produce different interface state distributions, for example, one peaked below the Si VBM and another above the Si CBM, so that the superposition of the tails from these two peaks would resemble a U-shaped curve in the thermal gap?

The explanation we propose is that there are Al impurities near the interface, some occupying the centers of AlO_4 tetrahedra, and others the fourfold coordinated positions at the interface where the Si and SiO_2 regions are joined. The latter Al impurities occupy the centers of distorted Si_2-Al-O_2 tetrahedra. Allowing for the different positions that neighboring interstitial M atoms would occupy at an actual interface, we would get two distinct distributions of interface levels, one offset in energy from the other. The distribution corresponding to AlO_4 + M would lie above the one corresponding to Si_2-Al-O_2+M, so the tail from the former would account for the upper part of the U, and the tail from the latter for the lower part. The individual pieces of this explanation are consistent with our calculations. Finally, we note that the exposure of the interface to H can be regarded not only as a means for saturating the dangling Si bonds, but also as a means for forming the AlO_4 + H and Si_2-Al-O_2 + M complexes that could be responsible (in whole or in part) for the U-shaped continuum.

9. Concluding Remarks

We have shown how idealized Si/SiO_2 superlattices can be used to describe the Si/SiO_2 interface and to construct simple models of defects such as Si and O vacancies and interstitials. The calculations we have discussed represent early attempts at dealing with complex geometrical situations. The results to date have been sufficiently encouraging to suggest that a more detailed understanding of these and other defects can be achieved by carrying out first-principles calculations self-consistently, taking lattice relaxation effects into account.

Acknowledgements. The author is grateful to many individuals for stimulating discussions and correspondence, particularly R.S. Bauer, F. Casula, D.J. Henderson, B.E. Hobbs, N.M. Johnson, R.V. Kasowski, A.G. Revesz, and W.E. Rudge.

References

1. Supported in part by ONR Contract Number N00014-79-C-0814
2. S.T. Pantelides (ed.): *The Physics and Chemistry of SiO_2 and its Interfaces* (Pergamon, New York 1978)
3. G. Lucovsky, S.T. Pantelides, F.L. Galeener (eds.): *The Physics of MOS Insulators* (Pergamon, New York 1980)
4. A.G. Revesz: J. Non-Cryst. Solids *11*, 309 (1973); Phys. Status Solidi a *57*, 235, 657 (1980); a *58*, 107 (1980);
   A.R. Revesz, G.V. Gibbs: In Ref.[3], p.92
5. B.E. Deal: J. Electrochem. Soc. *121*, 198C (1974)
6. Y.C. Cheng: Prog. Surf. Sci. *8*, 181 (1977)
7. M. Pepper: Contemp. Phys. *18*, 423 (1977)
8. C.R. Helms: J. Vac. Sci. Technol. *16*, 608 (1979)
9. N.F. Mott: Adv. Phys. *26*, 363 (1977); in Ref.[2], p.1; J. Non-Cryst. Solids *40*, 1 (1980)
10. G.N. Greaves: Philos. Mag. B*37*, 447 (1977); in Ref.[2], p.268; C.M. Svensson: In Ref.[2], p.328; G. Lucovsky, D.J. Chadi: In Ref.[3], p.301; R.H.D. Nuttall, J.A. Weil: Solid State Commun. *33*, 99 (1980)
11. D.L. Griscom: J. Non-Cryst. Solids *40*, 211 (1980); in Ref.[2], p.232; Phys. Rev. B*20*, 1823 (1979)
12. F. Herman, I.P. Batra, R.V. Kasowski: In Ref.[2], p.333
13. F. Herman, D.J. Henderson, R.V. Kasowski: In Ref.[3], p.107
14. F. Herman, R.V. Kasowski: J. Vac. Sci. Technol., in press
15. N.F. Mott, E.A. Davis: *Electronic Processes in Non-Crystalline Materials* (Clarendon Press, Oxford 1979), 2nd ed.;
    N.F. Mott: J. Phys. C: Solid State Phys. *13*, 5433 (1980)
16. J.M. Ziman: *Models of Disorder* (Cambridge Univ. Press 1979)
17. M.H. Brodsky (ed.): *Amorphous Semiconductors*, in Topics in Applied Physics, Vol.36 (Springer, Berlin, Heidelberg, New York 1979)
18. D.R. Uhlmann: J. Non-Cryst. Solids *42*, 119 (1980); D. Adler: ibid *42*, 315 (1980); A.C. Wright, G.A.N. Connell, J.W. Allen: ibid *42*, 509 (1980)
19. J.C. Phillips: Phys. Status Solidi b *101*, 473 (1980); Comments Solid State Phys. *9*, 191 (1980)
20. R.B. Laughlin, J.D. Joannopoulos, D.J. Chadi: In Ref.[2], p.321; Phys. Rev. B*20*, 5228 (1979); Phys. Rev. B*21*, 5733 (1980)
21. T. Sakurai, T. Sugano: In Ref.[3], p.241
22. F.L. Galeener: J. Non-Cryst. Solids *40*, 527 (1980)
23. W.Y. Ching: Phys. Rev. Lett. *46*, 607 (1980)
24. S.T. Pantelides, M. Long: In Ref.[2], p.339

25. G.H. Dohler, R.T. Dandoloff, H. Bilz: J. Non-Cryst. Solids *40*, 87 (1980)
26. J.D. Bernal: Proc. R. Inst. London *37*, 355 (1959); Sci. Am., Aug., 1960
27. R.J. Bell, P. Dean: Philos. Mag. *25*, 1381 (1972);
    P.H. Gaskell, I.D. Tarrant: Philos. Mag. *42*, 265 (1980)
28. G.S. Cargill III, P. Chaudhari (eds.): *Atomic Scale Structure of Amorphous Solids* (North-Holland, Amsterdam 1979); also published in J. Non-Cryst. Solids *31*, 1-286 (1978)
29. S.T. Pantelides, W.A. Harrison: Phys. Rev. B*13*, 2667 (1976);
    S.T. Pantelides: Comments Solid State Phys. *8*, 55 (1977)
30. F. Herman: J. Vac. Sci. Technol. *16*, 1101 (1979)
31. R.W.G. Wyckoff: *Crystal Structures*, Vol.1, 2nd ed. (Interscience, New York 1963)
32. L.E. Sutton (ed.): *Interatomic Distances Supplement* (The Chemical Society, London 1965) p. S 12S
33. K.F. Purcell, J.C. Kotz: *Inorganic Chemistry* (W.B. Saunders, Philadelphia 1977) p.318ff.
34. R.V. Kasowski, E. Caruthers: Phys. Rev. B*21*, 3200 (1980)
35. F. Herman: *Atomic Structure Calculations* (Prentice-Hall, Englewood Cliffs, NJ 1963)
36. R.V. Kasowski: to be published
37. A.J. Bennett, L.M. Roth: J. Phys. Chem. Solids *32*, 1251 (1971);
    T. Iizuka, T. Sugano: Japan. J. Appl. Phys. *12*, 73 (1973)
38. A. Goetzberger, E. Klausmann, M. Schulz: CRC Crit. Rev. Solid State Sci. *6*, 1 (1976);
    E.H. Nicollian: J. Vac. Sci. Technol. *14*, 1112 (1977);
    N.M. Johnson, D.K. Biegelsen, M.D. Moyer: In Ref.[3], p.311

Morphology of the Si-SiO$_2$ Interface

C.R. Helms

Stanford Electronics Laboratories, Department of Electrical Engineering
Stanford University
Stanford, CA 94305, USA

Introduction

The interface between silicon and thermally grown silicon dioxide is almost certainly the best studied solid-solid interface known. For silicon devices, we must produce oxides on silicon that are highly stable, present a diffusion barrier to undesired impurities, and form an interface with the silicon substrate that has a low density of fixed charge and interface traps. In addition, a precise control of the thickness of the oxide layer and its dielectric properties is necessary for the successful fabrication of silicon devices.

In this paper [1], our current understanding of the chemistry and morphology of the Si-SiO$_2$ interface will be reviewed with particular emphasis placed on those aspects which are particularly important in device processing and electrical characteristics.

These ideas will be compared to an ideal model of the interface made by matching β cristobalite to a 100 silicon surface. In contrast to this model, it appears that to accurately describe a real SiSiO$_2$ system it may be necessary to consider up to five separate regions: bulk Si, disordered Si near the metallurgical interface, nonstoichiometric SiO$_x$ at the interface, strained SiO$_2$ near the interface, and bulk SiO$_2$. The evidence for the existence of these regions will be presented and some implications with regard to electrical properties discussed.

The paper will be divided into sections relating to the various regions of the Si-SiO$_2$ interface believed to exist. Before that discussion, a description of a perfect interface structure will be presented.

Ideal Interface for (100) Si

The simplest possible picture for the Si-SiO$_2$ interface is shown in Fig. 1 [2]. This is obtained by matching the silicon atoms of an unreconstructed silicon (100) surface with a (100) surface of β cristobalite rotated by 45°. In addition, an extra half plane of oxygen atoms has been added to terminate the remaining silicon dangling bonds. The actual mismatch (not shown) between the silicon atom positions of 100 Si and 100 β cristobalite is ~6%. This interface structure has never been observed and may well be thermodynamically unstable. However, it forms a good basis for understanding what the actual interface might look like. In addition, the Si-O-Si bond angles are shown as 180°; the actual bond angles of β cristobalites are somewhat less.

The interface shown is probably the narrowest that can be constructed (2.4 Å). It is ideal in that it has the following properties.

(1) It is ideally flat.
(2) It spearates two perfectly ordered systems.
(3) It contains no defects.

It does, however, contain 1.04×10^{15} cm^{-2} silicon atoms whose nearest neighbor configurations neither appear as in SiO_2 or Si. 3.4×10^{14} cm^{-2} silicon atoms are bonded to two silicons and two oxygens as in SiO; 7×10^{14} cm^{-2} silicon atoms are bonded to three silicon atoms and one oxygen atom, as would be expected for Si_2O.

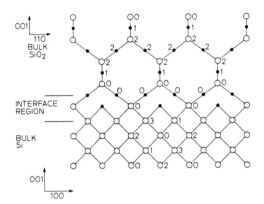

Fig. 1 Model of an Ordered Si-SiO$_2$ interface after Reference 2

The actual interface is far from the one pictured in Fig. 1. It is not ideally flat, with roughness observed by TEM measurements of up to 10 Å [3,4,5]. The SiO_2 is not a well-ordered form but amorphous and even though stoichiometric has a structure that is possibly graded in passing from the interface into bulk SiO_2 [6]. In fact, PANTELIDES and LONG [7] have succeeded in constructing an abrupt interface between amorphous SiO_2 and Si (100) by using slightly strained 144° bond angles.

The silicon atom positions on the silicon side of the interface also need not lie in their normal bulk positions. Indeed, silicon-free surfaces exist in reconstructed configurations with measurable bond strain present at least five layers below the surface [8,9]. There is evidence that similar effects may occur near the Si-SiO$_2$ interface [10].

The question of interface defects may be the most critical when considering device electrical characteristics. The most important defects of this type are vacancies leaving interface atoms with various dangling-bond configurations. It is well established that these interface vacancy complexes can lead to interface fixed charge or interface traps. It is difficult to determine the intrinsic concentration of such defects since they are very sensitive to time temperature history as well as small concentrations of impurities (especially hydrogen present).

Chemical Structure of the Nonstoichiometric Interface Region

The region referred to here includes the silicon atoms which are bonded neither to four other silicon atoms as in the substrate or to four oxygen atoms as in SiO_2. This region must exist even for an idealized interface, as discussed in the previous section. The properties of this region are probably most critical in determining the electrical properties of an MOS structure due to the possibility of defects leading to sites of fixed charge or interface traps.

The questions we would ask about this region include:

(1) How thick is it?
(2) How many Si atoms does it contain?
(3) What are their bonding configurations?
(4) What is the density of vacancy-type defects?

The first question is somewhat ambiguous since the question of the physical extent of the interface depends on how we decide to cut off the potential fluctuations present. The answer really lies in the number of silicon atoms per square centimeter that actually exist in nonstoichiometrically bonded configurations.

Results from various studies of the interface are shown in Table 1, along with comments regarding the validity of the actual values. The average over these experiments is $\sim 1.5 \times 10^{15}$ cm^{-2} (~ 2 monolayers for 100 Si with 7×10^{14} cm^{-2} atoms per monolayer). The lower values for the concentration of these configurations is obtained by electron spectroscopy [6, 11]. The experiments fall into two categories: one using chemical etching and one using sputtering, both for profiling. It can be argued that for either type of profiling the interface chemistry can be significantly perturbed by the profiling technique. Since electron spectroscopy is used, however, we can take advantage of the finite depth resolution (5 to 50 Å) to observe the interface region before damage due to the profiling technique is significant. This can be done by studying very thin oxides where the interface is still protected by the thin oxide present. The value of the concentration observed is then dependent on the accuracy with which the probing depth of the spectroscopy is known.

Table 1 Concentrations of silicon at the Si-SiO$_2$ interface in non-stoichiometric configurations

Technique	Value	Comment
AES [11]	$1 \pm 0.5 \times 10^{15}$ cm^{-2}	uncertain due to effects of ion beam etching employed
XPS [6]	6×10^{14} cm^{-2}	uncertain due to effects of chemical etching employed
RBS [10]	2×10^{15} cm^{-2} maximum	some contribution may be due to nonregistered near-interface substrate atoms
Ellipsometry [12]	$3 \pm 1 \times 10^{15}$ cm^{-2} maximum	some contribution may be due to nonregistered near-interface substrate atoms

The value obtained by channeling in Rutherford backscattering experiments [10] not only includes contributions from the nonstoichiometric region but also from substrate silicon atoms that are not in registry with substrate. Therefore the value obtained would represent an upper limit.

The ellipsometrically determined interface properties [12] were obtained from a theoretical fit to experimental data taken over a range of oxide thicknesses and wavelength. In this analysis, the thickness of the oxide, void fraction of the bulk SiO_2, thickness of the interface region, and its composition (which was used to estimate the interface optical constants) were allowed to vary. A value of 3×10^{15} cm^{-2} nonstoichiometrically bonded silicon atoms was obtained, somewhat larger than the other values. One possible source of this discrepancy (if indeed it is a discrepancy) is the sensitivity of the technique to nonregistered Si substrate atoms as was the case for RBS.

The question of the actual bonding configurations of the silicon atoms present in this region is much more difficult to obtain. The model of Fig. 1 shows two bonding configurations as in SiO and Si_2O. It would seem reasonable that a configuration like Si_2O_3 might also be possible.

Of critical importance for device properties is the existence of defects in this interface structure. It is believed, for example, that an SiO_3 configuration on the SiO_2 side of the interface with one "dangling" Si bond can lead to a center of fixed positive charge at the interface whereas a $Si(Si)_3$ configuration with one unsaturated bond on the silicon side of the interface can lead to an interface trap. Direct measurements of the latter type of center has been accomplished by POINDEXTER and coworkers [13,14] using EPR. A definite relationship between this center and interface trap density has been established. The SiO_3 center has not been observed directly, so its existence can only be confirmed from theoretical calculations.

Evidence for Strained SiO_2 Near the Interface

Although we normally think of the Si-SiO_2 interface terminating after stoichiometric SiO_2 is reached from the substrate, there is considerable evidence that the oxide within ~15 to 30 Å of the interface, although stoichiometric, is chemically different from so-called bulk SiO_2. Although the existence of this region has significant implications with regard to the oxidation process, it will also be important in determining the dielectric properties of very thin gate oxides. These effects stem from the different dielectric constant of this region compared to the bulk SiO_2. The importance of this region is due to the form of the relationship for flat-band and device threshold voltages involving terms in the capacitance of the oxide as C_{ox}^{-1}. For thick oxides, C_{ox} will scale as the inverse of oxide thickness. For thin oxides where the strained region is an appreciable fraction of the oxide thickness C_{ox} will no longer scale as the inverse of oxide thickness but as

$$C_{ox} \cong C_0 \left[1 + \frac{d}{x} (1 - \Delta\epsilon)\right]$$

where C_0 is the capacitance of the bulk oxide of thickness x (the total oxide thickness), d is the extent of the thin region near the interface, and $\Delta\epsilon$ is the ratio of the bulk dielectric constant to the near-interface dielectric constant.

Considerable evidence for such a region has been obtained by GRUNTHANNER and coworkers using XPS [6] in conjunction with theoretical calculations of MADHUKAR [15]. They have suggested that the broad silicon 2p core level peak in the XPS spectra associated with SiO_2 is made up of various components coming from different Si-O-Si bond angle configurations. By careful analysis of how this XPS peak (or set of peaks) vary with oxide thickness, they concluded that the distribution of bond angles is changing as the interface is approached, this change observed over a 16 Å distance.

In addition, WILLIAMS and GOODMAN [16] measured the contact angle of water on thin layers of SiO_2 on Si. They observed, for thicknesses less than 30 Å, that the contact angle varied monotonically from that characteristic of SiO_2 to that of Si. This indicates a significant change in the surface-free energy of the oxide within 30 Å of the interface that may be related to nonstoichiometry or changes in the chemical nature of the SiO_2 itself. The XPS results definitely rule out the nonstoichiometry argument. Additional support for a region of differing Si-O-Si bond angle come from calculations of PANTELIDES and LONG [7]. They performed simple calculations to show that an abrupt interface in stoichiometry could be accommodated if bond angle strain were allowed in the near interface SiO_2.

Evidence for Interface Roughness

Up to this point, we have been dealing with an interface assumed to be perfectly flat. Interface roughness almost certainly exists on some scale and has been suggested by numerous authors. The most clear-cut evidence for roughness present comes from TEM results [3,4,5], where roughness as large as 10 Å in height has been observed. The values obtained for roughness height and wavelength from the TEM studies are quite close to values estimated from studies of the variation in carrier mobility for high fields obtained in the early 1970's [17].

In addition, investigators using sputter profiling have suggested the presence of long wavelength roughness which would account for the large (~25 Å) interface extent observed in sputter profiling experiments [11, 18-22]. Recent calculations by FRENZEL and BALK [23] and measurements by HELMS [24] would suggest that the interface widths observed in these experiments are due to details of the sputtering process or knock-on mixing and damage-enhanced diffusion.

Evidence for Nonregistered Silicon Atoms in the Substrate

As mentioned above, silicon-free surfaces reconstruct so that the uppermost atom layers are out of registry with the substrate. There is also evidence that a similar effect may be present for the silicon atoms near the interface on the silicon side. This evidence comes from channeling results from Rutherford backscattering experiments [10,25]. In recent work by FELDMAN and coworkers in particular, they have found ~2 x 10^{15} cm^{-2} excess silicon atoms at the interface. The ellipsometric results of ASPENES and THEETEN [12] would also be sensitive to a layer of this type possible accounting for the 3 x 10^{15} cm^{-2} interface atoms observed. These atoms could exist in nonstoichiometric oxide configurations or as unregistered atoms near the interface but in the substrate. At present, it seems clear that at least 10^{15} cm^{-2} of these excess silicon atoms do exist in nonstoichiometric oxide configurations. The state of the remaining excess silicon atoms is, at this point, not certain.

Acknowledgement.

This research would not be possible without the long-term support of Dick Reynolds at DARPA, present contract MDA 903-79-C-0257.

23

References

1. A previous review of much of this material by this author appeared in J. Vac. Sci Technol. 16, 608(1979).

2. F. Herman, I. P. Batra, and R. V. Kasowski, in Proc. Intl. Conf. on the Physics of SiO_2 and Its Interfaces (Pergamon, New York, 1978), p. 333.

3. J. Blanc, C. J. Buiocchi, M. S. Abrams, and M. E. Han, Appl. Phys. Lett. 30, 120(1977).

4. O. L. Krivanek, T. T. Sheng, and D. C. Tsui, Appl. Phys. Lett. 32, 439 (1978).

5. O. L. Krivanek and J. H. Mazur, Appl. Phys. Lett. 37, 392(1980).

6. F. J. Grunthaner, P. J. Grunthaner, R. P. Vaszuez, B. F. Lewis, J. Maserjian, and A. Madhukar, J. Vac. Sci. Technol. 16, 1443(1979).

7. S. Pantelides and M. Long, in Proc. Intl. Conf. on the Physics of SiO_2 and Its Interfaces (Pergamon, New York, 1978), p. 339.

8. R. E. Schlier and H. E. Farnsworth, J. Chem. Phys. 30, 917(1959)

9. J. A. Applebaum and D. R. Hamann, Rev. Mod. Phys. 48, 479(1976), and references therein.

10. L. C. Feldman, P. J. Silverman, J. S. Williams, T. E. Jackman, and I. Stensgaard, Phys. Rev. Lett. 41, 1396(1978).

11. C. R. Helms, Y. E. Strausser, and W. E. Spicer, Appl. Phys. Lett. 33, 767(1978).

12. D. E. Aspenes and J. B. Theeten, J. Electrochem. Soc. 127, 1359(1980).

13. E. H. Poindexter, E. R. Ahlstrom, and P. J. Caplan, in Proc. Intl. Conf. on the Physics of SiO_2 and Its Interfaces (Pergamon, New York, 1978), p. 227.

14. F. J. Caplan, E. H. Poindexter, B. E. Deal, and R. R. Razouk, in Proc. Intl. Conf. on the Physics of MOS Insulators (Pergamon, New York, 1980), p. 306.

15. R. N. Nucho and A. Madhukar, in Proc. Intl. Conf. on the Physics of SiO_2 and Its Interfaces (Pergamon, New York, 1978), p. 55.

16. R. Williams and A. M. Goodman, Appl. Phys. Lett. 25, 531(1974).

17. Y. C. Cheng and A. Sullican, Surf. Sci. 34, 717(1973).

18. J. S. Johannessen, W. E. Spicer, and Y. E. Strausser, J. Vac. Sci. Technol. 13, 849(1976).

19. J. S. Johannessen, W. E. Spicer, and Y. E. Strausser, J. Appl. Phys. 47, 3028(1976).

20. C. R. Helms, W. E. Spicer, and N. M. Johnson, Solid State Commun. 25, 673(1978).

21. C. R. Helms, N. M. Johnson, S. A. Schwarz, and W. E. Spicer, J. Appl. Phys. $\underline{50}$, 7007(1979).

22. J. F. Wager and C. W. Wilmsen, J. Appl. Phys. $\underline{50}$, 874(1979).

23. H. Frenzel and P. Balk, in Proc. Intl. Conf. on the Physics of MOS Insulators (Pergamon, New York, 1980), p. 246.

24. C. R. Helms, to be published.

25. T. W. Sigmon, W. K. Chu, E. Lugujjo, and J. W. Mayer, Appl. Phys. Lett. $\underline{24}$, 105(1974).

Influence of Oxidation Parameters on Atomic Roughness at the Si-SiO$_2$ Interface

P.O. Hahn and M. Henzler

Institut B für Experimentalphysik, Universität Hannover
D-3000 Hannover, Fed. Rep. of Germany

1. Introduction

The quality of MOS-devices depends strongly on the quality of the interface between Si and SiO$_2$. Whereas the chemical composition and electrical properties have long been studied, the structural properties had been neglected due to the lack of appropriate measuring techniques. Only AES-depth-profiling and the transmission electron microscope have been applied to investigate the interface. A novel technique for studying the interface uses "low energy electron diffraction" (LEED) [1-7], which enables us for the first time to obtain quantitative results about the atomic roughness and its dependance of the oxidation parameters. If the surface of a crystal shows different levels due to atomic steps, the diffracted electron beam is modified by interference of the electrons at adjacent terraces. Varying the energy the interference changes periodically between constructive and destructive interference. Therefore the spot shape changes between sharp and broadened profiles. From the period the step height is derived with high precision, from the spot shape (relative half width) the distribution of terraces with respect to orientation, width and regularity is obtained [3, 7]. With the help of model calculations the spot profile gives us therefore a direct and quantitative measure of the step density.

2. Experimental

Silicon (111) and (100) samples have been oxidized at atmospheric pressure. After removal of the oxide by etching in HF the samples were transfered into an uhv system avoiding reoxidation by a coverage of methanol. Details of this method and their reliability are described elsewhere [7].

3. Influence of the procedure on the results

Fig. 1 shows the relative half width immediately after removal of the oxide (crosses: xxx). The half width is oscillating with energy. The upper solid curve is calculated for an edge atom density of 11.7%. Now the sample is transfered back out of the system, etched in HF, and transfered again into the system (plus: +++). A repetition of the same procedure (∅∅∅) does not change the roughness either.
Now the same sample was well annealed until the pattern showed the usual 7x7 structure. The measured relative half width is completely determined by the instrumental function. This well annealed sample again was

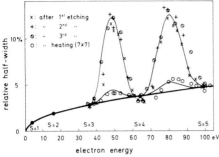

Fig.1: Relative half width versus electron energy for a sample after multiple etching: 1st etching (xxx), 2nd etching (+++), 3rd etching ($\emptyset\emptyset\emptyset$) (upper curve). The lower curve indicates the relative half width directly after heating to the 7×7 structure (step free surface). The center curve shows the residual roughness after etching of the step free surface in HF. S indicates the energies for in-phase scattering

transfered out of the system to be etched in HF. We obtain only a very small broadening (center curves: ooo) due to the oxidation in air during the transfer from the uhv system to the HF etching. These and further experiments [7] support the conclusion that the procedure does not alter the roughness at the interface appreciably and that the surface under investigation is wanted with respect to roughness.

4. Results
4.1 Oxidation at 800° C and 1000° C
(111) samples had been oxidized in dry atmosphere as a function of oxide thickness. The pretreatment was always an annealing to the 7x7 structure. The main result is (see Fig. 4 and [7]), that for low oxide thickness (< 20 nm) the edge atom density is appreciable higher (27%) than for higher thickness (15%). Oxidation at 1000° C reduces the roughness to 13%, while oxidation in wet atmosphere increases the roughness for both temperatures.

4.2 Variation of the treatment before and after oxidation
Four different starting conditions had been realized. All samples were then oxidized simultaneously at 1000° C for about two hours (see Fig. 2 in ref. [7]). We measure a strong dependence. To see, if a non oxidizing posttreatment would vary the results, the oxidized samples had been heated in nitrogen for 4 hours. Fig. 2 represents the result for a systematic variation of N_2 annealing temperature. It is seen that both a proper pretreatment and posttreatment is decisive for a low roughness at the interface.

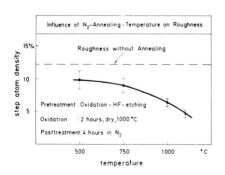

Fig. 2: Step atom density versus N_2 annealing temperature.

4.3 Roughness at the (100) surface

A (100)-orientated sample with 100 nm oxide but without N_2 annealing has been measured (Fig. 3, crosses xxx). Only between 65 eV - 110 eV could the profile be measured. The upper oscillating curves, are calculated for a relative half width of 16%. The solid curve is calculated only for monoatomic steps, the dashed one for steps with doubled height. Heating at about 1200° C we obtain nearly the instrumental function (solid lower curve). Finally a (100) sample was oxidized according to procedures which give at the (111) surface the smallest roughness observed. We see a very small roughness (circles ooo). The calculation was done for a step atom density of 5%.

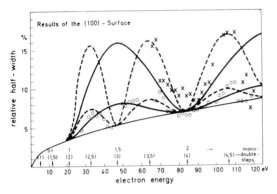

Fig.3. Relative half width versus electron energy for (100) samples.

(xxx) sample without N_2 annealing, 100-nm oxide. (ooo) sample, annealed 4 hours in nitrogen at 100 °C, 100-nm oxide. Oscillating curve: calculated for a relative half-width of 16% (upper curves) and 5% (lower curves). (—) for only monoatomic steps. (---) for only steps with doubled heigh. Curve below: instrumental function

5. Discussion

The results demonstrate, that the interface shows an appreciable number of edge atoms. A model including a roughening and a smoothing factor can explain the results summarized in fig. 4.

Summary of the results			
variation of the oxidation parameters ⟶ step atom density/%			
A. (111) - surface			
oxidation temperature	atmosphere	oxide thickness	
800 °C	dry	d < 20 nm	high 27%
		d > 20 nm	low 15%
1000 °C		d > 20 nm	low 13%
	wet		highest 35%
posttreatment: annealing in N_2 at various temperatures			
1100 °C	dry	100 nm	lowest 3%
↓			↓
500 °C			low 7%
pretreatment			
1. long storage in dir	dry	100 nm	high 27%
2. chemically etched in CP 6			↓
3. oxidation - HF - etching			
4. annealing in UHV to 7×7 structure			low 12%
B. (100) - surface			
1. oxide thickness: 100 nm; without N_2 — annealing			high > 16%
2: twice oxidized : 100 nm; annealed in N_2 at 1000 °C			low ~ 6%

Fig. 4

Random fluctuations of diffusion and reaction of oxygen is the roughening effect piling up more and more roughness with growing oxide thickness. On the other hand in thermal equilibrium a smooth, step free surface is energetically favoured similar to the free surface of silicon. Therefore diffusion of oxygen and silicon smoothens the interface depending on temperature, ambient and duration of heat treatment. The balance between roughening and smoothing determines the final roughness. Therefore high oxidation rate (at low thickness or with wet oxygen) increases roughness whereas high temperature and low oxidation rate (at high oxide thickness or in non-oxidizing ambient) lowers the roughness. The remarkable step (Fig. 2) may be related to growth kinetics. At low thickness (d < 20 nm) the thickness increases linear with time, at higher thickness with the square root in accordance with literature [8] where it is explained as reaction controlled and diffusion controlled respectively.

The dependance on pretreatment shows the influence of chemical differences. Contamination (e.g. metal or carbon) should alter the electronic properties and therefore the chemical reactivity leading to different step atom densities. First experiments with the (100) surface shows on the one hand the existence of mono- and double steps on the other hand this surface has the same dependance on the oxidation parameters.

A comparison with the results of the other structural techniques [7, 9] shows roughly an agreement. AES depth profiling found a transition layer of 3 - 5 nm independent of oxide thickness (25 - 100 nm), which increases slightly with growth temperature, however, in contrary to the present results. Here the influence of the ion beam, needed for depth profiling, is most likely the main cause for the difference. A LEED-study of ion bombarded germanium surfaces [6] yields an appreciably higher roughness than reported here. Therefore any technique using ion bombardment cannot reveal edge atom densities at the interface as found here.

TEM-measurements of the oxide after chemical removal of the silicon shows comparable step densities [10]. With that technique, however, it seems difficult to obtain quantitative information.

The presented quantitative results now enable a direct check of a correlation between electron mobility and step atom density which has so far been indirectly concluded from electrical measurements. Those measurements are in progress. In this way all kinds of procedures used for fabrication of MOS-devices may be checked with respect to interface roughness by use of the presented technique of LEED spot profiling.

Acknowledgements: The investigation has been supported by the US Army, European Research Office. The silicon crystals have been kindly provided by Wacker Chemitronic, Burghausen.

References

1. M. Henzler: in *Electron Spectroscopy for Surface Analysis*, ed. H. Ibach, Topics in Current Physics, Vol. 4 (Springer, Berlin, Heidelberg, New York 1977) p. 117
2. M. Henzler: in Advances in *Solid State Physics*, ed. by J. Treusch (Fest-körperprobleme XIX, 1979), p. 193
3. M. Henzler, Surf. Sci. *73*, 240 (1978)
4. P.O. Han, J. Clabes: M. Henzler, J. Appl. Phys. *51*, 4, 2079, (19807
5. M. Henzler, F.W. Wulfert: Proc. VIII. Conf. Physics of Semiconductors, Rome, 1976, p. 669
6. G. Schulze, M. Henzler, Surf. Sci. *73*, 553 (1978)
7. P.O. Hahn, M. Henzler: J. Appl. Phys. (in press)
8. B.E. Deal, A.S. Grove, J. Appl. Phys. *36*, 3770 (1965)
9. C.R. Helms, N.M. Johnson, S.A. Schwartz,W.E. Spicer: J. Appl. Phys. *50*, 7007 (1979)
10.T. Sugano, J.J. Chen, T. Hamano: Surf. Sci. *98*, 154 (1980)

Electronic and Optical Properties of SiO$_x$

K. Hübner, E. Rogmann, and G. Zuther

Sektion Physik der Wilhelm-Pieck-Universität
2500 Rostock, DDR

1. Introduction

Recently, it was shown that even in SiO$_x$ forms of the random bonding type
the Si-Si and Si-O bonds retain essentially those properties, which they have
in the limiting materials, i.e., in Si and SiO$_2$[1]. This concerns, for example, the values of the bond length and the valence charge density and is
in contrast to the behaviour of semiconducting mixed crystals. Therefore, it
can be assumed that SiO$_x$ is an inhomogeneous medium on a bonding scale, in
which the Si-Si and Si-O bonds are the corresponding subsystems. The electronic and optical properties of inhomogeneous media can be described by
means of the effective medium approximation. This approach leads to the fol-
lowing quadratic relation for the dielectric function $\varepsilon_x = \varepsilon_{1x} + i\varepsilon_{2x}$ of the
inhomogeneous medium [2]

$$\varepsilon_x^2 + \varepsilon_x[A(x)\varepsilon_a - B(x)\varepsilon_b] - [A(x)-B(x)+1]\varepsilon_a\varepsilon_b = 0 \quad . \tag{1}$$

The magnitudes A and B are functions of geometrical and depolarization factors, i.e., they involve local field effects. They depend on the average
chemical composition x of the medium with the two different contributions ε_a
and ε_b to the total electronic dielectric function ε_x. Inhomogeneous media
may reveal percolation effects. In agreement with the threshold in the dependence of various electronic properties of SiO$_x$ on x at x$\approx$1.25 we have
found the percolation threshold of the network of chemical bonds to be at
x = 1.25 [1].

2. Coexistence Approximation for the Determination of the Dependence of Optical Properties on x

As discussed above, there is a coexistence of Si-Si and Si-O bonds in SiO_x. On the other hand, total electronic properties of SiO_x show an ensemble behaviour, which is characterized by the fact that they vary quantitatively *and* qualitatively with the average chemical composition x of SiO_x. This means that not only the relative numbers of Si-Si and Si-O bonds, but also their mutual arrangement (e.g., in the form of special sequences) determine the macroscopic properties of SiO_x. Based on this observation the following approach is established.

The dielectric functions ε_a and ε_b of the Si-Si and Si-O bonds are expressed by the corresponding dielectric functions $\varepsilon_{Si}(E)$ and $\varepsilon_{SiO_2}(E)$ of the limiting materials Si and SiO_2, respectively (E means the energy). For numerical calculations the experimental values from PHILIPP [3] are used. The parameters A and B can be calculated exclusively from the x dependence of the static electronic dielectric constant $\varepsilon_x(E \approx 0) = \varepsilon_{1x}(E \approx 0)$ or from Sellmeier plots of the refractive index $n(E)\left\{[n^2(E)-1]^{-1} \text{ vs } E^2\right\}$ [4]. Analytical expressions for A(x) and B(x) determined in this way are given in [5]. With the help of these expressions the unknown values of $\varepsilon_x = \varepsilon_{SiO_x}(E)$ can immediately be determined from (1).

3. Results

The real and the imaginary part of the dielectric function of SiO_x in the energy range from E = 0 to E = 9 eV are shown in Fig.1. The percolation threshold mentioned above is revealed by the fact that the influence of Si-Si bonds on the lineshape of $\varepsilon_1(E)$ and $\varepsilon_2(E)$ vanishes only for $x \gtrsim 1.25$. The same is valid for the real [n(E)] and the imaginary part of the refractive index.

To check the agreement with experiment we compare in Fig.2 the theoretical energy variation of ε_1 and ε_2 for x = 1 with corresponding experimental values for silicon oxide ($SiO_1 \pm 0.1$) [3]. Considering the fact that SiO has a special intermediate "position" between the limiting materials Si and SiO_2, from which the energy dependences of ε_1 and ε_2 are taken over the total intermediate range of chemical composition, the agreement is surprisingly good.

The x dependence of the absorption coefficient $\alpha(E) = 4\pi\lambda^{-1}k(E)$ is shown in Fig.3 (λ is the wavelength). Although there is a strong influence of dis-

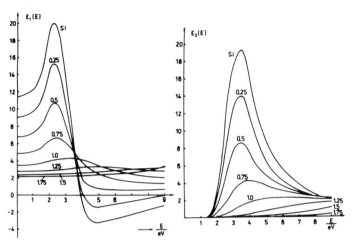

Fig. 1. Re$\{\varepsilon_1(E)\}$ and Im$\{\varepsilon_2(E)\}$ part of the dielectric function of SiO_x in dependence on energy (E)

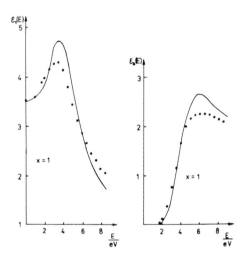

Fig. 2. Energy dependence of the real and imaginary part of the dielectric function of SiO [(——) experimental [3]; (...) theoretical, this work]

order effects on α due to small deviations from the value x = 2 in SiO_2, this magnitude can be used for a simple optical determination of the average chemical composition of SiO_x.

The energy E_{04} at which $\alpha = 10^4$ cm^{-1} is compared in Fig.4 with the x dependence of the corresponding optical gap measured by HOLZENKÄMPFER et al. [6].

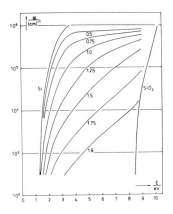

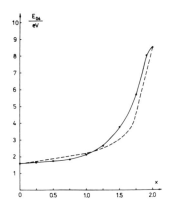

Fig. 3. Theoretical absorption coefficient of SiO$_x$

Fig. 4. Optical gap of SiO$_x$ [(---)
experimental [6]; (--o--) theoretical,
this work]

The theoretical and the experimental curves agree within the range of the
error for the chemical composition of the SiO$_x$ samples investigated by
HOLZENKÄMPFER et al.

4. Discussion

The parameters A and B of (1) are independent of energy if the Sellmeier plots
mentioned above are straight lines. Therefore, our approach to the x dependence of optical properties of SiO$_x$ can be extended also to the IR region,
where another ensemble of Sellmeier plots is found [7]. Furthermore, relation
(1) seems to have a more general meaning than expressed by the effective
medium theory. It can be shown, for example, that the application of the theory of dielectric screening of atomic potentials to the Si-Si and Si-O bonding units provides an analogous relation for ε_x [8].

   The comparison of the x dependence of E_{04} and of the fundamental gap E_g
of the electronic density of states of SiO$_x$ leads to the following observation. The optical gap shows a weak variation in the range from x = 0 to x ≈ 1.25
and a strong variation within the remaining range of x, whereas E_g as well as
many other electronic properties reveal a quite opposite tendency and are
nearly identical with those of SiO$_2$ already at x ≈ 1.5 [1]. The reason for
this behaviour of E_{04} is that the relative high polarizability of the Si-Si
bonds is dominating for the formation of total electronic properties up to

33

the state where the relative number of Si-Si bonds reaches the percolation threshold. This can be seen from the approximative condition $(2-x)\varepsilon_{Si} = 2x\varepsilon_{SiO_2}$, where $(2-x)/(2+x)$ and $2x/(2+x)$ are the probabilities that an SiO_x bond is an Si-Si and an Si-O bond, respectively.

Finally, it should be mentioned that similar results for ε_x of SiO_x were recently obtained by ASPNES and THEETEN [9]. Their assumptions and approximations, however, are much more complicated than those presented in this paper.

References

1. K. Hübner: Phys. Stat. Sol. (a)*61*, 665 (1980); J. Non-Cryst. Solids *35,36*, 1011 (1980)
2. C.G. Granqvist, O. Hunderi: Phys. Rev. B*18*, 2897 (1978)
3. H.R. Philipp: J. Phys. Chem. Solids *32*, 1935 (1971) and private communication
4. G. Zuther, K. Hübner, E. Rogmann: Thin Solid Films *61*, 391 (1979)
5. G. Zuther: Phys. Stat. Sol. (a)*59*, K109 (1980)
6. E. Holzenkämpfer, F.-W. Richter, J. Stuke, U. Voget-Grote: J. Non-Cryst. Solids *32*, 327 (1979)
7. K. Hübner et al.: to be published
8. G. Zuther, K. Hübner: Phys. Semiconductor Surface *11*, 61 (1980)
9. D.E. Aspnes, J.B. Theeten: J. Appl. Phys. *50*, 4928 (1979)

Hydrogenation of Defects at the Si–SiO$_2$ Interface

N.M. Johnson, D.K. Biegelsen, and M.D. Moyer

Xerox Palo Alto Research Centers
Palo Alto, CA 94304, USA

1. Introduction

The Si-SiO$_2$ interface possesses electronic defects which may be considered characteristic of the thermal oxidation process. These defects have three well-documented manifestations: (1) a broad peak in the interface-state distribution [1,2], (2) a paramagnetic center in electron spin resonance (ESR), which has been identified as trivalent silicon [3]. and (3) fixed positive space charge located in the oxide immediately adjacent to the interface. Hydrogen is generally believed to be involved in the removal of interface states, although the microscopic processes have not been established. For aluminum on SiO$_2$ it has been proposed that a low-temperature sinter (e.g., 450 C) releases residual hydrogen at the metal-oxide interface which subsequently diffuses to the Si-SiO$_2$ interface where it reacts with silicon dangling bonds and removes interface states [4,5]. However, this mechanism has been considered speculative because the chemical nature of the interface defect and its relation to interface states have not been established. Presented here are results from experimental studies of the mechanisms and effects of low-temperature processing and hydrogenation on defects at the Si-SiO$_2$ interface. Low-temperature anneals of the Al-SiO$_2$-Si system were used to compare the relative responses of the interface spin center and characteristic interface states.   To more directly probe the process of hydrogenation, thermal oxides on silicon were annealed in atomic deuterium. This isotope simulates hydrogen chemistry and is readily traceable even in high background concentrations of hydrogen.

2. Experimental Results

Interface-state distributions are shown in Fig. 1 for as-oxidized and post-metallization annealed metal-oxide-semiconductor (MOS) capacitors on (111)-oriented silicon. The devices consisted of SiO$_2$ layers, grown at 1000 C in dry O$_2$ on semiconducting p-type epitaxial silicon, with aluminum electrodes; the silicon substrate was degenerately doped (p$^+$) for ohmic back contact.   The distributions were obtained from deep-level transient spectroscopy (DLTS) performed in the current-transient mode [6,7], and the analysis assumed the value of 5x10^{-14} cm^2 for the hole capture cross section as determined previously for hole emission from characteristic interface states on (111)-oriented silicon [2]. In the as-oxidized sample the distribution displays the characteristic peak, which is located approximately 0.3 eV above the silicon valence-band maximum. Annealing the MOS capacitor in vacuum at 400 C for 15 min produced a ~60%

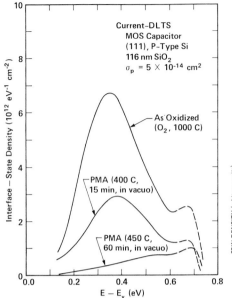

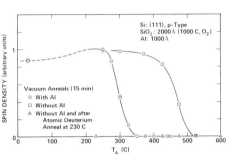

Fig. 1 Interface–state distributions in as-oxidized and post-metallization annealed (PMA) MOS capacitors

Fig. 2 Isochronal annealing of the electron spin center at the Si-SiO$_2$ Interface in as-oxidized and deuterated oxides

reduction in the peak density. The more conventional integrated-circuit anneal schedule of 450 C/60 min completely removed the characteristic peak. Near the silicon midgap the DLTS measurement is influenced by minority-carrier processes and experimental conditions (e.g., applied bias) and therefore does not yield the interface-state distribution; this is signified by the dashed-line segments and further discussed below.

The effects of low-temperature anneals on the interface spin center are summarized in Fig. 2 [8]. The specimens consisted of thermal oxides (1000 C, O$_2$) grown on (111)-oriented silicon wafers with aluminum overlayers. The MOS specimens were annealed in vacuum for 15 min, after which the aluminum was removed for ESR measurement of the spin density. In as-oxidized samples the density was ~6x10^{11} spins/cm^2. In specimens annealed with aluminum, the spin density decreases rapidly for isochronal anneal temperatures above 250 C. Without aluminum, temperatures above 400 C were required to significantly affect the density, and temperatures above 500 C were required to completely annihilate the spin signal. On the other hand, the spin signal could be completely removed by annealing specimens without aluminum in atomic deuterium at 230 C; and the spin signal was not regained with subsequent vacuum anneals up to 600 C.

The physical presence of deuterium in the SiO$_2$ layer of deuterated specimens was established with secondary ion mass spectrometry (SIMS). In Fig. 3 is

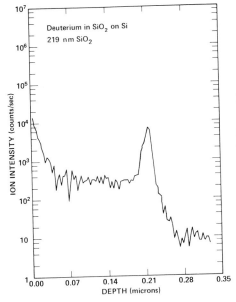

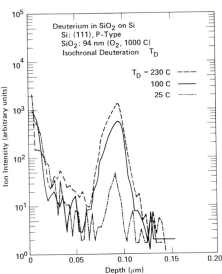

Fig. 3 Depth profile of deuterium in SiO_2 on Si from secondary ion mass spectrometry

Fig. 4 Depth profiles of deuterium in SiO_2 on Si for isochronal deuterations at different temperatures

shown the deuterium profile for an SiO_2 layer which was deuterated under the identical conditions required to annihilate the spin center (Fig. 2). Details of the deuteration technique, which is MOS process compatible, and SIMS measurement are presented elsewhere [9]. The profile reveals an accumulation of deuterium at the Si-SiO_2 interface. It has been demonstrated that this is not a consequence of ion migration ("snow plowing") during SIMS sputter profiling [9]. In Fig. 4 are shown deuterium profiles in oxides which were deuterated at different temperatures. Even at room temperature deuterium accumulates at the interface.

3. Discussion and Conclusions

The results in Fig. 1 demonstrate that the characteristic peak in the interface-state distribution can be removed by a post-metallization anneal. These results, obtained by current-transient DLTS on (111)-oriented silicon, are in agreement with previous results from constant-capacitance DLTS on (100)-oriented material [8]. Both transient techniques yield distributions in fully processed devices (e.g., bottom curve in Fig. 1) which qualitatively differ from those obtained by steady-state techniques, such as the quasi-static capacitance-voltage method [1,8]. Specifically, these latter techniques yield a U-shaped continuum of interface states with the minimum density near midgap, while from transient measurements of hole (majority-carrier) emission in MOS capacitors on p-type silicon the density increases monotonically with energy from the valence band toward midgap. This discrepancy has been discussed previously [10]. The distributions in Fig. 1 also display peaks at energies near midgap. The appearance of such peaks and their dependence on measurement parameters are to be expected from the participation of minority carriers in the transient response. The DLTS analysis of interface states is applicable when majority-carrier emission dominates [1]. Near midgap, minority-carrier emission from interface states becomes a competitive process; and the complete analysis of the recovery to equilibrium, after trap filling, must include emission and capture of both charge carriers [11].

It is concluded that characteristic defects at the Si-SiO$_2$ interface are strongly influenced by specific low-temperature treatments. Sintering aluminum thin films on thermal oxides removes both characteristic interface states and the interfacial paramagnetic defect center. However, the anneal processes occur at significantly different temperatures, which suggests that these two manifestations do not correspond to the same interface defect. The spin center has been identified with trivalent silicon [3]. Deuteration annihilated the spin center, which may be explained by the conversion of trivalent silicon defects to non-paramagnetic Si-D bonds. Secondary ion mass spectrometry revealed an accumulation of deuterium at the Si-SiO$_2$ interface on material with high spin density. However, deuterium accumulation also has been observed in thermal oxides on (100)-oriented silicon in which the spin density was negligible. It therefore appears that silicon dangling bonds are not the primary vehicle for deuterium accumulation at the interface. It is hypothesized that the accumulation arises from the breaking of strained Si-Si and/or Si-O bonds at the interface to form unstrained Si-D and O-D bonds.

The authors express their appreciation to E. H. Poindexter and W. Meuli for helpful discussions and to R. Lujan and W. Mosby for assistance with sample preparation. The SIMS measurements where performed at Charles Evans and Associates (San Mateo, California). The research was supported by the U. S. Army ERADCOM.

References

1. N. M. Johnson, D. J. Bartelink, and M. Schulz, The Physics of SiO$_2$ and its Interfaces, edited by S. T. Pantelides (Pergamon, New York, 1978), pp. 421-427.
2. N. M. Johnson, D. J. Bartelink, and J. P. McVittie, J. Vac. Sci. & Technol. 16, 1407 (1979).
3. P. J. Caplan, E. H. Poindexter, B. E. Deal, and R. R. Razouk, J. Appl. Phys. 50, 5847 (1979).
4. P. Balk, Ext. Abstr., Electronics Div., Electrochem. Soc. 14, 237 (1965).
5. E. Kooi, Phillips Res. Repts. 20, 578 (1965).
6. N. M. Johnson, M. J. Thompson, and R. A. Street, AIP Conf. Proc., in press.
7. T. G. Simmons and L. S. Wei, Solid-State Electron. 17, 117 (1974).
8. N. M. Johnson, D. K. Biegelsen, and M. D. Moyer, Physics of MOS Insulators, eds. G. Lucovsky, S. T. Pantelides, and F. L. Galeener (Pergamon, New York, 1980), pp. 311-315.
9. N. M. Johnson, D. K. Biegelsen, and M. D. Moyer, Appl. Phys. Lett., in press.
10. M. Schulz and N. M. Johnson, Appl. Phys. Lett. 31, 622 (1977).
11. N. M. Johnson, D. K. Biegelsen, and M. D. Moyer, J. Vac. Sci. & Technol., in press.

Stress Behaviour of Hydrogen Annealed Interface States

L. Risch

Siemens Research Laboratories, Otto-Hahn-Ring 6
D-8000 München 83, Fed. Rep. of Germany

Abstract

Si-SiO$_2$ interfaces, with a very low fast interface state (N_{IT})
density have been realized through strong annealing with
hydrogen. The enhanced sensitivity of these hydrogenated defects under negative bias temperature stress upon the electric
field is explained with the aid of a second oxide defect. In
the range of low N_{IT} the rise in N_{IT} is determined by this
second oxide defect, whereas at higher N_{IT} densities and for
oxides without strong hydrogen annealing the previous NBS model
is confirmed.

1. Introduction

Besides simulating the long-term behaviour of MOS devices, bias
temperature stress measurements give insight into the defect
structure of the Si-SiO$_2$ interface. According to JEPPSON and
SVENSSON [1] hydrogen that saturates dangling bonds [2], [3]
dissociates under stress conditions and initiates the simultaneous creation of fast interface states and charged oxide
defects (N_F).

With improved hydrogen annealing using a layer that re-
leases hydrogen, devices with extremely low $N_{IT} < 1 \times 10^9$ cm^{-2}eV^{-1}
can be fabricated [4]. The behaviour of this hydrogen-rich
interface under stress conditions should therefore show more
clearly the typical properties attributed to the hydrogen in the
NBS model in [1].

2. Experimental Details

The samples were MOS-capacitors on a p-Si (100) substrate with
$N_A = 2 \times 10^{15}$ cm^{-3}. Gate oxidation was performed dry at 1000 °C,
the oxide thickness being 80 nm. Oxidation was immediately
followed by N$_2$ annealing at 1000 °C (15 min) and at the end
of the process, by low temperature N$_2$ annealing at 475 °C
(30 min). Oxide type I was annealed with H$_2$ at 475 °C (30 min),
whereas oxide type II was additionally annealed with a layer
of amorphous Si at 450 °C (30 min). The preparation of the
layer has been described in [5]. The corresponding N_{IT} densities
at midgap measured by the conductance method are shown in Fig1.

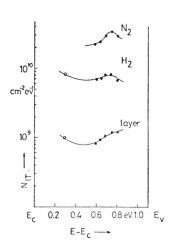

Fig.1 N_{IT} after N_2 and hydrogen annealing

3. Measurements

3.1 Positive Bias Stress

Figure 2 shows the N_{IT} density after 15 min bias stress at 25°C,
150 °C and 250 °C with electric field strength between 0 to
7 MV/cm. Whereas for oxide I the rise in N_{IT} is negligible small,
the hydrogenated oxide II already indicates that N_{IT} rises
for electric fields above 4 MV/cm at temperatures greater than
150 °C. The magnitude of the rise is similar for both types of
oxides and follows a t 1/4 law with the stress time interval [6].
Maybe due to ionic contaminations sometimes large differences
are observed from specimen to specimen.

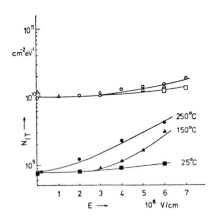

Fig.2 Rise in N_{IT} as a function
of the positive electric field
(upper curve: oxide type I,
lower curve: oxide type II)

3.2. Negative Bias Stress

N_{IT} of oxide I rises in the temperature range of 25 °C to 250 °C
exponentially with the electric field strength (Fig.3).

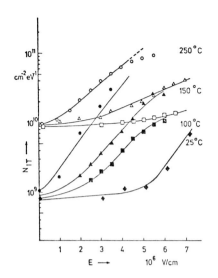

Fig.3 Rise in N_{IT} as a function of the negative electric field (upper curve: oxide type I, lower curve: oxide type II)

This process is thermally activated, at higher temperatures the rise shifts to lower electric fields. Oxide II behaves similarly but the influence of the electric field is greater in the range $N_{IT} = 1 \times 10^9$ to 2×10^{10} cm^{-2}eV^{-1} [7]. At higher N_{IT} densities this strong rise levels off to approach the smaller rise typical for oxide I.

According to JEPPSON and SVENSSON [1] N_{IT} increases with t 1/4 with the stress time interval. The more complicated rise in N_{IT} of oxide II is governed by two t 1/4 laws, the first dominating at short stress times up to $\approx$ 16 seconds and the second at longer times and corresponding approximately to the rise in N_{IT} of oxide I. Rising N_{IT} is accompanied by a rise in N_F which shifts the flat band voltage to negative values.

4. Discussion

According to the NBS model in [1], N_{IT} of oxide II (with a large number of SiH groups available) rises under NBS exponentially with the electric field strength, is thermally activated and follows a t 1/4 law. N_{IT} of oxide I (with less SiH groups available) behaves similarly, but with a smaller rate constant. For both types of oxides the rise in N_{IT} can be described by

$$N_{IT} = N_{ITO} + R_1 \cdot t\ 1/4 \tag{1}$$
$$R_1 = R_0 \cdot exp\ [-q/KT\ (\emptyset_0 - 1/2\ aE)] \tag{2}$$

but for the hydrogenated oxide the rise consists of two reactions (R_0 is a constant, a the distance at which Si-O-Si bonds are broken, $\emptyset_0$ is the activation energy of OH diffusion, t is the stress time interval in seconds, E the electric field strength in V/cm.

In Table 1 the four constants N_{ITO}, R_0 and $\emptyset_0$ and a, fitted to the experimental values, are given. Also positive bias stress can be described with (1). For all cases the activation energy is ≈ 0.3 eV. With these constants the approximate behaviour of N_{IT} under various stress conditions may be predicted.

Table 1

	Oxide I		Oxide II		+ Bias
	- Bias		-Bias		+ Bias
$\emptyset_0$ [eV]	0.3	0.3 [1]	0.28	0.3	0.28
a [nm]	0.3	0.32 [1]	0.6	0.3	0.3
R_0 [cm^{-2}eV^{-1}s$^{-1/4}$]	1×10^{12}		$1,5 \times 10^{11}$	2.5×10^{11}	4×10^{10}
N_{ITO} [cm^{-2}eV^{-1}]	7×10^9		5×10^8	1×10^{10}	5×10^8
			16 s		

It follows from Table 1 that the enhanced rate constant $R = f(E)$ of oxide II under NBS depends on different trap distances from the interface. In the N_{IT} range below 2×10^{10} cm^{-2}eV^{-1} hydrogen liberated from the interface reacts preferentially with a defect deeper in the oxide. At $N_{IT} \approx 2 \times 10^{10}$ cm^{-2}eV^{-1} the rise in N_{IT} of oxide II changes to the rise of oxide I, therefore it is concluded that the second trap is $\approx 2 \times 10^{10}$ cm^{-2}. One known trap in this range is the fixed oxide charge. With an assumed rate constant for the reaction with a charged defect

$$R^* = R_0 \exp\left[-q/KT(\emptyset - aE)\right]$$

the dependence of the electric field is twice as great and the trap distance is 0.3 nm for all N_{IT} reactions in Table 1. At higher N_{IT} densities the rise of the hydrogenated oxide thus will approach the normal oxide, where only the reaction with $N_{IT} > 2 \times 10^{10}$ cm^{-2}eV^{-1} yields can be detected.

The trap mechanism according to (1) is valid for E=0 to 6 MV/cm and T=25 to 250 °C. At higher electric fields a completely different behaviour is observed [8].

References

1 K.O. Jeppson, C.M. Svensson, J. Appl. Phys. 48, 2004 (1977)
2 F. Montillo, P. Balk, J. Electrochem. Soc. 118, 1463 (1971)
3 C.M. Svensson, K.O. Jeppson, The Physics of SiO$_2$ and ITS Interfaces, ed. S.T. Pantelides (New York, Pergamon 1978)
4 L. Risch, E. Pammer, K. Friedrich, Insulating Films on Semiconductors, Inst. Phys. Conf. Ser. No. 50, 114 (1979) The Institute of Physics Bristol and London
5 I. Weitzel, R. Primig, K. Kempter, Thin Solid Films, 75, 143 (1981)
6 W. Mohr, private communication
7 L. Risch, publication pending in Solid State Electronics
8 K.R. Hofmann, G. Dorda, this issue

On the Si/SiO$_2$ Interface Recombination Velocity

M.W. Hillen, J. Holsbrink, and J.F. Verwey

Department of Applied Physics, Groningen State University, Nijenborgh 18
9747 AG Groningen, The Netherlands

1. Introduction

The recombination of minority carriers at the Si/SiO$_2$ interface
plays an important role in modeling the base current of bipolar
devices. For instance in Integrated Injection Logic (I^2 L) gates,
up to 40% [1,2] of the base current of the upward-down operated
npn switching transistors is caused by this recombination and
severely affects the current gain.

The recombination process at the Si/SiO$_2$ interface takes place
via interface traps. A quantity used to characterize this
process is the surface recombination velocity S. It is mostly
defined as [3, 4]

$$S = U_s / \Delta n_s \qquad (1)$$

where Δn_s is the excess electron concentration at the Si/SiO$_2$
interface and U_s is the surface recombination rate. Under
certain conditions U_s is directly proportional to the electron
current I_{no} flowing into the Si/SiO$_2$ interface. However, this
quantity is a strong function of the surface potential. So,
the actual value of S will strongly depend on the incidentally
fluctuating amounts of oxide charge, etc. This is the reason
why we will focus our attention on S_0, the fundamental
recombination velocity. And we will use it for the characterization of the Si/SiO$_2$ interface with respect to recombination.

2. Test Structures

First we describe the test structures used to study the interface recombination in bipolar transistors and to determine S_0.
Fig. 1a shows schematically a transistor structure that can be
used to measure the interface recombination in the proper way.
In this structure the injection is from the bottom n-p junction
(I_{no}) in order to obtain a laterally homogenous concentration
of charge carriers at the Si/SiO$_2$ interface. Moreover, the injection mode is then similar to that in the switching transistors
in I^{2}L cells. In the measurement of the base current I_B, a
voltage V_{eb} is applied to the emitter-base junction of the
upward-down operated transistor while the surface potential of
the base surface is varied in a controllable way by the voltage
ramp Vg applied to a gate electrode. Actually we used a modification of this structure in which for ease of the analysis, the
base region was homogeneously doped (Fig. Ib). This was obtained
by using an epitaxial layer for the base region and V-groove
etching (followed by oxidation) for isolation of the base region.

The gate G and contacts to the emitter (E), base (B) and col-
lector (C) are indicated in the figure.

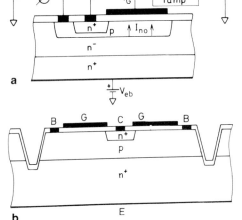

Fig. 1

(a) Cross-section of a n-p-n
transistor operated in the upward mode. The arrows indicate
the electron flow in the base
region perpendicular to the
gated Si/SiO_2 interface.

(b) Cross-section of gated
transistor with epitaxially
grown base region, as used in
the present research. The
surface and the emitter-base
space-charge layers (scl) are
indicated in the figure.

3. The surface recombination rate U_s

As already mentioned in the introduction, the quantity U_s is a
strong function of the surface potential φ_s. Calculated values
of U_s as a function of φ_s (or V_g) are shown in Fig. 2. It is
calculated with usual approximations like low level injection
(but $V_{eb} >> kT/q$),etc. Some of the parameters used in the calculation are indicated in the figure. The three curves are for
three values of V_{eb}. It may be seen that U_s is strongly peaked.

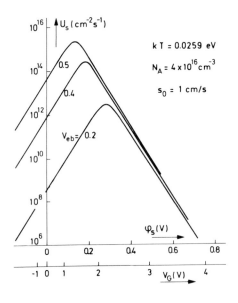

Fig. 2
Surface recombination rate as
a function of the surface
potential and the gate voltage
(data of the wafer JH6,
$V_{FB} = 0$).

We draw expecially the attention to the point where $V_g=0$ (and $\varphi_s=0$). This is the situation that exists approximately in transistors without a gate. One may notice that for increasing injection levels (increasing V_{eb}) the value of U_S strongly increases and nearly reaches its maximum. A small increase of φ_s by some positive charge in the oxide may actually bring U_S to its maximum. Possibly this explains the large scatter in the value of S reported in the literature (in the range $1-10^5$cm/s). The maximum $I_{no,m}$ in the base current as a function of V_g (see below) is proportional to the maximum in U_S.

4. Results on base current

Typical results for measurements of the base current I_B as a function of V_g are shown in Fig. 3. The base current is recorded from accumulation ($I_{B,acc}$) of the base surface to inversion ($I_{B,inv}$). The measurements were carried out in devices with different geometries. The minor differences between the two curves at the top of the figure are presumably due to the differences in geometry. A discussion of these differences is out of the scope of this paper. In the figure the current I_{no} is indicated that is stronly dependent on V_g. This current is ascribed to interface recombination. The value of I_{no} at its maximum $I_{no,m}$ is used to calculate S_o. For the calculation of this quantity we have

$$I_{no,m} = qA_g\, S_o n_i\; (\frac{qV_{eb}}{kT})\; \exp\; (\frac{qV_{eb}}{kT})\; . \qquad (2)$$

Where Ag is the gate area and n_i is the intrinsic carrier concentration. For S_o we found values ranging from 0.5 to 2 cm/s. A remarkable effect should be noticed that occurs at very low

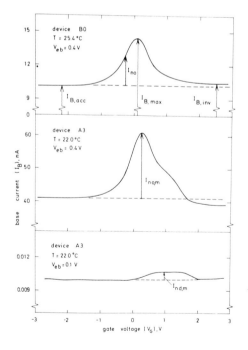

Fig. 3

Typical base current vs gate voltage characteristics, obained on wafer JH6.

injection (V_{eb}=0.1V). In the bottom curve of Fig. 3 one may observe that I_{no} is in a region of V_g independent of φ_s, i.e. a flattened peak $I_{nd,m}$ is observed. This means that the surface recombination is relatively high compared to the current in the neutral base region. In other words the current is diffusion limited. This gives the possibility to determine the base acceptor concentration. One may use {5}

$$I_{nd,m} = qA_g \frac{D_n n_i^2}{LN_A} \left\{ \exp \left(\frac{qV_{eb}}{kT} \right) - 1 \right\} \qquad (3)$$

where D_n is the diffusion coefficient of the electrons in the base region, L its width, and N_A the acceptor concentration. This flattened peak for low injection levels was also oberserved for double-diffused bipolar transistors. Then, for the analysis, the quantity LN_A in eq.(3) must be replaced by the Gummel number.

5. Separate determination of S_0

The expression we use for S_0 is

$$S_0 = \sqrt{\sigma_n \sigma_p} \quad \upsilon_{th} \, kTD_{it} \quad . \qquad (4)$$

Where σ_n and σ_p are the capture cross-section for electrons and holes, respectively, υ_{th} is the thermal velocity of the carriers, and D_{it} the interface state density. It should be noted that S_0 was introduced by Grove and Fitzgerald {6} as the surface recombination velocity of a depleted surface. The difference between their and our definition of s_0 is a factor $\pi/2$.
It can be seen in eq. (4) that a separate determination of D_{it} and the capture cross-sections will lead to an independent value of S_0. These quantities were meas ed on the same chip because we had a low-doped epitaxial base region. Results on D_{it} for wafer JH 6 are given in Fig. 4. It shows the interface state density as a function of energy in the bandgap. For the recom-

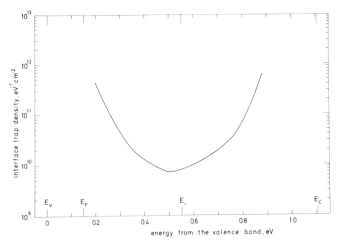

Fig. 4 Interface trap density as determined from a quasi-static C-V curve at V_{eb}= 0 V.

bination process the value at mid-gap is most important. There-
fore, we estimate from Fig. 4 that $D_{it}\approx10^{10}cm^{-2}eV^{-1}$. This value
was confirmed by conductance measurements ($D_{it}=5.8\times10^9cm^{-2}eV^{-1}$).
With the aid of the conductance technique we determined the
capture cross-section of holes at mid-gap and found $\sigma_p=3.1\times10^{-15}$
cm^2. We did not yet succeed in determining capture cross-sections
at other energies. This is probably caused by the fact that
conductance measurements on p-type silicon are hard to perform
because of the presence of surface potential fluctuations. In
literature a vast range of σ_p values has been reported, from which
no definite conclusion can be drawn. Assuming equal values of
σ_n and σ_p, we compute for our devices $S_0=7.5cm/s$, which is larger
than the values found in the base current experiments. Usually
the asumption $\sigma_n\approx\sigma_p$ is made at the start of the theoretical
treatment. However, for the above mentioned energy region, no
values of σ_n in p-type material have been reported. We too did
not succeed in determining σ_n on our wafers. Recently, values of
$\sigma_n\approx10^{-16}cm^2$ bave been measured on p-type silicon in the upper
part of the band-gap {7,8}. Assuming these values also to be
valid in the lower part of the band-gap S_0 can be computed to be
1.4cm/s, which is in much better agreement with the values found
previously.

6. Conclusions
The base current component recombining at the oxide covered base
surface can be used to characterize the Si/SiO_2 interface.
The only suited parameter for this characterization is the
fundamental recombination velocity $S_0 = \sqrt{\sigma_n\sigma_p}\ \upsilon_{th}\ kTD_{it}$.
Measurements of S_0 have been carried out, using special test
transistors (epitaxially grown, low-doped base and a gate on
top of the base oxide). Typical values of S_0 in the range 0.5-
2cm/s have been obtained. Independent measurements (QSCV and
conductance) have been carried out to find σ_p and D_{it}. Using
values for σ_n and σ_p of 10^{-16} and $3\times10^{-15}cm^2$ respectively, a
good correspondence to the base current measurements is
obtained.

References

1 H.E.J. Wulms, IEEE J.Solid-St.Circ. SC-12, 143 (1977)
2 L. Halbo and J. Haraldsen, 9th Europ.Solid-St.Device Research
   Conf., Munich, Sept.10-14, 1979
3 D.J.Fitzgerald and A.S. Grove, Surf.Sci. 9, 347 (1968)
4 M.V. Whelan, thesis, tech.univ.of Eindhoven, 1970
5 C.A. Grimbergen, thesis, univ. of Groningen, 1977
6 A.S. Grove and D.J. Fitzgerald, Solid-St.Electron, 9,783(1963)
7 R.J.Kriegler, T.F. Devenyi, K.D. Chik, and J. Shappir,
   J.Appl.Phys. 50, 398(1979)
8 T.C.Poon and H.C. Card, J.Appl.Phys. 51, 5880(1980)

Optical Excitation of MOS-Interface-States

K. Blumenstock and M. Schulz

Institut für Angewandte Physik, Universität Erlangen-Nürnberg, Glückstraße 9
D-8520 Erlangen, Fed. Rep. of Germany

1. Introduction

The general properties of interface states in n-type MOS-capacitors as analysed by electrical techniques are frequently reviewed /1,2/ . A continuum of states is usually observed in the silicon band gap with an increase in density towards the conduction band edge. The electron capture cross-section shows a strong decay near the conduction band.

The properties of MOS-interface-states are explained by a new, the variable-energy (VE) model /2,3/ . The VE-model assumes a discrete trap level in SiO_2, whose energy in the interface region is dependent on the distance from the interface. This variation of the trap energy leads to a continuum of interface states observed in electrical measurements. The decrease of the electron capture cross-section for shallow interface states is explained by an increasing tunneling depth.

In this paper, we test the VE-model by an experiment which excites electrons in shallow interface states by monochromatic infrared light. For states near the conduction band, a decrease of the absorption cross-section is expected if a tunneling process is involved to transfer the electron from deep in the SiO_2 to the silicon.

2. Measurement Principle

The measurement principle is explained in figure 1.

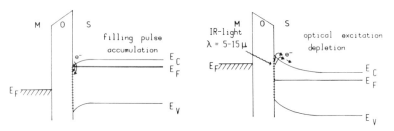

Fig.1 Sketch of the band structure of the MOS-capacitor during the filling pulse and the photoexcitation.

In order to fill interface states with electrons, the MOS-capacitor is biased into accumulation by a filling pulse. For the optical excitation of trapped electrons, the capacitor is biased into depletion. Monochromatic infrared light in the wavelength region $\lambda=5\mu m$ to $\lambda=15\mu m$ is used to excite only shallow states at energies less than $E_C-E_T=0.24eV$ below the conduction band edge. The photoexcited charge carriers are removed from the interface by the electric field in the space charge layer.

The capacitance transient induced by the charge transfer is sensed by a capacitance bridge. In the experiment, we keep the capacitance and thus the width of the space charge layer constant by a feedback to the bias voltage. The transient interface charge is then monitored by the transient observed in the bias voltage.

The transient of the photoexcitation is dependent on the photon flux and the absorption cross-section of the interface traps. In order to reduce thermal emission of electrons from interface states, the sample has to be kept at temperatures as low as possible. The lowest temperature at which capacitance measurements are still possible was determined to $T=45K$.

The time constants τ for thermal emission are shown in figure 2 for different sample temperatures as a function of the energy position of the interface trap.

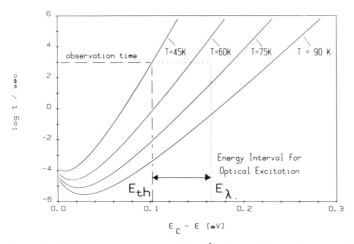

Fig.2 Time constant variation for thermal emission. Parameter is the sample temperature.

The time constants are calculated. A typical electron capture cross-section with a decay near the conduction band edge was assumed.

Interface states of an energy less than $E_C-E_T=0.1eV$ empty out within 1000 sec. after the filling pulse by thermal emission at the sample temperature $T=45K$. For infrared light with energy E_λ, an energy interval between E_{th} and E_λ is available for photoexcitation. The lower energy limit E_{th} can be varied by the sample temperature, and the upper energy limit E_λ by the wavelength of the IR light used.

The absorption cross-section σ_n^0 in the energy interval $E_{th}...E_\lambda$ depends on the bias transient $\Delta V(t)$ by equation (1).

$$\Delta V(t) = \frac{A \cdot q}{C_{ox}} \cdot \int_{E_{th}}^{E_\lambda} \{N_{IT}(E) \cdot (1 - \exp\{-\sigma_n^0 \cdot \phi_\lambda \cdot t\})\} \, dE \, , \qquad (1)$$

$N_{IT}(E)$ is the interface state density, ϕ_λ the photon flux at the interface, A the area of the gate contact and C_{ox} the oxide capacitance.

3. Experimental

The experiments are performed on samples with n-type Si sub-strate. The silicon material is 4-6Ωcm phosphorous doped, and 100 oriented. An oxide layer of approx. 0.1µm is grown by thermal oxidation. Infrared transparent gate contacts are obtained by a polysilicon layer, 0.5µm thick, and nominally un-doped. A helium flow cryostat is used to cool the sample to a variable temperature in the range 4K-300K. A cross-section of the cryostat is shown in figure 3.

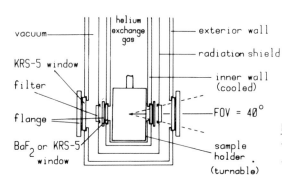

Fig.3 Cross-section of the sample chamber in the He flow cryostat.

The sample is mounted in the inner chamber which is filled with dry He exchange gas and can be illuminated through two IR-trans-parent windows (BaF$_2$ or KRS-5 windows) in the inner and outer wall with a field of view of 40°.
The wavelength selection is achieved by four different inter-ference filters which are mounted in the cold radiation shield. Four wavelength intervals are therefore separately available when the sample holder is rotated. The half width passbands of the filters are 10% of the center wavelength.

Infrared light source is the 300K background radiation and a globar. The relative emission spectrum of the globar was deter-mined by an infrared spectrometer. The value of the total photon flux reaching the Si-SiO$_2$ interface is obtained by using the

sample itself as a detector and the 300K background radiation
as a calibrated light source. The photon flux from the thermal
background (FOV=40°) is calculated by the data of the filters,and
the transmissivities of the windows and the polysilicon- and
SiO_2-layer /4/.

The photon flux at the interface is in the order of $10^{15} cm^{-2} sec^{-1}$.
The irradiation on the sample can be interrupted by a cold
shutter which is mounted in the sample holder.

4. Results

Typical transients of the photoexcitation of MOS-interface-
 states are shown in figure 4. The feedback voltage is plotted
as a function of time; the calibration of the scale to the
density of states is given in the insert.

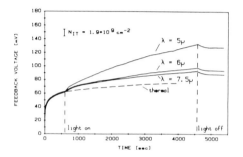
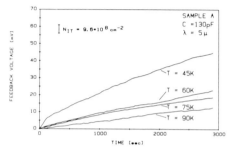

Fig 4a. Typical transient of   Fig 4b. Typical transient of
the photoexcitation. The   the photoexcitation. The para-
parameter is the wavelength.   meter is the sample temperature

In fig.4a, the irradiation starts 600 sec. after the filling
pulse. The amplitude of the photoexcitation signal decreases
with increasing wavelength, because the energy interval $E_{th}..E_\lambda$
and thus the number of states available for photoexcitation is
reduced. The dashed line is the dark transient due to pure
thermal emission. When the irradiation is stopped by closing
the cold shutter, a constant level is reached after a short
transient.

The number of states available for photoexcitation can also be
varied by increasing the sample temperature. The photoexcitation
transients with the thermal emission substracted are shown in
fig.4b at constant wavelength but at different sample temperatures. Because of the increased thermal emission at high sample
temperatures, the energy interval for photoexcitation is reduced.

The measured transients as shown in fig.4 were fitted by equ.(1)
to determine the absorption cross-section and the interface
state density. The photoexcitation measurement has been performed
for two samples having different interface state densities.

The general results are :

For a sample having a low density of states ($\sim 10^{10} cm^{-2} eV^{-1}$)
- The optically determined interface state density is consistent with CC-DLTS and CV results.
- The absorption cross-section for interface states ~ 0.13 eV below the conduction band edge is in the order of 10^{-19} cm^2
- The absorption cross-section decreases with increasing wavelength.

For a sample having a high density of states ($\sim 10^{12} cm^{-2} eV^{-1}$) shows a complicated behaviour. The photoexcitation signal is dependent on the bias voltage. The optically determined interface state density is inconsistent with CC-DLTS results. These effects are not yet understood. The absorption cross-section determined in the energy interval 0.1eV to 0.24eV, however, is of the same low order of magnitude, below 10^{-18} cm^2.

In fig.5, the absorption cross-section determined for the sample having a low interface state density is shown together with results taken from Greve et al./5/ . The absorption cross-section is plotted as a function of the excitation energy above the conduction band edge $E_\lambda - E_T$.

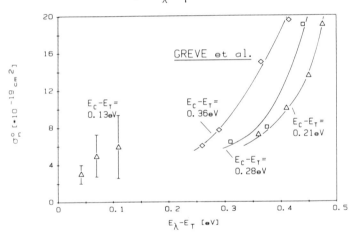

Fig.5 Comparison of the absorption cross-section with results from Greve et al. /5/.

For a constant $E_\lambda - E_T$, the absorption cross-section decreases with decreasing energy depth $E_C - E_T$ of the state.
For a fixed trap depth $E_C - E_T$, the absorption cross-section decreases with decreasing energy of the infrared light.
Although taken for a different sample, our results are generally consistent with the observations by Greve and extend the measurement range to low excitation energies.

5. Conclusions

By infrared excitation of MOS-interface-states, we are able to determine the absorption cross-section and the density of

states. The properties of the absorption cross-section qualitatively support the VE-model for interface states. For a
quantitative comparison , better understanding of the excitation process is necessary.

Acknowledgment

Financial support of the ERO London is gratefully acknowledged.
The authors wish to express their appreciation to G. Sixt at
AEG-Telefunken for the assistance with the sample preparation.

References

1. A.Goetzberger, E.Klausmann, M.Schulz : Critical Review in
    Solid State Sciences, Vol.6, 1 (1976)

2. M.Schulz : Inst. Phys. Conf. Ser., Vol.50, 87, (1979)

3. M.Schulz, K.Blumenstock, E.Klausmann : Proc. of the fourth
    Int. Symposium on Silicon Materials,Science and Technology,
    to be published

4. K.Schroder, R.N.Thomas, J.C.Schwartz : IEEE Trans. on
    Electr. Dev., Vol.ED 25, 254 (1978)

5. D.W.Greve, W.E.Dahlke : Inst. Phys. Conf. Ser., Vol.50,
    107 (1979)

Part II

Thin Insulating Films

Langmuir-Blodgett Films on Semiconductors

G.G. Roberts

Department of Applied Physics and Electronics, University of Durham
Durham, England

1. Historical Introduction

Over two hundred years ago BENJAMIN FRANKLIN described to the
Royal Society the results of some experiments he had carried out
on the spreading of oil films on the pond at Clapham Common in
London. He reported that
  "...the oil, though not more than a teaspoonful, produced
  an instant calm over a space several yards square, which
  spread amazingly, and extended itself until it reached
  the lee side, making all the quarter of the pond, perhaps
  half an acre, as smooth as a looking glass".

It was many years before any methods for the control and
manipulation of these films were proposed; FRAULEIN POCKELS {1}
in 1891 is credited with first using the form of trough subsequently developed by Langmuir. At the turn of the century,
LORD RAYLEIGH {2} began to consider the structure of these films
and concluded they were only one molecule thick. However,
IRVING LANGMUIR is accepted as the pioneer of scientific research
in the field. While working in the General Electric Company
Laboratories shortly after the first World War, he developed the
theoretical and experimental concepts which form the basis of
our current understanding of the behaviour of molecules in
insoluble monolayers. Later, with KATHLEEN BLODGETT, he devised
a process to control and transfer these monomolecular layers
onto solid substrates.

At one time the Langmuir trough technique was used principally
as a method of determining the molecular weights of large
molecules. However, Langmuir-Blodgett (LB) films are now the
subject of broad and intense study in both the physical and
biological sciences. The rekindling of interest in the field is
due in part to HANS KUHN {3} who published some classic papers
about ten years ago; his elegant work showed how one could
utilize monomolecular layers to assemble precise supermolecular
structures for optical and energy transfer investigations. It
is only during the last four or five years that the extensive
scientific and technological possibilities have been appreciated
{4}. This stage in the development of LB films has required an
improvement in the design of the deposition system {5} and
the synthesis of novel materials {6,7} with the required
structure to form oriented monolayers on a substrate. Progress
has only been achieved by strong collaboration between solid
state physicists and innovative synthetic chemists.

Films of great perfection and controlled thickness can be
assembled using the Langmuir trough. The technique itself is
described in the next two sections while the rest of the
article focusses on the properties of junctions prepared by
depositing LB films on the semiconductor substrates.

2. The Langmuir Trough-Technique

Langmuir-Blodgett films are prepared by depositing a small
quantity of a solution of a suitable organic material onto a
liquid surface and waiting for the solvent to evaporate; the
floating molecules may then be compressed until a quasi-solid
one molecule thick is formed. To assemble multilayer structures,
a suitably prepared substrate is dipped and raised through the
surface of the subphase (usually highly purified water); the
thickness of the film then depends on the number of monolayers
deposited and the molecular size of the material used.

The container which holds the liquid subphase upon which the
monolayer floats is termed a Langmuir trough. In the sophisticated
system at the University of Durham the working area is defined
by a constant perimeter plastic coated glass fibre barrier and
the working area is controlled by a low geared motor. The
compression system is coupled to a microbalance which
continually monitors the surface pressure via a sensor in the
liquid surface. This differential feedback arrangement carefully
maintains the monolayers at a desired surface pressure, a
feature which is critically important when dipping is in progress and molecules are thus being removed from the subphase.
Figure 1 shows a schematic diagram of a monolayer under study in
a Langmuir trough; the inset shows a sketch of the constant
perimeter barrier. The automated Durham troughs have been carefully engineered to avoid vibration effects and are housed in
a Class 1000 microelectronics clean room. Their dimensions

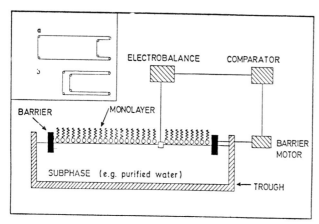

Fig. 1: Schematic diagram of Langmuir trough; the monolayer molecules
have their polar heads (circles) immersed while their non-polar
hydrocarbon chains (zig-zag lines) stick out of the subphase.
Also shown is the constant perimenter PTFE tape defining (a)
maximum area (b) minimum area.

are determined by the size of the substrates to be coated but
are typically 0.5 to 1.0m in length and 0.3m wide.

In order to investigate the structure of the monolayer and
establish the correct dipping conditions, isotherms must first
be recorded. The temperature and pH of the subphase are important
variables which must be optimized for each molecule studied. The
surface pressure where dipping occurs also affects the quality
of the film produced. If conditions are carefully controlled
and if appropriate molecules are used, one monolayer is transferred on each excursion of a substrate through the water surface.
A motor powered micrometer screw system is used to carry out
this delicate operation. For normal deposition conditions the
two molecules of successive layers are oriented in opposite
directions so that the molecules stack in a polar-head to polar-
head and non-polar-chain to non-polar-chain arrangement. Sketches
of other possible configurations are given in the book by
GAINES {8}.

3. Molecular Engineering

Various organic materials have been deposited using the Langmuir
trough technique but most data have been obtained for fatty
acids and their salts. This is because these molecules possess
hydrophilic and hydrophobic terminations which ensure that they
float and orient naturally on a water surface. A representative
molecule of this type is stearic acid ($C_{17}H_{35}COOH$) whose
molecular size is approximately 2.5nm. Although it is a good
insulator, this material and related dielectrics are relatively
unstable both thermally and structurally and are unlikely to
form the basis of a practical device. Recent activity {9} has
therefore concentrated upon producing an improved material

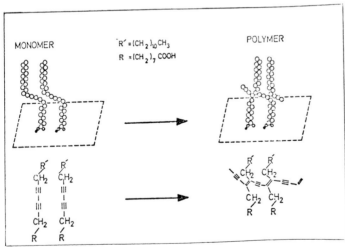

Fig. 2: Schematic diagram showing the relationship between the diacetylene
fatty acid monolayer (left) and the corresponding polymer produced
on UV irradiation (right). In the case shown the diacetylene polymer
is derived from penta-cosa-10, 12-diynoic acid.

without sacrificing any of the advantages of this long chain open structure. One particularly promising material is di-acetylene, a u.v. polymerizable organic compound to which various long chain hydrophobic and hydrophilic groups can be attached. An example of diacetylene suitably substituted for Langmuir trough deposition is given in Fig. 2. After coating the required number of monomer monolayers the film can be cross-polymerized by exposure to an intense 300nm ultra-violet source - this can be achieved without losing the integrity of the film which is then stable up to 230°C. An even more stable material is phthalocyanine; this highly polarizable molecule with its extensive delocalized electron system can be cycled to temper-atures in excess of 400°C, a high enough value for us to entertain using conventional lithographic techniques to define required electrode patterns. Good film adhesion to a substrate is obviously of extreme importance in a reliable device structure. The phthalocyanine molecule excels in this regard; it appears to be able to distort its polarization in such a way as to adhere well to virtually any substrate.

In the early stages of our research it was recognized that the scope of the Langmuir trough technique would be considerably widened if aromatic materials with only very short aliphatic chains could be formed into high quality LB multilayers. The motivation for the work was the realization that conjugated systems with their extended π-electron orbitals might show more useful electrical properties than conventional long chain aliphatic materials such as stearic acid. Scientists at Imperial Chemical Industries Ltd. synthesized a range of lightly substituted anthracene derivatives with side chains as short as four CH_2 units {6}. Because of the layered structure of these films their conduction properties turn out to be very different to those of crystalline anthracene; for example, the in-plane conductivity exceeds that in a direction perpendicular to the layers by a factor of over a million {10}. In the example shown in the inset to Fig. 7 the hydrocarbon substitution is C_4H_9 and the material is therefore termed C 4 anthracene. This substituted organic material of molecular size 1.2 nm has been used effectively as a tunnelling spacer in some devices and also displays interesting optical properties {11}.

The LB technique provides a means of building up organized super-molecular functional units which would be difficult or impos-sible to achieve by other means. This capability is likely to be capitalized upon in future investigations of high temperature superconductors, organic semiconductors and conducting polymers, Esaki—type multilayer structures etc. {4}.

4. Preparation of Field Effect Devices

To obtain LB layers that are well ordered and structurally stable requires scrupulous cleanliness and meticulous attention to experimental detail. Structural analyses using X-ray and electron diffraction confirm that carefully prepared films do assemble as predicted classically. Some appreciation of the quality and reproducibility of multilayers can also be gained by studying the capacitance of either MIM or MIS

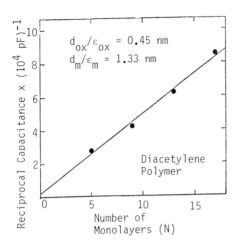

Fig. 3
Reciprocal capacitance at
1 kHz versus number of
monolayers N for LB films
of the polymer shown in
Fig. 2.

structures as a function of the number of deposited layers.
Data obtained at a frequency of 10 kHz are presented in Fig.3
for a diacetylene polymer deposited on to an aluminium coated
glass slide with an evaporated gold top contact. The linear
dependence of the reciprocal capacitance versus number of
monolayers N demonstrates clearly the reproducibility of the
monolayer capacitance and hence dielectric thickness from one
monolayer to the next. The slope of this straight line plot
gives the dielectric thickness of each monolayer while the
intercept yields similar information about the interfacial
oxide layer. Direct current measurements (not shown) confirm
that the films are highly resistive; Poole-Frenkel conduction
is observed through the multilayer sandwich structure. These
substituted diacetylene films have breakdown strengths
$\sim 2 \times 10^6 Vcm^{-1}$ and can suitably be incorporated in field effect
devices. The capacitance shows virtually no dispersion
throughout the frequency range studied. Another feature is
the small degree of hysteresis observed when they are placed
in MIS structures. Previous efforts using polymers deposited
in conventional ways have always been plagued by this problem.

Experience has shown that when an energetic process such as
sputtering is used to deposit an insulating layer on a
semiconductor, a surface damaged layer is produced which in-
variably dominates the electrical characteristics of the
junction so formed. The Langmuir trough technique, being a
low temperature deposition process provides an opportunity
to circumvent this particular difficulty. It does mean,
however, that substrate preparation prior to dipping is of
paramount importance in determining the quality of the interface
produced. That is, the thin natural 'oxide' layer formed during
the etching procedure is likely to remain undisturbed and thus
plays a significant role even after it has been coated with
a LB monolayer. For this reason it is important to first carry
out a systematic study of the surface chemistry of the substrates.
We have recently reported the results of an ESCA survey of the
effectiveness of several types of etchant on InP and have
correlated these with subsequent device performance {12, 13}.

The etchants employed could be categorized into two groups.
Those that produced de-oxidized surfaces (etchants such as
HF and all inorganic/peroxide mixtures) led to MIS structures
that exhibited large clockwise hysteresis but devices prepared
with oxidized surfaces (using etchants such as aqueous bromine
or bromine-methanol) displayed far less hysteresis. The latter,
however, were more prone to show electron injection effects
in forward bias. By using a suitable combination of the two
types of etchant, optimum characteristics could be achieved.

5. Langmuir-Blodgett MIS Structures

During the past few years we have examined both MIS and transistor structures based on several different insulator-semiconductor junctions. A selection of typical results is now presented.

5.1 Silicon

The native oxide layer on Si, whose excellent insulating
properties makes this semiconductor such a good material for
field effect devices, is not entirely dissimilar to the
commonly used aluminium oxide as a surface on to which to
deposit LB films. In fact, all the early MIS work involving
monomolecular layers utilized Si as the substrate. TANGUY {14}
reported capacitance data as a function of bias voltage for
junctions prepared using orthophenanthroline and LUNDSTROM {15}
described similar experiments using fatty acid salts. The latter
subsequently extended his charge injection studies to thin films
of chlorophyll on silicon {16} .

Figure 4 shows a set of admittance data for p-type Si coated with
cadmium stearate. Similar curves for other thicknesses of LB
film confirm that accumulation and inversion values are obtained

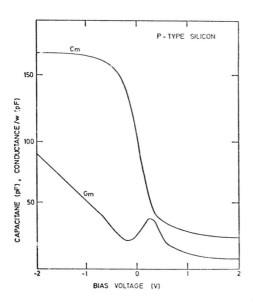

Fig.4 Measured values of
capacitance and conductance
at 10kHz for a p-type Si-
$CdSt_2$ MIS structure. Insulator thickness = 31 monolayers
(80nm); electrode area =
$5.5 \times 10^{-7} m^2$

in forward and reverse bias, respectively. In this case the interface is essentially governed by the thin natural oxide layer and the orangic film provides the required insulation. HICKMOTT {17} has recently reported that the presence of charge within an insulator or dipoles at its surface can be established by measuring the flatband voltage as a function of insulator thickness. This experiment has been carried out for several LB film thicknesses deposited on the same substrate; The V_{FB} versus thickness plot is linear as predicted by theory but interestingly it possesses the opposite slope to that normally obtained for MOS devices on Si; the positive slope we find is indicative of negative charge in the insulator a result reinforced by the fact that the admittance characteristics display structure slightly to the right of the zero volts axis in Fig. 4. LB films do not generally adhere as well on to Si as they do on to other slightly oxidized substrates; poor stacking of the multilayers can then occur resulting in a higher than normal defect induced conductivity. This accounts for the non-ideal conductance characteristic.

The properties of hydrogenated amorphous Si have been discussed extensively in the literature. This material is thought to be useful in Schottky barrier solar cells and field effect transistors. LB films have a large area capability and might therefore be used as passivating layers in these applications. Preliminary measurements made in Durham have looked most promising. Results obtained for devices based on glow discharge produced n-type αSi and diacetylene polymer show good accumulation and depletion characteristics not too dissimilar from their crystalline counterparts.

5.2 Indium Phosphide

The success of integrated circuit devices based on Si is due largely to the impressive qualities of its native oxide. The group III-V semiconductors do not possess this inherent advantage and therefore a great deal of effort has been expended in finding a suitable technology for the formation of fully compatible insulating films on these materials {18}. Current theories and experimental evidence suggest that Fermi level pinning occurs at the surfaces of III-V compounds such as GaAs in InP. This arises due to extrinsic surface states localized at surface imperfections and stoichiometric defects; their energy levels are thought to be independent of the characteristics of the adatoms so that if an energetic process is used to deposit the insulator on the semiconductor, the natural oxide might be expected to play only a minor role. It is instructive therefore to compare admittance data for MIS structures on III-V compounds prepard using sputtered oxides and insulators deposited using the low temperature Langmuir method. In the latter case, as has been discussed in the previous section, the etchant used to prepare the semiconductor substrate is very important.

The published work on p-type InP passivated with oxides shows that the semiconductor surface is already inverted even at zero gate bias and there are encouraging reports of n-channel

inversion-mode devices based on this material. The contrasting
admittance data for the same semiconductor coated with
diacetylene polymer films is shown in Fig. 5. The G/V data are
distincitive in that they exhibit structure dependent on the
surface preparation. The corresponding C-V curve is not plotted
but shows both accumulation and depletion regions, features
which contrast markedly with reports of inversion behaviour
and failure to accumulate using conventional oxides. The inset
in Fig. 5. confirms the linear relationship between C^{-2} and V
expected in a depletion situation.

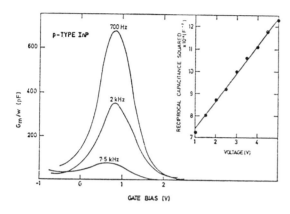

Fig. 5 Measured conductance versus voltage for p-type InP coated
with 25 monolayers of diacetylene polymer; the inset
shows the measured capacitance at 70 kHz.

The above example highlights the differences in the characteristics of p-type InP MIS devices prepared with insulators
deposited using low and high temperature methods. Similar
conclusions can be reached if one examines results for n-type
InP. Comprehensive data for LB films on this material have
already been published {19}. They illustrate that a relatively
clean interface is present between the LB film and the semiconductor substrate; moreover, the surface state density is
sufficiently   low for field effect transistor devices to be
fabricated using our organic layers.

5.3 Cadmium Telluride

It has also been possible to deposit LB films successfully
on to group II-VI semiconductor compounds including CdS, ZnS
and CdTe. Double dielectric structures are sometimes involved
but in this case it is possible to avoid the natural 'oxide'
layer by using a suitable etchant. Data are presented in Fig. 6
for p-type CdTe coated with 31 monolayers of cadmium stearate.
The classical results shown cannot be obtained using conventional
insulators. LB films adhere equally well to Cd:Hg:Te and work
is in progress in Durham to passivate this important alloy for
infra-red detector applications.

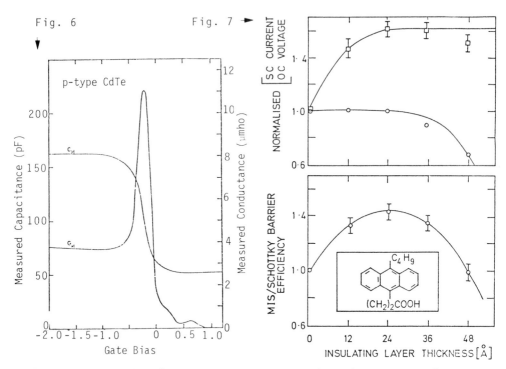

Fig. 6   Fig. 7 ➤

Fig.6   Measured values of capacitance and conductance at 100 kHz for a p-type CdTe-CdSt$_2$ MIS structure Insulator thickness = 80nm, electrode area = 5.5x10^{-7}m^2

Fig. 7 The upper diagram shows the open circuit voltage (upper curve) and short circuit current (lower curve) data for an n-type CdTe single crystal coated with LB films of C4 anthracene. The molecular size of this molecule is 1.2nm and therefore successive points refer to 1,2,3 and 4 monolayer thicknesses. The lower diagram shows the ratio of the efficiencies of the MIS and simple Schottky barrier devices allowing for fill factor changes. The inset shows the C4 anthracene molecule.

There are several reasons for introducing an ultra-thin film of tunnelling dimensions between a metal and a semiconductor. For example, the conversion efficiency of a photovoltaic Schottky barrier device can be enhanced in this way {20}. The presence of an interfacial layer can also enhance the tunnel injection of minority carriers into a semiconductor resulting in increased recombination electroluminescence. In both cases, theoretical studies emphasize the important influence of surface states and the critical control of insulator thickness required to optimize the effect. The Langmuir trough technique is ideally suited to this problem. Figure 7 illustrates the effect of incorporating one, two, three and four monolayers of C4 anthracene (molecular structure shown in inset) between CdTe and a metal electrode. The upper diagram displays the open circuit voltage of the barrier and

the short circuit current through the device as a function of
thickness. The lower curve shows that the solar cell efficiency
can be increased by approximately 50% if a 2.5nm layer is
coated on to the semiconductor (this optimized value takes
into account the relevant changes in the fill factor of the
cell).

LB films can be cooled to liquid helium temperatures without
degrading their quality and therefore have been used for
conventional tunnelling measurements between different metals.
The ability of single monolayers to withstand high electric
fields is likely to be capitalized upon in applications such
as SQUIDS and inelastic tunnelling spectroscopy.

6. Special Transducers

This article has described how LB films can be used to good
purpose to passivate the surfaces of several types of semiconductor. The degree of control afforded by the technique is
likely to be important. However, some practical applications
may also capitalize on the fact that the films are organic
and therefore normally respond more positively than inorganic
materials such as SiO_2 to external stimuli such as pressure,
temperature and ambient gas changes. For example, we have
examined the potential of LB field effect devices as microelectronic sensors and have observed definite changes in the
surface state spectrum of devices exposed to several gases
including ammonia, carbon monoxide and hydrogen. That is, gases
can be made to penetrate to the interface between the organic
insulator and the semiconductor and have a selective effect
which is evaluated with microprocessor based instrumentation
{21}. One can also foresee pyro-FETs or piezo-FETs based on LB
molecules with the required symmetry.

Biophysicists and biochemists have long been aware of the usefulness of the Langmuir trough to form several species into an
accurately defined supermolecular structure{22}. Lipid molecules
with their fatty and polar regions are well suited to the technique and have therefore received widespread attention.
Langmuir trough developments and related methods have enabled
more representative bilayer structures of lipids found in vivo
to be assembled. The incorporation of proteins and other essential
cell membrane constituents into lipid layers is essential in the
modelling of biological systems. This type of research also has
implications in the transducer field {23}. That is, by assembling
biological molecules such as antibodies and enzymes within
insulating LB films, it is possible to propose field effect
devices for monitoring immunological response (IMFETs), etc.
The successful incorporation of LB films of specially substituted
molecules into conventional MIS and FET structures augers well
for the longer term development of these more complex structures.

7. Other Applications of LB Films on Semiconductors

There are clearly many variants of some of the devices already
mentioned e.g. the equivalent of MNOS structures, floating gate
thin film transistors and a.c. electroluminescent devices.
However, this section is reserved for two quite distinct
applications.

7.1 Integrated Optics

The Langmuir trough technique is attractive in thin film guided
wave optics applications because the thickness and refractive
index of the film can be selected with great precision; the
molecular length is known to a fraction of an angstrom, better
than that required for any conceivable optical device. PITT
and WALPITA {24} and COLOMBINI and YIP {25} have demonstrated
waveguiding effects in fatty acid LB films; the average
attenuation losses in these films is relatively low ($5dBcm^{-1}$).
A particularly effective use of multi-layers in integrated
optics may be as post-fabrication claddings on conventional
waveguides to enable fine-tuning to be achieved. The work also
has immediate application to several device concepts such as
planar couplers and acoustic surface wave devices which can also
capitalize on the fine control of refractive index. Dyes can be
incorporated into LB films and this has led to the suggestion
of planar laser structures using suitable monolayers as the
inert matrix.

7.2 Electron Beam Lithography

LB films can also assist in the fabrication of conventional
elctronic devices based on Si. BARRAUD et al {26} have reported
improved electron beam polymerizable lithographic resists based
on multilayers of ω-tricosenoic acid which is a C24 fatty acid
with one double bond C = C linkage at its end. Films of this
material, approximately 50nm thick, have all the desired
characteristics of a resist i.e. they are compact and pin-hole
free, they have a constant and controlled thickness, are highly
resistant to chemical reagents after electron beam exposure,
are tenaceously adherent to the substrate and lead to high
resolution.

8. Conclusions

This review has touched on only those applications of LB layers
that involve a semiconductor substrate, and represents therefore only a narrow selection of the numerous situations where
it may be possible to capitalize on the close control over complex
multilayer structures that is available with the Langmuir trough
approach. Much more research work is required into the
synthesis of suitable materials and the deposition technique
itself before any firm conclusions can be reached concerning
its commercial attractiveness. The research described here
should only be regarded as a first step towards more ambitious
endeavours which will require the cooperation of physicists,
biologists, electronic engineers and chemists. The progress
that has already been achieved does, however, firmly establish
a role for Langmuir-Blodgett films in thin film technology.

Acknowledgements

My thanks are due to Dr.M.C.Petty who helps supervise much of
the Langmuir trough work at the University of Durham. Some of
the data presented here are the work of our graduate students,
P.J.Martin, R.W. Sykes, K.K.Kan, J.P.Lloyd and I.M.Dharmadasa.
The valuable assistance of colleagues at the I.C.I. Corporate
Laboratory is also gratefully acknowledged.

References
1. A. Pockels, Nature 43, 437 (1891).
2. Lord Rayleigh, Phil Mag. 48, 321 (1899).
3. H.Kuhn and D.Möbius, Angew.Chem.Int.Ed.Engl.10,620(1971).
4. P.S.Vincett and G.G.Roberts, Thin Solid Films 68,135(1980).
5. G.G.Roberts, P.S. Vincett and W.A.Barlow,Phys.Technol.12
            69(1981).
6. P.S.Vincett, W.A.Barlow,F.T.Boyle, J.A.Finney and G.G.Roberts,
            Thin Solid Films, 60, 265(1979).
7. J.P.Fouassier, B.Tieke and G. Wegner, Israel J.Chem. 18,
            227(1979).
8. G.L.Gaines, Insoluble Monolayers at Liquid-Gas Interfaces
            (Interscience, New York,1966).
9. K.K.Kan, M.C.Petty and G.G. Roberts, Proc.Raleigh Conf.
            on the Physics of MOS Insulators (Pergamon,
            New York, 1980) p. 344.
10. G.G.Roberts, T.M.McGinnity, W.A.Barlow and P.S. Vincett,
            Thin Solid Films, 68, 223(1980).
11. G.G.Roberts, T.M.McGinnity, W.A.Barlow and P.S.Vincett,
            Solid St.Comm. 32, 683(1979).
12. D.T.Clark, T.Fok, G.G.Roberts and R.W.Sykes, Thin Solid
            Films, 70, 261(1980).
13. R.W.Sykes, G.G.Roberts, T.Fok and D.T.Clark, Proc.IEE pt.1,
            127, 137 (1980).
14. J.Tanguy, Thin Solid Films 13, 33 (1972).
15. I.Lundstrom and D.McQueen, Chem.Phys.Lipids 10, 181(1973)
16. I.Lundstrom, Physica Scripta 18, 424(1978)
17. T.W.Hickmott, Proc.Raleigh Conf.on the Physics of MOS
            Insulators.(Pergamon New York, 1980) p227.
18. H.H.Wieder, Proc.Durham Conf.on Insulating Films on
            Semiconductors (Inst.of Physics, Bristol,1980),p234.
19. G.G.Roberts, K.P.Pande and W.A.Barlow, Solid St. and
            Electron Devices 2, 169 (1978)
20. H.C.Card,Proc.Durham Conf.on Insulating Films on Semicon-
            ductors (Inst.of Physics. Bristol. 1980) p140
21. P.J.Martin and G.G.Roberts. Proc.I.E.E. pt 1, 127,133
            (1980).
22. R.H.Tredgold, Adv. in Physics 26, 79 (1977).
23. P. Bergveld, N.F. DeRooij and J.N.Zemel, Nature 273,
            438 (1978).
24. C.W.Pitt and L.M.Walpita, Electrocomp.Sci and Tech. 3,
            191 (1977).
25. E. Colombini and G.L.Yip, Trans. IECE Jap. E61, 154 (1978).
26. A. Barraud, C.Rosilio and A.Ruaudel-Teixier, Solid St.
            Tech. 22, 120 (1979).

A Study of MIS Structures Prepared Under Ultra-High-Vacuum Conditions [1]

M. Commandré[2], J. Derrien[2], A. Cros[3], and F. Salvan[3]

Faculté des Sciences de Luminy, Département de Physique, Case 901
F-13288 Marseille Cedex 9, France

G. Sarrabayrouse and J. Buxo

LAAS, 7 avenue du Colonel Roche
F-31400 Toulouse Cedex, France

1. Introduction

In present semiconductor device technology (MIS, MISS ...) the $Si-SiO_2$ interface is a vital part. The realization of very thin and pure oxide films (less than 30 Å for some applications) is very difficult to control under classical ways due to

i) the presence of a native and unknown composition oxide film at the semiconductor surface extending sometimes to > 10 Å,
ii) the very rapid growth rate of the early stages of oxidation in the oxygen pressure range usually used (several Torrs to atmosphere).

To avoid these difficulties, we have studied the kinetics of thermal oxidation of Si (111) surfaces prepared under UHV conditions. Once the oxide formed, aluminium was then evaporated onto the sample "in situ" and the MIS structure was then electrically tested. We present in this paper some preliminary results.

2. Experiment

All experiments were performed in a stainless steel ion pumped vessel equipped with various surface techniques as Auger spectroscopy (AES), electron loss spectroscopy (ELS), low energy electron diffraction (LEED), sputter ion gun, gas entries and metal evaporators. Samples were mounted into a sample holder able to be placed in front of every probe. Thermal heatings of samples (1100°C), in a base pressure of $5 \cdot 10^{-10}$ Torr, removed all contaminants as checked by AES. Moreover LEED showed sharp spots of a (7x7) ordered Si surface. Temperature was checked by thermocouple attached near a side of the sample and also by a I.R. pyrometer. Samples were heated by Joule effect, using a regulated electrical current passing through the sample. The oxidation process was performed "in situ" with high purity oxygen at low pressure (10^{-3} Torr to 1 Torr), the dynamic equilibrium was maintained by using a throttle-valve for O_2 gas introduction and an additional turbomolecular pump for its evacuation. O_2 pressure was measured with a capacitance gauge (Baratron type).

1 Work supported by "Groupement circuit intégré au Silicium"
2 ERA CNRS n° 899
3 ERA CNRS n° 373

3. Kinetics results

3.1 AES measurements

AES spectra were recorded in the first derivative mode $\frac{dN}{dE}$ of the N(E) electron distribution, always under the same electronic conditions (incident electron energy 2 kV, beam current 0.5 µA/mm^2, 2 V peak to peak modulation). Some typical spectra recorded during the oxidation process are shown in Fig. 1. Curve a displays an Auger spectrum of a clean Si surface.

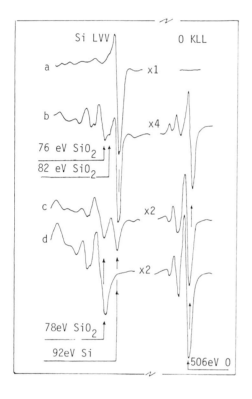

Fig. 1 Auger spectra of a Si surface at various stages of oxidation.
a) clean surface. No contaminant is found. Only a Si line at 92 eV is observed.
b) a slightly oxidized surface. Peaks, due to SiO_2 (∿ 72 ev), SiO_x (∿ 82 eV) and 0 (∿ 510 ev), appear.
c) The SiO_2 is growing now on top of the SiO_x transition layer. Peaks of SiO_2 and 0 are observed and the substrate is still not completely screened.
d) The SiO_2 film is now sufficiently thick to screen the substrate.

oxidation   conditions

10^{-2} Torr - 1000°C

Only Si lines are observed (Si LVV transition at 92 eV). No contaminants are recorded. Fig. 1b shows the early stages of an oxidized surface. One can observe the appearance of a oxygen peak (0 KLL transition ∿ 510 eV), a SiO_x peak (∿ 82 eV) and a SiO_2 peak (∿ 76 eV) not yet fully developed (estimated thickness ∿ 5 Å - see section 3.3). The interface is not abrupt and there is a transition layer of SiO_x, x < 2, rich in silicon, located between the Si substrate and the growing silica film. Fig. 1c shows the next stage of oxidation where the SiO_2 film is developing on top of this transition layer. The SiO_x peak is now screened by a higher peak of SiO_2 (∿ 78 eV) and concurrently the 0 peak shifts towards a lower energy (506 eV). All these AES features were already discussed in literature [1]. Fig. 1d corresponds to a SiO$_2$ film of about 30 Å thick. It displays all features of a bulk and clean SiO$_2$ sample where only the SiO$_2$ (78 eV) and O (506 eV) peaks are observed. The film is amorphous as tested by LEED.

3.2 ELS measurements

We also recorded ELS spectra of sequential oxidation stages (incident electron energy $\sim$ 100 eV, beam current $\sim$ 0.1 µA/mm^2, spectrum recorded in the second derivative mode d^{2}N/dE^2 with a 1 V peak to peak modulation). Fig. 2a shows a clean Si surface spectrum with characteristic bulk and surface plasmons ($\hbar w_p$ $\sim$ 17 eV and $\hbar w_s$ $\sim$ 11 eV). The S$_3$ and S$_2$ peaks involve clean surface states and E$_2$ is an interband transition. During oxidation, the ELS features change and can be explained within a molecular energy scheme [2]. In particular Fig. 2b displays the features of a O$_2$ monolayer chemisorbed on the Si surface. Fig. 2c, d and e show the following stages of oxidation and one can observe the development of SiO$_2$ loss transitions (peaks from 10 eV to 25 eV in Fig. 2e) [2]. Again we can see the gradual evolution from a pure Si ELS spectrum to that of a SiO$_2$ film. The presence of a transition layer is revealed on top of the substrate prior to the SiO$_2$ growth.

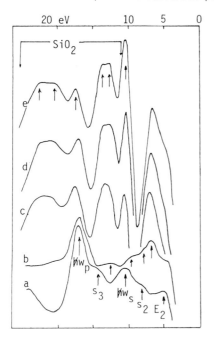

Fig. 2   Energy loss spectra of a Si surface at various stages of oxidation. Only loss peaks are shown and referenced in energy versus the elastic peak not shown (located at 0 eV).
a) Clean surface, see |2|.
b) A surface covered with a O$_2$ monolayer.
c) d) and e) Several stages of oxidation. All features characteristic of a thick SiO$_2$ film are fully developed in e).

3.3 Oxidation kinetics plots

Using the AES peak to peak heights measured "in situ" and during the oxidation process, we have plotted some curves of thermal oxidation. The thickness estimation is based on the fact that Auger electrons of the Si substrate are attenuated when traversing a silica film of thickness d, according to an exponential law

$$I_{Si}(d) = I_{Si}(0) \exp(- d/ \lambda) \qquad (1)$$

where $I_{Si}(0)$ and $I_{Si}(d)$ are respectively the Si (92 eV) Auger signal of a clean surface and a surface covered with an oxide layer of d thickness. λ is the escape depth of electrons of 92 eV energy. An accurate value of λ is necessary. We have used a value given by the "universal curve" of electron escape depth [3], taking into account the density of SiO_2. We find that λ for SiO_2 is then ~ 6.5 Å. The application of equation (1) is only valid when the SiO_2 layer is uniform.

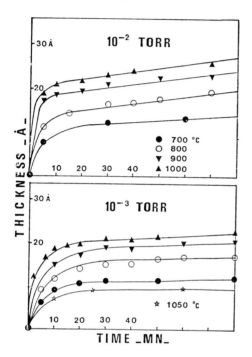

Fig. 3 Oxidation kinetics plots. The thickness is expressed in Å and oxidation time in minutes. Two regimes are observed. The rate is very rapid at the early stages of oxidation and a saturation is quickly reached depending on the temperature and the pressure.

All the kinetics plots display a very rapid rate at the beginning of oxidation, followed by a dynamic saturation, depending upon the substrate temperature and the O_2 pressure (Fig. 3). It is still premature to draw definitive conclusions and to build up a model based on these kinetics, but our results seem to follow the same trend of oxidation curves obtained at higher O_2 pressure mixed in nitrogen [4]. A remarkable result is to be mentioned. For a constant O_2 pressure (Fig. 3), the higher the temperature the higher the growth rate But a limit temperature is soon reached and then no more oxidation can be obtained. It seems that competition between the temperature and the pressure occurs. This phenomenon has not been observed at sufficiently high O_2 pressures, [4], and work is in progress to clarify this point.

4. Fabrication of MISS devices

In order to check the "quality" of these oxides, we have fabricated some MISS devices and measured their electrical properties. We used industrial

wafers with already patterned SS structures (P⁺N), composed with windows of
160 μm diameter, located at every 1 mm distance and separated with thick SiO_2
masks (> 5000 Å). The wafers were cleaned under UHV conditions and then thin
oxide layers were formed as previously explained (20 Å ∿ 100 Å). Pure aluminium was then evaporated "in situ" onto these oxide films and the samples
were then cleaved to be mounted in their housing. I(v), C(v) and G(v) curves
were then performed. The results show, at the present stage, that these "in
situ" MIS diodes have a too large conductance (∿ 10^{-5} A for a 5 V voltage
bias, MIS area ∿ 210^{-4} cm^2). Moreover this large conductance is probably
caused by conducting paths through the oxide because when high voltage pulses
are applied to the MIS diode, the oxide film sometimes displays a more
insulating behaviour, the pulses "burning" these conducting paths.

5. Conclusion

When working with UHV surface techniques, it is possible to form very thin
and clean Si oxides. Preliminary results show that the thickness and the
oxide growth rate can be controlled, with a subtle balance between the
temperature and the O_2 pressure. Although the oxide film seems to show all
ideal atomic features, as checked by AES and ELS, their dielectric properties
are still too bad and hence more studies are needed to clarify this point.
In particular, it is not impossible that aluminium atoms diffuse through
these very thin oxide films giving rise to the conducting paths discussed
previously.

References

1 See for example C.R. Helms, N.M. Johnson, S.A. Schwarz and W.E. Spicer,
  J. Appl. Phys. 50, 7007 (1979) and references cited therein.
2 See for example N. Lieske and R. Hezel, Thin Solid Films 61, 197 (1979)
  and references cited therein.
3 M.P. Seah and W.A. Dench, NPL Report. Chemic. 82 April 1978, U.K.
4 Y. Kamigaki and Y. Itoh, J. Appl. Phys. 48, 2891 (1977).

Electrical Properties of Ultrathin Oxide Layers Formed by DC Plasma Anodization

R.B. Beck, A. Jakubowski, J. Ruzyłło

Institute of Electron Technology, Technical University of Warsaw
Koszykowa 75
00662 Warszaw, Poland

1. Introduction

It appears that insufficient control over the ultrathin (25-50 Å) oxides growth process and stability still limits the further development of MIS tunnel devices such as solar cells, various transistor structures, and switching devices.

In silicon technology the ultrathin SiO_2 layers are grown mainly by thermal oxidation. The present work, however, exploits the possibility of using plasma anodization to grow ultrathin SiO_2 on silicon. This technique offers in some aspects considerable advantages over the thermal oxidation. This includes oxidation of only the top surface of the wafer, low-temperature process, possibility of silicon surface vacuum treatment prior to anodization, and also metallization of the formed ultrathin oxide without exposing it to the surrounding atmosphere.

The purpose of this work is to evaluate the basic electrical properties of the ultrathin oxide films grown on silicon by plasma anodization, and to discuss the usefulness of this technique to the ultrathin SiO_2 processing in the light of obtained results.

2. Experimental

The plasma anodization process was carried out in the vacuum system arranged in the way schematically presented in Fig.1. As a substrate n- and p-type silicon wafers of 2-Ωcm resistivity were used. Prior to the anodization they were chemically cleaned and etched, and then the aluminum was evaporated on the back surface of the wafer to provide the electric contact of the wafer to the wafer holder.

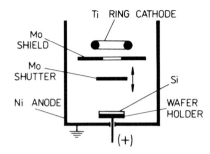

Fig. 1. Schematical view of the plasma anodization system used in this study

In the system shown in Fig.1 plasma is initiated by the DC glow discharge in the pure oxygen under the pressure of 80 mTorr [1] introduced to the system which was initially pumped down to a vacuum better than 10^{-6} Torr. While the discharge is stabilized, the wafer is shielded by the Mo shutter. Then, the shutter is lifted, and the silicon wafer is anodized at the constant current which was usually set at the value of 0.1 mA/cm^2 or less. The process is continued until the desired voltage drop on the wafer is obtained.

The thickness of the oxide was measured by ellipsometry. The vacuum-evaporated aluminum was used as the gate electrodes. These were in the form of dots 0.6 mm in diameter obtained by use of the shadow mask. The diodes were formed over the area of the wafer of up to 5 cm^2 which was exposed to the plasma during the anodization.

3.   Results and Discussion

The preliminary information on the impact of the plasma anodization on the surface properties of silicon was obtained from the ellipsometric measurements and from the I-V characteristics of the formed MIS tunnel diodes. The typical I-V characteristics obtained are shown in Fig.2a. The shape of both forward and reverse characteristics is similar to the typical I-V curves of the nonequilibrium MIS tunnel diodes [2]. Also, the results of the current-versus-temperature measurements shown in Fig.2b suggest that tunneling through the oxide dominates the current of the diode. These data and the ellipsometric measurements confirm the formation of the ultrathin oxide on the Si surface. The same conclusion emerges from results of C-V measurements shown in Fig.3. In addition, these results show rather good stability of the device against the bias-temperature treatment. This points out small influence of the slow-

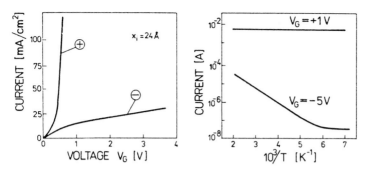

Fig. 2. (a) I-V characteristics of the obtained MIS tunnel diodes, (b) plots of current vs 1/T for the forward and reverse bias

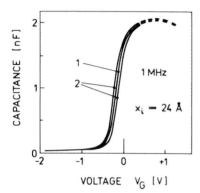

Fig. 3. C-V curves of the studied devices, (1) after processing, (2) after B-T treatment at 150°C, 30 min., $E_{ox} = \pm 106$ V/cm

trapping effects and mobile charges within the oxide on the surface potential of studied diodes.

In the course of this study it was found that the low-temperature annealing (300°C) in dry nitrogen has a profound influence on the device properties. It especially concerns the conduction through the oxide since as a result of annealing the reduction of current and its better stability were observed as shown in Fig.4a. At the same time the annealing usually resulted in the slightly decreased oxide thickness which suggests the oxide densification taking place during annealing.

The annealing was also proved to affect the interface properties of the studied devices as seen from C-V curves in Fig.4b. It was found that the density of the total uncompensated charge in the studied devices was in average equal to 1.5×10^{12} cm^{-2} and 6×10^{11} cm^{-2} before and after the annealing, respectively. The exact nature of these effects is at this point difficult to assure. However, very similar results obtained after the annealing in argon

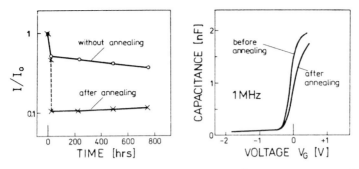

Fig. 4a,b. Influence of annealing at 300°C in nitrogen on (a) reverse current where I_0 stands for the initial current of the as formed device, (b) C-V curves

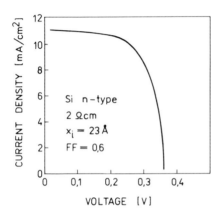

Fig. 5. Output characteristic of the MIS solar cell with ultrathin oxide grown by plasma anodization

suggest that nitrogen does not interact with the oxide, and the observed changes result from the heat treatment causing the structural rearrangements within the oxide (Si-O bonds formed with the participation of the "implanted" oxygen atoms).

In order to further evaluate properties of studied oxides the anodization process described in this work was applied to grow the ultrathin oxide in some practical devices such as the Lateral MIS Tunnel Transistor [3] and MIS solar cell. In Fig.5 the typical output characteristic obtained in the latter case is presented. In both cases, however, good device stability and performance comparable with those incorporating the ultrathin thermal SiO_2 were observed. Also, other photoelectric measurements performed on the MIS tunnel devices made in the way described in this work demonstrated good oxide quality [4].

In summary, the DC plasma anodization appears to be a suitable method of growing tunnel oxide on silicon. The oxide obtained in this way is comparable with the thermal oxide, but considering specific advantages of the plasma anodization this method seems to be more promising than the thermal oxidation in some applications in MIS tunnel devices.

Aknowledgement. The authors wish to acknowledge the participation of Mr. J. Żurawski in this work.

References

1. R.B. Beck, M. Patyra, J. Rużyłło, A. Jakubowski: Thin Solid Films *67*, 261 (1980)
2. R.A. Clarke, J. Shewchun: Solid State Electron. *14*, 957 (1971)
3. J. Rużyłło: IEEE Electron Dev Lett. EDL-*1*, 197 (1980)
4. S. Krawczyk: This volume

Influence of Different Technologies of Metal Deposition and of Oxide Growth on the Electronic Properties of MIS Tunnel Diodes

Georges Pananakakis, Georges Kamarinos

Laboratoire de Physique des Composants à Semiconducteurs, ERA-CNRS 659
ENSERG - 23, avenue des Martyrs
F-38031 Grenoble Cedex, France

Vincent Le Goascoz

THOMSON-EFCIS, Avenue des Martyrs, B.P. 217
F-38019 Grenoble Cedex, France

1. Introduction : Nature of interface states

We study the influence of different technological processes on the electronic properties of the Si-SiO$_2$ interface of Al-SiO$_2$-Si (N-type) MIS tunnel diodes. I(V) (in darkness or under illumination) and C(V), G(V) characteristics (100 Hz $\leq$ f $\leq$ 1 MHz) are plotted and analysed.

From C(V) and G(V) measurements ([1], [2]) the density and the energy distribution of interface states are deduced ; in addition, the surface potential of the semiconductor (ψ_S) is obtained.

The analysis of the I(V) characteristics is performed by comparison with a detailed and self-consistent numerical model [3] ; it can fit very tightly both I(V) and ψ_S(V) experimental curves. It takes into account the interfacial parameters of the structure : the insulator thickness δ, the barrier heights of the insulating layer for semiconductor electrons and holes, the semiconductor doping, the difference between the work functions of the metal and the semiconductor ϕ_{ms} and finally, the density and the nature of interface states that are supposed to interact with the three carrier reservoirs : the valence and the conduction bands of the semiconductor and the metal. In this way we can examine the kinetic and electrostatic action of the interface states.

Assuming the validity of thermionic theory [4] we can establish and solve (numerically) the exact equations for the transport of carriers across the MIS structure.

After comparison of experimental results with theoretical results given by our model, we deduce that the interface states present in all the devices tested behave as donor-like states and their occupancy is controlled by the semiconductor electrons Fermi level.

2. Influence of metal (Al) deposition

Two methods of metal deposition are tested :
 *Electron gun deposition
 *In-source deposition.

In these cases the oxide is thermally grown under atmospherical pressure at 750°C.

We show that the electronic properties (density N$_{SS}$, and energy distribution of interface states) does not depend on the metal deposition process (Fig.1).

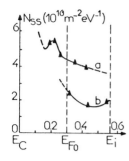

Fig.1 Interface states distribution into the silicon gap.
(a) Electron gun deposition
(b) In-source deposition (without annealing)

Nevertheless, the devices fabricated by the electron gun process, and before any annealing, exhibit the most stable I(V) characteristics [5]. The stability of these devices may be attributed to the influence of the high density of defects, into the oxide, induced by the electron gun X-rays [6].

3. Influence of the oxidation process

Three different oxidation processes are tested :
   Thermal oxidation under atmospheric pressure at 750°C.
   Thermal oxidation under low oxygen pressure at 970°C (L.P.O_2).
   Chemical deposition under low pressure at 900°C (L.P.C.V.D.).

The I(V) characteristics of the devices have the shape of that given in the fig.2 except for L.P.C.V.D. devices after annealing (400°C).

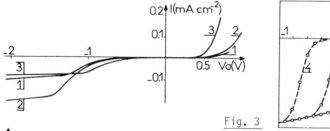
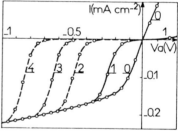

Fig. 3

Fig.2 I(V) characteristics of different MIS structures (L.P.O_2 oxidation) before annealing (oxide thickness $\delta \simeq 25$ Å, measured by ellipsometry)

Fig.3 Theoretical I(V) curves showing the "threshold effect" in the reverse characteristics of MIS structures (δ=30 Å ϕ_{ms}=0,2eV) provided with donor-like interface states (N_{ss}=2,5. $10^{16}m^{-2}$) ; their energy level corresponds to : 0) 0,1eV 1)0,3eV 2)0,55eV 3)0,7eV 4)0,9eV below the conduction band. (for the semiconductor and interface parameters used see [3]).

The saturation currents are, as expected, lower after annealing.

The I(V) curves are fitted by our model and they reveal donor-like interface states located near the midgap (cf. theoretical curves on fig.3).

Capacitance and conductance measurements give (fig.4) the energy distribution and the density of interface states. They confirm and complete the above I(V) analysis : existence of donor-like states in the midgap ; their density is lower after annealing and this independently on the technological process. Nevertheless the L.P.C.V.D. devices exhibit lower state densities.

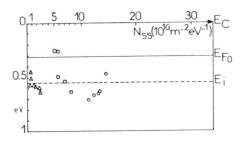

Fig.4 Interface state distribution in the silicon gap. Δ L.P.O$_2$ oxide after annealing. $\square$ LPCVD oxide before annealing (after annealing $N_{SS} < 10^{16} m^{-2} eV^{-1}$). $\circ$ L.P.O$_2$ oxide without annealing

From high frequency (1 MHz) C(V) measurements we deduce ψ_s vs V_a (fig.5). These results are in very good agreement with the theoretical calculation issued from our model (fig.6).

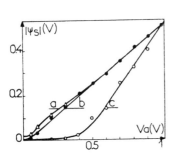

Fig.5 Variation of the Si surface potential ψ_s vs reverse applied bias V_a for LPCVD devices. Before annealing (c) ; after annealing (a,b)

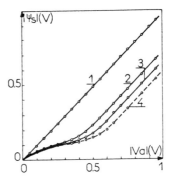

Fig.6 Computed ψ_s vs V_a curves ; $\overline{N_{SS}} = 0$; 2×10^{16} ; $2,5\times10^{16}$; $3\times10^{16} m^{-2}$ (curves 1,2,3,4)

Single energy level interface states located at the midgap.

6. CONCLUSION

Our results show that the LPCVD process provides the devices with a minimum density of states at the Si-SiO$_2$ interface ($8 \times 10^{16} m^{-2} eV^{-1}$ without annealing). On the contrary, we obtain a maximum semiconductor surface potential $\psi_s(0,15V)$ in the case of thermally grown oxide.

Our study makes therefore, easier the choice of the technological processes for the fabrication of thin oxide film devices.

References

1. E.H. Nicollian, A. Goetzberger, *The Bell Syst. J. 46,* n°6, 1055 (1967)
2. S.J.Fonash, *J. Appl. Phys. 48,* n°9, 3953 (1977)
3. G. Kamarinos, G. Pananakakis, P. Viktorovitch, INFOS 79 Durham, *Inst. Phys. Conf. Serv. n°50,* chap. 3, 166 (1979)
   see also : G. Pananakakis, G. Kamarinos, P. Viktorovitch, *Rev. Phys. Appl., 14,* 639 (1979)
4. E.H. Rhoderick, Metal Semiconductor contacts (Oxford-Clarendon) (1978)
5. H. Jaouen, mémoire DEA, ENSERG Grenoble (1979)
6. H.F. Ragaïe, Thèse INP Grenoble (1980)

Photoelectric Methods as a Tool for the Analysis of Current Flow Mechanism in MIS Tunnel Diodes

Stanislaw K. Krawczyk

Institute of Electron Technology, Technical University of Warsaw
Koszykowa 75

00662 Warszawa, Poland

1. Introduction

In Fig. 1 the energy band diagram of the illuminated MIS tunnel diode is shown.

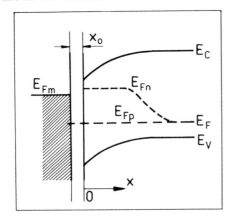

Fig. 1 Energy band diagram of the illuminated $Al-SiO_2-Si/p-$type/tunnel diode

Photoelectrical properties of such structures can be described by the following set of equations:

$$I = I_{ph} - I_{RDO}[exp(\Delta u_F) - 1] - I_{VTO}[exp(-u_G) - 1] \qquad (1)$$

$$I_{CTO}[exp(\Delta u_p) - 1] = I_{ph} - I_{RDO}[exp(\Delta u_F) - 1] \qquad (2)$$

$$u_G = \Delta u_p - \Delta u_F \qquad (3)$$

where I is the total MIS current, I_{ph} is the current of light generated carriers, u_G is the gate voltage in kT/q units,

$$\Delta u_F = [E_{Fn}(0) - E_{Fp}(0)]/kT$$

$$\Delta u_p = [E_{Fn}(0) - E_{Fm}]/kT$$

I_{CTO}, I_{VTO} and I_{RDO} stand for "zero currents" for electron tunneling, hole tunneling and generation-recombination-diffusion processes, respectively.

The attempt to determine the I_{CTO}, I_{VTO} and I_{RDO} currents from photoelectric measurements is the purpose of this work. The principal idea of the proposed methods is based on the fact that the gate bias and illumination induce to some extent independent changes of the quasi-Fermi levels displacement near the semiconductor-oxide interface. It results in the changes of the current contribution to the total MIS current and makes it possible to evaluate some of them separately.

For the experiments the $Al-SiO_2-Si/p-type/$ tunnel diodes with various oxide thickness /15Å-30Å/ were manufactured by DC plasma anodization [1]. The illumination of $\lambda=0.9\ \mu m$ was provided by a CQYP20 LED.

2. Measurements of I vs. I_{ph}

On the basis of (1-3) we can write the following equations:

$$\frac{\Delta I}{\Delta I_{ph}} = \frac{1}{1 + (I_{RDO}/I_{CTO})\exp(-u_G)} \tag{4}$$

$$\frac{\Delta I}{\Delta I_{ph}} = \frac{I_{CTO} + I_{CT}}{I_{CTO} + I_{ph}} \tag{5}$$

where I_{CT} stand for the tunnel current of electrons.

It is easy to notice that by inserting into (4) the value of $\Delta I/\Delta I_{ph}$ measured at the given gate voltage the value of I_{RDO}/I_{CTO} can be calculated.

Fig.2 shows the procedure of the I_{RDO}/I_{CTO} calculations from the $\Delta I/\Delta I_{ph}$ measurements, performed at various open-circuit voltages.

Furthermore, the measurements of the $\Delta I/\Delta I_{ph}$ at $u_G=0$ or the measurements of the I_{sc}/I_{ph} can be used as a tool for the determination of the barrier height in MIS tunnel diode. On the basis of the well known theory [2], the following relation can be formulated:

$$I_{RDO}/I_{CTO} \sim T^{-1/2}\exp\left[E_C(0) - E_{Fm} - E_G/2\right]/kT . \tag{6}$$

By the determination of the I_{RDO}/I_{CTO} at various temperatures the barrier height can be found. Fig.3 illustrates such a procedure and shows the band diagram of the semiconductor surface region resulting from this measuring procedure.

Eq.(5) can be used to determine the type of the MIS tunnel diode. For example, if the examined diode is a minority carrier one, thus, under strong illumination /so strong that short-circuit current is placed above the plateau on the I-V characteristics/ and at the open-circuit conditions, the following simplification can be assumed:

$$\Delta I/\Delta I_{ph} \cong I_{CTO}/I_{ph} . \tag{7}$$

As an example of (7) utilization, the following results have been obtained for one of the samples:

I_{ph}=93 μA, V_{oc}=0.250 V, $\Delta I/\Delta I_{ph}$=0.67, calculated I_{CTO}=62 μA

I_{ph}=157.5 μA, V_{oc}=0.306 V, $\Delta I/\Delta I_{ph}$=0.40, calculated I_{CTO}=63 μA.

Very good agreement between the two calculated values suggest that the examined diode is an almost ideal minority carrier. The obtained value of I_{CTO} is marked on the I-V characteristic and seems to be very reasonable /Fig.4/.

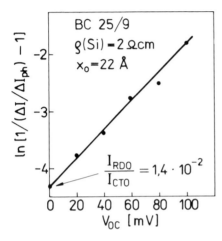

Fig.2   The procedure of the I_{RDO}/I_{CTO} determination

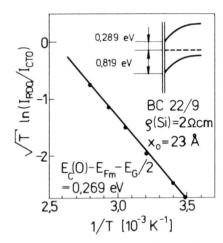

Fig.3   The procedure of the barrier height determination

3.   Measurement of I vs. u_G

Another method for the I_{RDO}/I_{CTO} determination can be suggested. In this procedure we have to measure the $\Delta I/\Delta u_G$ when the gate voltage equals zero, first in the dark and then under illumination. The I_{RDO}/I_{CTO} can be calculated from the following equation:

$$\frac{\Delta I}{\Delta u_G}\bigg|_{I_{ph}\neq 0,\, u_G=0} - \frac{\Delta I}{\Delta u_G}\bigg|_{I_{ph}=0,\, u_G=0} = \frac{I_{RDO}\, I_{sc}}{I_{CTO}} \quad . \tag{8}$$

The obtained values were close to the values obtained by the previously described method.

4.   Comparison of the I-V and I_{sc}-V_{oc} Characteristics

Some essential differences between I-V and I_{sc}-V_{oc} characteristics can be observed. For example, in the minority diode, tunnel current of minority carriers saturates under the forward bias condtions. It results in plateaus on the I-V curves. At open-circuit conditions the electron quasi-Fermi level lies at the same level as the Fermi level in metal or above it. In such cases the minority carrier current cannot saturate. This causes a difference between the I-V and I_{sc}-V_{oc} curves. The I-V and I_{sc}-V_{oc} curves comparison can give qualitative information concerning the role of current components. Furthermore, from the linear approximation of the I_{sc}-V_{oc} curve, for I_{sc} above the plateau, the value of $I_{RDO}I_{VTO}/I_{CTO}$ can be found /Fig. 5/. If the ratio I_{RDO}/I_{CTO} has been already determined, value of I_{VTO} can be now estimated.

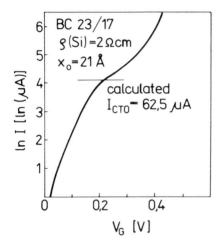

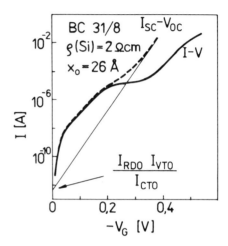

Fig. 4 Measured I-V curve of the Al-SiO_2-Si/p-type/ diode

Fig. 5 Comparison of the I-V and I_{sc}-V_{oc} characteristics

5. Conclusions

The proposed methods have been discussed here on the basis of a simple theory. However, good agreement between theoretical and experimental relations has been observed. Therefore, it is apparent that the photoelectric methods can be considered as an effective instrument for the analysis of current flow mechanisms in MIS tunnel diodes

References

1 R.B. Beck, A. Jákubowski, R. Ruzyllo; This volume
2 K.K. Ng, H.C.Card; IEEE Trans. ED-27, 716 (1980)

Part III

Charge Injection into Insulators

Charge Injection into Wide Energy Band-Gap Insulators

D.J. DiMaria

I.B.M. Thomas J. Watson Research Center
Yorktown Heights, NY 10598, USA

Abstract. Charge injection into wide energy band-gap insulators such as silicon dioxide, which is one of the building blocks of the solid-state industry, is very difficult because of the large energy band-gap of SiO_2 ($\sim$ 9 eV) and the large energy barriers ($>$ 2 eV) it forms with contacting metals and semiconductors. High current injection into insulators at moderate to low electric fields can be obtained by modifying the insulator-contact interfaces by either insulator energy band-gap grading, surface roughening, or through the use of multiphase materials such as Si-rich SiO_2. Experimental examples of each of these three cases will be discussed and novel applications in the area of electronic devices using these phenomena will be presented.

1. Introduction

Insulators such as SiO_2, Al_2O_3, or Si_3N_4 with large energy band gaps of approximately 9 eV [1], 9 eV [2], and 5 eV [3], respectively, have been used extensively in the electronics industry for passivation or as part of the active region of devices. With the advent of planar Si technology, SiO_2 (which can be easily thermally grown from Si or deposited from a chemical vapor phase) has become the most important insulator, particularly in the gates of insulated-gate field-effect-transistors (IGFETs) which are used extensively for memory and logic operations. SiO_2 has also been shown to be superior to Al_2O_3 and Si_3N_4 in terms of its low trap densities in the forbidden gap for either electrons in the conduction band or holes in the valence band [4] and in terms of the number of interface states formed when contacted to semiconductors such as silicon [5]. Al_2O_3 and Si_3N_4 have been used particularly as charge storage layers in non-volatile memories where the carriers are tunneled into and out of these insulators by means of a thin ($\simeq$ 25 Å) tunnel oxide region grown on a Si substrate [6]. Also SiO_2 has a high mobility of 20-40 cm^2/V-sec as compared to other insulators for electrons once they are injected from the contacts into the SiO_2 bandgap [4]. Effective hole mobilities are much less in SiO_2 [4].

Large energy bandgap insulators form large energy barriers with contacting metals or semiconductors. This property causes the good insulating properties usually observed. Many solid-state devices require this blocking contact behavior. For instance in an n-channel IGFET, channel electrons flowing from the source to drain regions are kept confined to the Si surface layer by using SiO_2 between the channel and the gate electrode.

There is, however, a class of devices where a dual nature (both insulating and conducting) of the insulator is required. These devices are used in non-volatile memories where information storage is needed for very long periods of time which requires the use of good insulators. The stored information usually takes the form of electrons or holes trapped on a metal-like layer buried in the SiO_2 gate insulator of an IGFET [7]. This metal-like layer usually is degenerately doped n-type polycrystalline Si (poly-Si) [7]. The charge state of the storage layer affects the operation of the transistor by its internal field and therefore performs a memory function. To write or erase the charge storage layer, carriers must be exchanged through the insulating layers with the metal or Si contacts. This is difficult because of the large energy barriers of the materials used for the insulating layer (such as SiO_2) at the contacts. Injection into the insulator usually requires heating of the carriers in the metal or semiconductor using photons or electric fields [4] so that the excited carriers can surmount the energy barriers. Also if very large electric fields are applied, carriers can tunnel from near the top of the Fermi level for metals or from near the bottom of the conduction band (electrons) and the top of the valence band (holes) for semi-conductors into the SiO_2 energy bands [4,8]. An alternative approach would be to devise an insulator system with a dual nature where the material is highly blocking to carrier injection from the contacts at low electric fields where the device would be normally read (that is, the charge state of the storage layer is sensed) and memory retention is required. However, the material must allow high current injection into the insulator from the contacts at higher electric fields when writing or erasing the charge storage layer is required. This review will be concerned with such a class of insulating systems, the way to achieve this dual physical nature in practice, and experimental structures using these concepts.

2. High Current Injection

In this section, high current injection into insulators at low to moderate electric fields will be discussed for insulator-contact systems where the insulator energy band gap is graded [9,10], the contact forms a rough surface with the insulator [11,12], or the insulating material at the interfaces with the metal or semiconductor contacts has at least two phases [13,14]. In the following section 3, an experimental example of an electrically alterable read-only-memory (EAROM) using the two phase system approach [15,16,17] will be presented in detail.

  A. Graded Energy Band gap

The idea of grading the energy band gap of materials to obtain novel devices has been prevalent in the area of semiconductors [18]. Recently, this idea has been extended to insulators [9,10] as depicted schematically in Figs.1

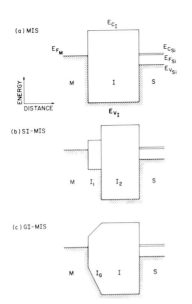

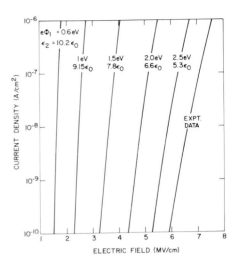

Fig.1 Zero applied field energy band diagram for (a) MIS, (b) stepped insulator MIS, and (c) graded insulator MIS structures. Taken from Reference 9.

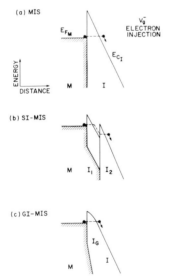

Fig.2 Energy band diagram for negative gate bias (electron injection from the gate electrode) for (a) MIS, (b) stepped insulator MIS, and (c) graded insulator MIS structures. Taken from Reference 9.

Fig.3 Magnitude of the areal particle current density injected into the conduction band of a graded band-gap insulating system as a function of the magnitude of the average applied electric field. The energy barrier at the contact-insulator interface is graded linearly from the value assumed for SiO_2, which is 3.1 eV, to the values indicated on the various curves by $e\Phi_1$. Also the low frequency dielectric constant is increased linearly from that of SiO_2 which is 3.9 to the values assumed at the contact-insulator interface which are also indicated on the curves by the numerical factor in the expression for ε_2 with ε_0 being the permittivity of free space. The grading is assumed to be done over 500 Å of insulating material. The curve labeled "experimental data" which is taken from Reference 38 is the actual characteristic observed for electron injection from near the bottom of the Si conduction band (contact) into the SiO_2 conduction band (insulator) where the energy barrier is 3.1 eV.

and 2 for a metal-insulator-semiconductor (MIS) system. Fig.1 shows several insulating systems with no applied voltage dropped across the insulators. Fig.1a shows a homogeneous insulator, Fig.1b shows a stepped insulator system with two insulators with different energy band gaps, and Fig.1c shows a graded insulator system which is just that of Fig.1b in the limit of a large number of very thin stepped insulators with sequentially decreasing energy band gaps from the bulk insulator to the metal contact. Fig.2 depicts the band bending for the insulating systems in Fig.1 that occurs when a negative voltage is applied to the metal contact. At low electric fields, as depicted in Fig.1, a tunneling electron from near the top of the Fermi level of the metal electrode "sees" effectively the largest energy barrier of the insulating system (namely, that of the bulk insulator). However, when enough voltage is applied to bring about the band bending shown in Fig.2, a tunneling electron "sees" the lowered energy barrier at the metal contact with the graded energy band-gap system. In addition, the tunneling distance is smaller in the graded system (comparing Fig.2a and 2c) for the same bulk insulator electric fields. Clearly, a large tunneling current enhancement into the insulator conduction band from the metal contact is expected for the graded system as compared to a homogeneous insulator. This is demonstrated in Fig.3 where calculated FOWLER-NORDHEIM tunneling current density [19] as a function of the average electric field is shown for several insulating systems linearly graded from bulk SiO_2 to various energy band gaps with respectively different contact-insulator energy barriers as indicated in this figure. Also, the calculations shown in this figure assume that the dielectric constant of the insulators increases linearly to the final value indicated with decreasing energy band gap which is consistent with experimental observations for most materials of interest in the semiconductor industry.

Several experimental attempts have been made to obtain insulating systems graded from SiO_2 to Si_3N_4 [9,10]. DI MARIA tried to do this using an As implant to smooth the SiO_2–Si_3N_4 interface energy step [9] (see Figs.1b and 2b), while HIJIYA et al. have reported that they obtained grading through oxynitride phases from Si_3N_4 to SiO_2 by oxidizing Si_3N_4 that had been grown by nitridizing Si with NH_3 at temperatures $\geq 1000°C$ [10]. Both of these attempts at band-gap grading were incorporated into EAROM devices which worked similarly to what was expected if band-gap grading had occurred [9,10,19]; that is, either electron injection to charge or hole injection to discharge (depending on applied gate voltage bias) a storage layer (discrete traps [9] or a floating poly-Si layer [10]) in the bulk of the SiO_2 via a graded energy band-gap insulator system near one of the contacting electrodes. However, no direct measure of the different interfacial energy barriers, using internal photoemission [13], has been reported.

B. Surface Roughness

Another way to get high current injection into an insulator at low to moderate average applied electric fields from a contacting electrode is by using surface roughness at the contact-insulator interface [11,12]. A rough semiconductor or

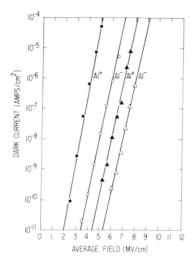

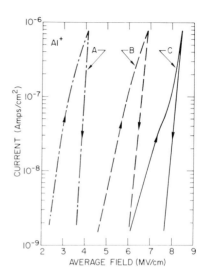

Fig.4 Point by point magnitude of the areal dark current density as a function of the magnitude of the average electric field in the oxide for polycrystalline and single crystal Si-SiO_2-Al metal-oxide-semiconductor (MOS) structures with 500 Å oxide layers grown at $1000°C$ in O_2. The data were generated by measuring the current in the external circuit 2 min. after each 0.5 MV/cm step increase in the magnitude of the average electric field starting at 0 V. Solid and open circles are data for positive and negative gate voltage biases, respectively, on a polycrystalline Si MOS. Solid and open triangles are data for positive and negative gate voltage biases, respectively, on a comparable single crystal Si MOS. Taken from Reference 11.

Fig.5 Ramped magnitude of the areal dark current density measured in the external circuit as a function of the magnitude of the average electric field for positive gate voltage bias and a ramp rate of 9.5×10^{-3} MV/cm-sec. Composition of MOS samples A, B, and C are:

A: Al - thermal SiO_2 (450 Å) - poly Si

B: Al - CVD SiO_2 (520 Å) - thermal SiO_2 (70 Å) - poly Si

C: Al - CVD SiO_2 (520 Å) - W ($\simeq 10^{14}$ atoms/cm^2) - thermal SiO_2 (70 Å) - poly Si.

Taken from Reference 21.

metal surface with many hillocks or asperities will locally produce large electric fields near the tips of the asperities which decrease to the average applied field value in the bulk of the insulator. The local field enhancement must extend out at least a tunneling distance into the insulator; for example, 20-50 Å in SiO_2. The size, shape, and density of the asperities will affect the amount of local electric field enhancement which in turn controls the injected current density [20]. The integrated effect of the local current injection over the entire contact area can easily produce an increase in the average areal current density by many orders of magnitude as compared to a similar structure with a planar interface. This is due to the strongly field-dependent tunneling current mechanism which is operative at all contact interfaces with wide energy band-gap insulators, at least initially.

The top surface of poly-Si films, which are used extensively in the semiconductor electronics industry, is always naturally rough due to the many randomly oriented crystalline grains. If a thermal SiO_2 layer is grown on top of the poly-Si or a chemically vapor-deposited (CVD) SiO_2 film is put down on this surface, high current injection from the asperities is always seen as shown in Fig.4. The electric field enhancement for applied voltage biases favoring injection from the rough poly-Si surface is always larger than injection from the other insulator interface (see Fig.4) which usually has some roughness (particularly on thin films) due to partial replication of the rough poly-Si surface by the insulator [11]. When deposited or grown on rough surfaces, insulators with many trapping states in their forbidden band gap (such as CVD Si_3N_4) will usually screen the effect of the locally high electric fields by building up trapped space charge near the injecting tips of the asperities [21]. An experimental example of this phenomena is shown in Fig.5 where a discrete tungsten trapping layer is sandwiched between a $\approx$ 70 Å thermal SiO_2 layer grown over a rough poly-Si surface and a 520 Å CVD SiO_2 capping layer [21]. This tungsten trapping layer readily captures electrons into energetically deep trapping sites with respect to the bottom of the SiO_2 conduction band, and screens the locally high fields of the asperities as shown in Fig.5 where the areal current density is shifted to larger average electric fields (compare structures A and C). The CVD SiO_2 capping layer by itself also contains trapping sites due to H_2O incorporated into the film during deposition [4,14,21], and it will also screen the high local fields. These traps however have a lower electron capture probability and are further away from the asperities than the tungsten sites, and therefore will not be as effective as is seen in Fig.5.

The surface roughness of the top surface of poly-Si layers used in Si-gated devices has been used to advantage in some devices (particularly EAROMs [22,23]) and reduced in others to minimize current leakage [12,24]. Many floating poly-Si gate memory type devices which are variations of the original floating-gate avalanche-injection metal-oxide-semiconductor (FAMOS) device introduced by FROHMAN-BENTCHKOWSKY [7] use the rough poly-Si asperities to "erase" the floating poly-Si gate by removing the electrons, via the high local electric fields, to a top metal or poly-Si control gate electrode [22,23]. One novel device used in a non-volatile memory operation has three levels of poly-Si (two rough top surfaces are used) where both writing (putting electrons on) and erasing (taking electrons off) the floating poly-Si storage layer is achieved by electric field distortion caused by poly-Si surface roughness [25].

However, other types of devices using buried poly-Si gate electrodes which require low current leakage near the poly-Si surfaces have these surfaces fabricated as smooth as possible. A reduction in asperity size and probably aspect ratio (sharpness of the asperity) has been demonstrated by ANDERSON and KERR [12] for oxides grown thermally at 1150°C above poly-Si surfaces. They also showed a reduction in the enhanced current injection consistent with the surface smoothing directly observed with scanning-electron-microscopy (SEM) [12].

C. Two Phase Materials

The final means to be discussed for achieving high current injection into an insulator at low to moderate applied fields involves incorporating a region inbetween the contact and insulator which is composed of a two phase material having semiconductor or metal regions mixed into an insulator matrix. CVD Si-rich SiO_2 [26] or semi-insulating polycrystalline silicon (SIPOS) doped with oxygen atoms [27] which is a two phase mixture of Si and SiO_2 [28-31] that can be deposited from gaseous mixtures of N_2O and SiH_4 is one example. Cermets or granular metal films composed of insulating regions of such materials as SiO_2, Al_2O_3, or Si_3N_4 and metallic grains of such materials as Al, Ni, W, Au, Pt, Pd, or Mo which can be co-deposited by electron beam evaporation or sputtering systems [32] are other examples. With this type of insulator system, the cermet or Si-rich SiO_2 layer is fairly insulating at very low applied fields, but at larger electric fields it becomes very conductive compared to the underlying wide band-gap insulator and readily moves carriers from one metal or semiconductor island to another via Poole-Frenkel conduction, direct tunneling, or percolation until the interface of the cermet or Si-rich SiO_2 layer with the wide band-gap insulator is reached [13]. At this interface, electric field distortion will occur if the metal or semiconductor islands there have any curvature [13]. Fig.6 depicts this physical mechanism using energy band diagrams.

Recently, several publications from the IBM T.J. Watson Research Center [13,14] have demonstrated this concept using a two phase mixture of Si and SiO_2 (Si-rich SiO_2). Examples of high electronic injection currents into a wide band-gap insulator of SiO_2 achieved with this system for various mixtures of Si-rich SiO_2 materials are shown in Fig.7. The region in Fig.7 from 0 V to −20 V for these structures is due to electron injection from the Al contact into the Si-rich SiO_2 layer and subsequent charging of the Si islands. The region beyond −20 V is due to field-enhanced electronic Fowler-Nordheim tunneling of the last layer or layers of Si islands in the Si-rich SiO_2 material into the underlying thicker SiO_2 layer [13,14].

Further experiments by the IBM group demonstrated that stacked vertical layers composed of Si-rich SiO_2, SiO_2, and Si-rich SiO_2 could be formed using CVD

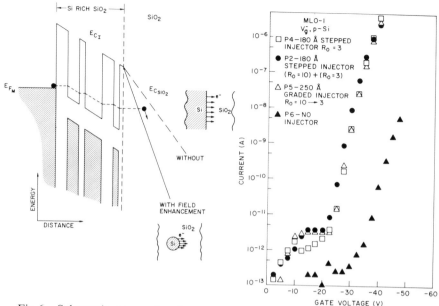

Fig.6 Schematic energy band representation of conduction in the Si-rich SiO_2 layer via direct tunneling between isolated Si regions in the SiO_2 matrix of this two phase system and subsequent high field injection into the underlying SiO_2 region due to local electric field enhancement caused by the curved surfaces of the Si regions. Electronic Fowler-Nordheim tunneling into SiO_2 from a planar Si surface is shown for comparison. Taken from Reference 17.

Fig.7 Point by point magnitude of the dark current as a function of negative gate voltage on MIS structures with various stacks of Si-rich SiO_2 on top of thermal SiO_2. In this measurement, the gate voltage was stepped by -2.5 V starting from 0 V every 20 sec with the dark current being measured in the external circuit 18 sec after each voltage step increase. The Si-rich SiO_2 layer was either stepped or graded with R_0 defined as $[N_2O]/[SiH_4]$ as an indicator of the Si content of this layer. R_0 from 10 (40% atomic Si) to 3 (46% atomic Si) was used with Si content increasing towards the top metal gate electrode when several layers were stacked on top of the underlying 550 Å thick thermal SiO_2 layer ($R_0 = 10 + 3$) or when a graded layer was used ($R_0 = 10 \rightarrow 3$). Taken from Reference 13.

techniques [15]. Enhanced electron injection from either the bottom contact (Si substrate) or the top contact (Al metal gate) for positive and negative gate voltage biases, respectively, could be obtained by means of the appropriate Si-rich-SiO_2–SiO_2 interfacial region in these structures [15]. The measured current-voltage characteristics for these dual electron injector structures, called DEISs, were also shown to be equivalent to the appropriate enhanced injection current-voltage characteristic observed on similar structures with just one Si-rich SiO_2 layer present [15].

These DEISs also exhibit an asymmetry in the enhanced electron injection from the Si-rich-SiO_2–SiO_2 interfaces (the bottom interface gives larger currents than the top interface for the same average electric fields) even though the CVD Si-rich SiO_2 regions of the DEIS stack were fabricated under identical conditions [15]. This asymmetry is believed to be due to the fact that the bottom Si-rich SiO_2 layer is deposited on Si which is either crystalline or polycrystalline, while the top layer is deposited on top of SiO_2 which is amorphous [15]. Cross-section transmission-electron-micrographs (CRTEM) have shown evidence for a "rougher" interface with the SiO_2 layer for the bottom Si-rich SiO_2 layer which also appears to have larger and denser numbers of Si islands than the top Si-rich SiO_2 layer [17,33,34].

In the very low field region, the cermets or Si-rich SiO_2 layers cause another interesting effect. The very small metal or semiconducting islands in these layers act as charge trapping sites for carriers injected at very low fields due to asperities or particulate on the contact, and they field screen these weak spots by the space charge they build-up [17,21,33] (see section 2-B). This space charge is not permanent, but it can move back to the contact when the electric fields are reversed [13,17]. Capacitor structures using Si-rich SiO_2 layers inbetween the contact and bulk SiO_2 have been shown to give outstanding breakdown histograms due to this phenomenon with very few low field breakdown events as compared to control structures processed identically but without the Si-rich SiO_2 layers present [13,33].

In the next section, experimental examples of EAROM structures using DEIS stacks inbetween floating and control gate poly-Si layers will be discussed in detail.

3. DEIS EAROMs

The DEIS EAROM is a non-volatile semiconductor memory which uses the high current injection from the Si-rich-SiO_2–SiO_2 interfaces of a DEIS stack to charge or discharge a floating degenerate n-type poly-Si storage layer as depicted in Fig.8 [15-17,33]. Negative control gate voltage biases are used to charge (write operation) the floating poly-Si by injecting electrons onto it while positive control gate voltage biases (erase operation) are used to discharge it by pulling electrons off of it. Over-erasure is also possible by leaving the poly-Si in a state with a net positive charge due to ionized donors.

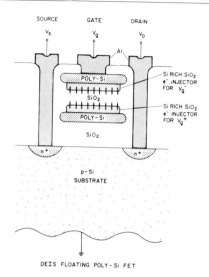

SOURCE   GATE   DRAIN

V_s V_g V_D

Al

POLY-Si

Si RICH SiO_2
e^- INJECTOR
FOR V_g^-

SiO_2

POLY-Si

Si RICH SiO_2
e^- INJECTOR
FOR V_g^+

SiO_2

n^+ n^+

p-Si
SUBSTRATE

DEIS FLOATING POLY-Si FET

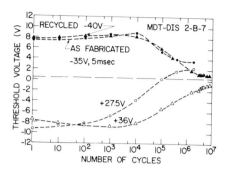

Fig.8 Schematic representation of a non-volatile n-channel field effect transistor memory using a dual electron injector stack between a control gate and a floating poly-Si layer. Writing (erasing) is performed by applying a negative (positive) voltage, V_g^- (V_g^+), to the control gate which injects electrons from the top (bottom) Si-rich SiO_2 injector to the floating poly-Si storage layer (back to the control gate). Structure is not drawn to scale. Taken from Reference 15.

Fig.9 Threshold voltage after writing and erasing as a function of the number of write/erase cycles on an n-channel DEIS FET from wafer MDT-DIS 2-B (290 Å gate oxide from the floating poly-Si layer to the Si substrate and a CVD DEIS stack of 150 Å Si-rich-SiO_2–150 Å SiO_2–150 Å Si-rich-SiO_2 from the floating poly-Si to control gate poly-Si with the Si-rich SiO_2 having 46% atomic Si). Solid and open symbols correspond to the threshold voltage after writing at −35 V and erasing at +27.5 V for 5 msec, respectively, the as-fabricated structure. Solid and open triangles correspond to the threshold voltage after writing at −40 V and erasing at +35 V for 5 msec, respectively, the same structure after threshold window collapse due to charge trapping in the intervening CVD SiO_2 layer of the DEIS stack. The horizontal dashed line indicates the initial threshold voltage of the as-fabricated FET before cycling.

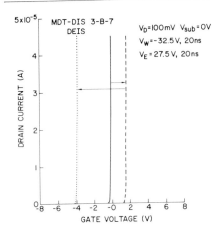

Fig. 10 Drain current as a function of the gate voltage with the drain biased at $V_D = 100$ mV and the source and substrate grounded $V_S = V_{sub} = 0$ V on an n-channel DEIS FET from wafer MDT-DIS 3-B (100 Å gate oxide from the floating poly-Si layer to the Si substrate and a CVD DEIS stack of 100 Å Si-rich-SiO$_2$–100 Å SiO$_2$–100 Å Si-rich-SiO$_2$ from the floating poly-Si to the control gate poly-Si with the Si-rich SiO$_2$ having 46% atomic Si). The device was first written with -32.5 V in 20 ns from its virgin uncharged state and then erased with $+27.5$ V in 20 ns as indicated by the arrows.

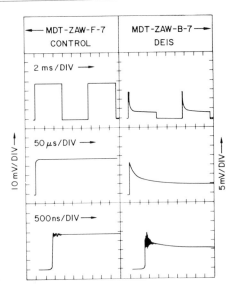

Fig. 11 Oscilloscope traces with 50 Ω shunted across the input of the oscilloscope of the drain current flowing due to a 5 V step applied to the control gate as a function of time with $V_D = 100$ mV and $V_S \approx V_{sub} = 0$ V on n-channel DEISs and control FETs from wafers MDT-ZAW-B,F. Wafer B has the CVD DEIS stack (100 Å Si-rich-SiO$_2$–500 Å SiO$_2$–100 Å Si-rich-SiO$_2$ with the Si-rich SiO$_2$ having 46% atomic Si) as the gate insulator between the Si substrate and the control gate poly-Si electrode, while the control wafer has just the CVD 500 Å thick layer of SiO$_2$ as the gate insulator. These traces show that the FET operation is not too severely deteriorated due to the Si islands in the Si-rich SiO$_2$ layer deposited on the substrate Si acting as surface states if the operating time is kept in the hundreds of nanoseconds regime.

The cycling (write/erase) characteristic of such a DEIS EAROM is shown in Fig.9 where positive threshold voltages of the device indicate stored electrons on the floating poly-Si while negative threshold voltages indicate ionized donors [35]. Clearly from this figure, the device can be written and erased approximately 10^4 times before the threshold voltage starts to degrade and collapse. This collapse is caused by electron capture on energetically deep trapping sites in the band gap of the intervening CVD SiO_2 layer [4,14,15,33]. These traps are believed to be related to H_2O impurities incorporated into the CVD SiO_2 layer during fabrication [4,36,37] and their number can be reduced by high temperature ($1000°C$) annealing in nitrogen or forming gas (N_2/H_2 mixtures) prior to the control gate deposition [15,17,37]. The cycling characteristic can be recovered by thermally discharging the electrons trapped on these sites by using mild heating in forming gas or room air with temperatures between $200°C$ and $400°C$ [33]. Also, recovery can be obtained by applying larger gate voltages to return the electric fields near the Si-rich-SiO_2–SiO_2 interfaces (which have been lowered by the internal fields of the trapped charge) to the values they had initially [33] as shown in Fig.9.

At the very low voltages that the device gate is held when not in use, the retention characteristics (perturbation of the stored information by charge leakage off of the floating gate) of the DEIS EAROM are excellent [16,33] because of the insulating behavior of the Si-rich SiO_2 regions in this voltage regime (see section 2-C). Also, since the voltages at which the device is read (that is, the stored information on the floating poly-Si layer is sensed by the conduction of the Si surface channel between the source and drain of the transistor) are low ($\lesssim 5$ V), read perturb effects due to electron removal off of the floating poly-Si by tunneling are negligible [33].

The DEIS EAROMs also have excellent breakdown characteristics, particularly in minimizing low voltage breakdown events [16,17,33], due to the field screening action of the Si-rich SiO_2 layers as was discussed in section 2-C. This ability of the DEIS stacks to pass very high electronic particle current densities allows them to be written or erased in very short times as shown in Fig.10. This figure demonstrates 20 ns operation which we believe is the fastest time ever reported for switching an EAROM in a non-destructive manner.

Other types of DEIS EAROMs can be fabricated with the DEIS stack over the channel region of the device rather than inbetween the control and floating gate electrodes. However, this structure has the disadvantage that the Si islands in the Si-rich SiO_2 layer will act similarly to surface states by capturing carriers from the channel region and building up an electric field which will degrade device performance. If the device is read in times that are short compared to those necessary for most of the nearest Si islands to capture electrons, then the device performance will not be as severely degraded as shown in Fig.11.

Another interesting phenomenon associated with the reversible trapped space charge which builds up on the Si islands in the Si-rich SiO_2 layers of DEIS stacks is the injection efficiency degradation shown in Fig.12. The effect

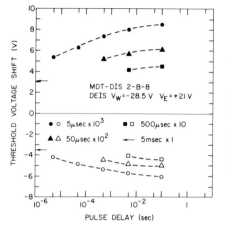

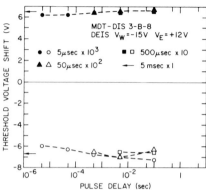

Fig.12 Threshold voltage shift as a function of the time delay between write or erase pulses for pulse trains of various pulse time duration but all adding to 5 msec on an n-channel DEIS FET from wafer MDT-DIS 2-B which was described in the caption of Fig.9. Each point was taken by either writing with a voltage of -28.5 V (as indicated by solid symbols) or erasing with a voltage of $+21$ V (as indicated by open symbols) using the number of pulses with the indicated pulse time duration listed opposite the appropriate symbols. Prior to each write or erase measurement, the device was reset into an opposite charge state (for example, an ionized donor state on the floating poly-Si prior to a write pulse train). The device was reset to the same opposite charge state for each write or erase measurement by cycling about 25 times with -28.5 V and $+21$ V and then stopping after the appropriate write or erase pulse.

Fig.13 Threshold voltage shift as a function of the time delay between write or erase pulses for pulse trains of various pulse time duration but all adding to 5 msec on an n-channel DEIS FET from wafer MDT-DIS 3-B which is described in the caption of Fig.10. Symbols have the same meanings as in Fig.12, but with write or erase voltages of -15 V or $+12$ V, respectively.

demonstrated in this figure is due to trapped electrons building up in the Si-rich SiO_2 region near the injecting contact, screening the electric field, and thereby lowering the injection efficiency from the contact. This phenomenon is more pronounced if the Si-rich SiO_2 layer is thick and/or has a small percentage of excess Si so that these trapped electrons can not move as easily through the Si-rich SiO_2 layer to the Si-rich-SiO_2–SiO_2 interface and then be injected into the SiO_2 layer (compare Figs.12 and 13). Clearly in Fig.12, a longer delay between each pulse of a string of write or erase pulses gives a larger threshold voltage shift magnitude since more electrons are put onto or removed from the floating poly-Si layer. This is because the trapped electrons on the Si islands which block contact injection have had sufficient time to move back to the appropriate contact possibly having this motion influenced by their own internal electric field. Fig.12 also shows that a large number of pulses of small duration are more efficient than one large pulse of the same total time duration. Again this is due to the trapped electronic charge build-up on the Si islands limiting contact injection particularly when there are few delays, delays of small duration, or no delays between each pulse of the write or erase pulse train. Another phenomenon closely related to that depicted in Figs.12 and 13 occurs during sequential write/erase cycling of DEIS EAROMs with single pulses [33]. In this case, if the delay time from the write to the erase pulse (or vice versa) is too short, the electric field generated by the trapped electronic space charge in the opposite Si-rich SiO_2 injector which was used during the previous write or erase operation decreases the electric field near the Si-rich-SiO_2–SiO_2 interface which is being used currently [33]. This phenomenon is more pronounced on DEIS EAROMs with thin SiO_2 regions and thick and/or low excess Si content Si-rich SiO_2 regions [33].

4. Conclusions

This review has been concerned with a means of obtaining high current injection into wide band-gap insulators such as SiO_2, Si_3N_4, or Al_2O_3 at low to moderate electric fields using a modification of the contact-insulator interface. This modification has been obtained by generating locally high electric fields or by grading the insulator energy band-gap. Other materials and means for doing this will probably evolve in the future because of the need for such insulating systems in the area of non-volatile semiconductor memory. Electrically alterable memories such as the one described in section 3 will find increasing use in electronic equipment requiring microprocessors which should rapidly develop in the 1980's. The EAROMs described here may evolve into pure non-volatile random-access-memories (NVRAMs) if the cycling degradation problem due to charge trapping in the SiO_2 layer can be solved and large numbers of cycles ($\gtrsim 10^{12}$) obtained.

Acknowledgements

The author would like to gratefully acknowledge helpful discussions with K.M. DeMeyer and D.R. Young; the calculations performed by P. Ko depicted in Fig.3 which he performed while at IBM during the summer of 1978; the sample fabrication and materials work of D.W. Dong, C.M. Osburn, and the Si Processing Facility at the IBM Thomas J. Watson Research Center; and the critical reading of this manuscript by M.H. Brodsky. This work was supported by the Defense Advance Research Projects Agency, Washington, DC and monitored by the Rome Air Development Center, Deputy for Electronic Technology, Solid State Sciences Division, Hanscom AFB, MA 01731 under contract No. MDA903-81-C-0100.

References

1.   Z.A. Weinberg, G.W. Rubloff, and E. Bassous, Phys. Rev. B19, 3107 (1979).
2.   E.T. Arakawa and M.W. Williams, J. Phys. Chem. Solids 29, 735 (1968).
3.   H.R. Philipp, J. Electrochem. Soc. 120, 296 (1973).
4.   D.J. DiMaria, in The Physics of SiO_2 and Its Interfaces, ed.   by S.T. Pantelides (Pergamon, New York, 1978), p. 160 and references contained therein.
5.   M. Schulz, in Insulating Films on Semiconductors, ed. by G. Roberts (Institute of Physics and Physical Society, London, 1980), p. 87.
6.   J.J. Chang, Proc. IEEE 64, 1039 (1976) and references contained therein.
7.   D. Frohman-Bentchkowsky, Solid State Electronics 17, 517 (1974).
8.   M. Lenzlinger and E.H. Snow, J. Appl. Phys. 40, 278 (1969).
9.   D.J. DiMaria, J. Appl. Phys. 50, 5826 (1979).
10.   S. Hijiya, T. Ito, T. Nakamura, N. Toyokura, and H. Ishikawa, presented at the IEDM, Washington, DC, 1980, paper no. 23.4.
11.   D.J. DiMaria and D.R. Kerr, Appl. Phys. Lett. 27, 505 (1975).
12.   R.M. Anderson and D.R. Kerr, J. Appl. Phys. 48, 4834 (1977).
13.   D.J. DiMaria and D.W. Dong, J. Appl. Phys. 51, 2722 (1980).
14.   D.J. DiMaria, R. Ghez, and D.W. Dong, J. Appl. Phys. 51, 4830 (1980).
15.   D.J. DiMaria and D.W. Dong, Appl. Phys. Lett. 37, 61 (1980).
16.   D.J. DiMaria, K.M. DeMeyer, and D.W. Dong, IEEE Electron Device Lett. EDL-1, 179 (1980).
17.   D.J. DiMaria, in The Physics of MOS Insulators, ed. by G. Lucovsky, S.T. Pantelides, and F.L. Galeener (Pergamon, New York, 1980), p. 1.
18.   A.G. Milnes and D.L. Feucht, Heterojunctions and Metal-Semiconductor Junctions (Academic Press, New York, 1972).
19.   P. Ko and D.J. DiMaria, unpublished.
20.   T.J. Lewis, J. Appl. Phys. 26, 1405 (1955).
21.   D.J. DiMaria, D.R. Young, and D.W. Ormond, Appl. Phys. Lett. 31, 680 (1977).

22. S.A. Abbas and C.A. Barile, 13th Annual Proceedings, Reliability Physics, Las Vegas, Nevada, 1975 (IEEE, New York, 1975), Vol. 13, p. 1.

23. H.S. Lee, Appl. Phys. Lett. 31, 475 (1977).

24. H.R. Huff, R.D. Halvorson, T.L. Chiu, and D. Guterman, J. Electrochem. Soc. 127, 2482 (1980).

25. R. Klein, W.H. Owen, R.T. Simko, and W.E. Tchon, Electronics 52, 111 (1979).

26. D. Dong, E.A. Irene, and D.R. Young, J. Electrochem. Soc. 125, 819 (1978).

27. T. Matsushita, T. Aoki, T. Otsu, H. Yamoto, H. Hayashi, M. Okayama, and Y. Kawana, IEEE Trans. on Electron Devices ED-23, 826 (1976).

28. M. Hamasaki, T. Adachi, S. Wakayama, and M. Kikuchi, J. Appl. Phys. 49, 3987 (1978).

29. A.M. Goodman, G. Harbeke, and E.F. Steigmeier, in Physics of Semiconductors 1978, ed. by B.L.H. Wilson (The Institute of Physics, Bristol and London, 1979), p. 805.

30. A. Harstein, J.C. Tsang, D.J. DiMaria, and D.W. Dong, Appl. Phys. Lett. 36, 836 (1980).

31. E.A. Irene, N.J. Chou, D.W. Dong, and E. Tierney, J. Electrochem. Soc. 127, 2518 (1980).

32. B. Abeles, RCA Rev. 36, 594 (1975).

33. D.J. DiMaria, K.M. DeMeyer, C.M. Serrano, and D.W. Dong, J. Appl. Phys., May 1981.

34. J.M. Gibson, N.J. Chou, and C.M. Serrano, *private communication*.

35. S.M. Sze, Physics of Semiconductor Devices (Wiley-Interscience, New York, 1969), Chapter 10.

36. E.H. Nicollian, C.N. Berglund, P.F. Schmidt, and J.M. Andrews, J. Appl. Phys. 42, 5654 (1971).

37. D.R. Young, E.A. Irene, D.J. DiMaria, R.F. DeKeersmaecker, and H.Z. Massoud, J. Appl. Phys. 50, 6366 (1979).

38. C.M. Osburn and E.J. Weitzman, J. Electrochem. Soc. 119, 603 (1972).

Oxide and Interface Charge Generation by Electron Injection in MOS Devices

F.J. Feigl*, D.R. Young, D.J. DiMaria, and S. Lai

IBM Thomas J. Watson Research Center
Yorktown Heights, NY 10598, USA

Introduction

The development of charge within the oxide layer of an MOS structure, during transport of an electron current across the oxide
layer, is illustrated in Figure 1.

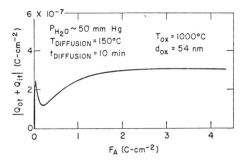

Fig. 1: MOS device charging
curve measured on a water
diffused oxide (see Text).
Data were obtained at room
temperature.

Q_{ot}, the oxide trapped charge, is negative, and is due to
capture of mobile electrons into localized states associated with
defects or impurities within the amorphous SiO_2 network. This process produces an increase in the ordinate of Figure 1, and is
the dominant process at very low and very high values of the
charge fluence F_A (the abcissa in Figure 1). At intermediate
values of the charge fluence (0.01 $C-cm^{-2} \leq F_A \leq 0.3$ $C-cm^{-2}$),
the ordinate decreases with increasing F_A. This is due to the
generation of electron states in the immediate vicinity of the
$Si-SiO_2$ interface. These states are unoccupied under the specific
conditions used to measure the net charge $Q_{ot} + Q_{it}$, and are
positively charged when unoccupied. We shall designate this
net positive charge as interface charge Q_{it}, in accordance with
the terminology recently proposed by Deal[1].

Negative Q_{ot} buildup produced by trapping of electrons injected
into oxide films was studied under carefully controlled
conditions by Nicollian and coworkers [2] These investigators

---

*Permanent Address: Department of Physics, Sherman Fairchild
Laboratory, Lehigh University, Bethlehem
Pennsylvania, 18018

showed that the trapping process obeyed simple first-order kinetics

$$Q_{ot}(F_A)=Q_{ot}(\infty) \ (1-e^{-\frac{\sigma_C}{e}F_A})$$ (1)

and scaled with the injected electron charge fluence $F_A = \int_0^{t_A} J_A dt$.

Nicollian and coworkers also studied the formation of interface states by passage of an injected electron current. It seems quite likely that this process, as described by these workers, was in fact the positive Q_{it} buildup shown in Figure 1. Nicollian and coworkers demonstrated that both Q_{ot} and Q_{it} buildup was strongly affected by the presence of water related impurities in the oxide film.

The studies just described were followed by numerous subsequent investigations. {3, 4} Of particular importance to the present investigation are the work of Gdula {5} and the very recent studies of Young and coworkers {6}, Miura and coworkers {7} and Lai and Young {8}. The accumulated experience of all of these investigations can be expressed in three summary points which are central to the present investigation.

First, both negative oxide charge Q_{ot} and positive interface charge Q_{it} are produced during injection of electrons into thermal SiO_2 films. Second, generation of both Q_{ot} and Q_{it} depends on the total charge fluence F_A transported across the oxide film. This point is not clearly established in the literature, and was addressed directly in our research program. Third, the kinetics of both Q_{ot} and Q_{it} generation are strongly affected by incorporation of water into the SiO_2 film. The fact that water-related impurities are involved in Q_{ot} and Q_{it} generation is clearly established in the literature. However it is not at all clear that it is the trapping kinetics which are predominantly affected, and this will be the main experimental point of the present investigation.

Experimental

Details of sample preparation have been described by Young and coworkers {6}. Very dry oxides were produced on (100), p-type silicon wafers with a nominal resistivity of 0.2 Ω-cm. Oxidations were carried out at 1000°C, under conditions designed to minimize incorporation of water related impurities. All oxide films were subjected to a post-oxidation annealing step (POA) for at least twenty minutes in a flowing, dry N_2 ambient at 1000°C.

After the oxidation step, most of the oxidized wafers were divided. Dry control samples were metallized (Al) immediately after oxidation. Some of the dry control samples were given a postmetallization annealing step (PMA) for thirty minutes in a forming gas ambient at 100°C.

Research samples were placed inside a diffusion furnace as soon as possible after oxidation. The diffusion was carried out in an ambient of flowing N_2 gas bubbled through heated deionized water. Water diffusion was done at diffusion (sample) temperatures in the range 135-300°C, and at bubbler temperatures in the range 30-60°C. For the bubbler temperatures quoted, nominal (saturated) water vapor pressures are approximately 50-150 mmHg. The water diffusion process was intended to be identical to that described by Nicollian and coworkers {6}.

The research samples were metallized (Al) immediately after the water diffusion step. No post-metallization annealing step was performed on the water-diffused research samples.

Electron trapping and electron transport induced positive charge buildup in both control oxide and water-diffused oxide MOS capacitors were studied using the constant-current rf avalanche method developed by Young {6}. Pulsed electron flow is induced through the SiO_2 film by rf avalanche in the p-silicon surface depletion layer, and the rf voltage level V_A is continuously and automatically adjusted to maintain a constant average dc current I_A throughout a given run. The high frequency C-V flat band voltage V_{FB} is automatically monitored and recorded at preset intervals throughout the run.

The experimental data reported in this paper is the flat band shift ΔV_{FB} as a function of cumulative avalanche injection time t_A. As in Figure 1, these data are most often presented as plots of $(Q_{ot} + Q_{it})$ or ΔV_{FB} vs. F_A, where $F_A = I_A t_A / A$ is the integrated current density or charge fluence over the electrode area A. Note that $(Q_{ot} + Q_{it}) = -(C_{ox}/A) \Delta V_{FB}$, where C_{ox} is the device oxide capacitance. Data runs were made with the device maintained at room temperature, and with the device maintained at approximately 100°C. These runs were made on separate electrodes on a given sample.

The flat band voltage shift measurements were augmented with extensive measurements of other MOS capacitor device character-istic shifts (for example, photo IV characteristics {3}). These experiments are described elsewhere {9}.

Results and Analysis

The generation of negative Q_{ot} and of positive Q_{it} scaled with the integrated current density F_A. This is shown in Figures 2 and 3, both of which represent the same experimental data. The several data runs in each figure were made on separate electrodes on a single half-wafer which had been subjected to a water diffusion step. The water diffusion conditions were as shown on the figures. The several data runs were performed at different current densities $J_A = I_A / A$, as indicated on the figures.

The data in Figure 2 is plotted as a function of the cumulative avalanche injection time t_A. In Figure 3, the same data is replotted as a function of the integrated current density or charge fluence F_A, which is proportional to the injection time t_A. It is clear that both the negative Q_{ot} buildup (positive flat band shift) and the

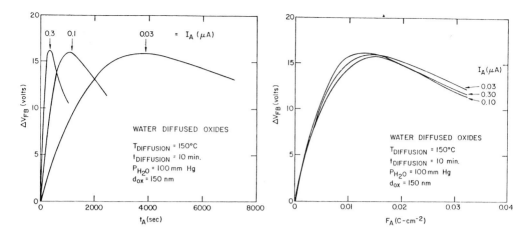

Fig. 2: Flat band voltage shifts resulting from avalanche in-jection of electrons into water diffused oxides, measured at $22°C$. These data runs were made on different electrodes on a single half-wafer.

Fig. 3: Flat band voltage shifts resulting from avalanche injection of electrons into water diffused oxides. These data runs are the same as those exhibited in Fig. 2 , replotted as a function of the avalanche injection electron fluence $F_A = I_A \cdot t_A$.

positive Q_{it} buildup (negative flat band shift at fluences greater than approximately 0.01 C-cm^{-2}) scale with the charge fluence. The deviations of individual data runs from a single normalized curve in Figure 3 are within the electrode-to-electrode variation of identical data runs on a single wafer.

We thus assert specifically that Q_{it} buildup scales with the $J_A t_A$ product. This same behavior is observed in dry control oxides.

The generation of Q_{ot} and Q_{it} are both strongly affected by the presence of water in the oxide film. This is shown in Figures 4 and 5. In these figures, the charge densities (including sign) are plotted as a function of the avalanche fluence for a dry control oxide (Figure 4) and a water-diffused oxide (Figure 5). In each of these figures, the curves marked "$Q_{ot}+Q_{it}$" and "Q_{ot}" are experimental data. The former curve ($Q_{ot}+Q_{it}$) is obtained from a flat band shift measurement at 22°C (room temperature RT). The latter curve (Q_{ot}) is obtained from a flat band shift meas-urement at approximately 100°C (high temperature HT).

Thus:

$$Q_{ot} + Q_{it} = - (C_{ox}/A) \; \Delta V_{FB}(RT) \qquad (2)$$

$$Q_{ot} = - (C_{ox}/A) \; \Delta V_{FB}(HT). \qquad (3)$$

107

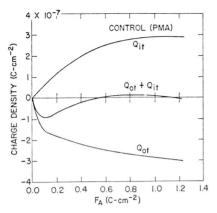

Fig. 4: MOS device charging curves obtained on a dry control oxide which had been given a postmetallization anneal (forming gas ambient at 400° C for thirty minutes). The two lower curves were obtained from data. The top curve was calculated (see text).

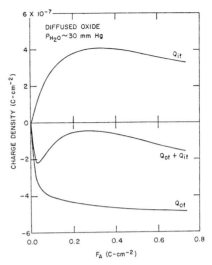

Fig. 5: MOS device charging curves obtained on a water diffused oxide. The two lower curves were obtained from data. The top curve was calculated (see text).

Young and coworkers {6} have shown that C-V flat band shift measurements under positive bias at elevated temperatures (≥100°C) can be interpreted in terms of bulk trapping only. This point is also discussed by Lai and Young{8} and by the present authors {9}.

It is clear from Figures 4 and 5 that the <u>dominant</u> effect of water diffusion into dry oxide films is to alter the kinetic rates for both Q_{ot} and Q_{it} generation. Specifically, the general shape of the experimental curves for dry control and water-diffused oxides are similar. Also, the total amount of oxide trapped charge differs by less than a factor of two. However, the rate of generation of Q_{ot}, as characterized by the dominant capture cross section σ_c in Equation (1), differs by approximately a factor of four. For dry control oxides {6} $\sigma_c \sim (2-4) \times 10^{-18} cm^2$, and for wet oxides {2,6} $\sigma_c \sim (1-2) \times 10^{-17} cm^2$. Also the characteristic rate of generation of Q_{it}, as indicated by the fluence at which the room temperature flat band curve - i.e., $|Q_{ot}+Q_{it}|$ - has a minimum, differs by a factor of three for dry and water-diffused oxides.

The curves marked "Q_{it}" in Figures 4 and 5 were obtained by calculation. Thus, from Equations 2 and 3,

$$Q_{it} = (C_{ox}/A) \cdot \left\{ \Delta V_{FB}(HT) - \Delta V_{FB}(RT) \right\} . \qquad (4)$$

These data are replotted in Figure 6.

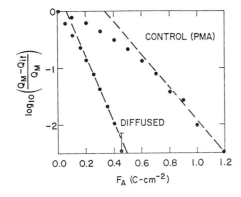

Fig. 6: Kinetics of positive interface charge buildup. The data points (solid circles) were obtained from Figs. 4 and 5 (see text). The dash lines are exponential functions fitted to the points at larger fluence values.

In Figure 6, Q_M is the saturated value of Q_{it} (i.e., $Q_{it}(\infty)$ as $t_A \to \infty$), corrected for the large-fluence linear decrease in Q_{it} apparent in Figure 5. We have no explanation for this linear decrease.

Discussion and Conclusions

We now propose a heuristic model to account for the production of Q_{ot} and Q_{it}. The following pair of coupled kinetic equations describe this process:

$$\frac{dQ_{ot}(t_A)}{dt_A} = \frac{\sigma_c}{e} J_A \left\{ Q_{ot}(\infty) - Q_{ot}(t_A) \right\} \qquad (5)$$

$$\frac{dQ_{it}(t_A)}{dt_A} = \frac{\sigma_e}{e} J_A \left\{ Q_{ot}(t_A) - Q_{it}(t_A) \right\} .$$

The first equation can be solved analytically to yield Equation (1), and describes direct electron capture by neutral trapping centers with capture cross section σ_c. We have used Equation (1) to generate numerical solutions to Equation (3). These solutions, with the two adjustable parameters σ_c and σ_e, reproduce the data variation exhibited in Figure 6 to within the overall accuracy of the experimental data. Furthermore, the two conditions required by Equations (5) and (6) are satisfied, within the overall experimental accuracy. These conditions are:

(1) $Q_M \equiv Q_{it}(\infty) \approx Q_{ot}(\infty)$, and (2) $F_0 \approx \sigma_c/e$,

where F_0 is the abcissa intercept of the dash lines in Figure 6.

We now discuss the physical implications of this kinetic model. For low fluences ($F_A \ll F_0$), the only process occurring in the oxide is bulk trapping of electrons to form negative oxide trapped charge Q_{ot}. This is described by Equation (5). For intermediate fluences ($F_A \sim F_0$), both bulk trapping and, also, a second process resulting in formation of donor states occur. The donor states are measured as positive interface charge Q_{it} in the experiment.

109

We suggest that this second process is initiated in the oxide bulk by interaction of a mobile electron with the trapped electron centers which constitute Q_{ot}. This interaction does not change the charge state of the trapped electron centers. The interaction does, however, result in the transport of some neutral species to the Si-SiO$_2$ interfacial region. The arrival of the mobile species at the interface ultimately results in the formation of the donor states. Note that σ_e is the effective cross section for interaction of mobile electrons with the previously formed trapped electron centers.

Finally, for high fluences ($F_A \gg F_0$), the electron trapping process has saturated, and only the second process, resulting in Q_{it} formation, is occurring. Since $Q_{ot}(t_A) \approx Q_{ot}(\infty)$, a constant in this limit, Equation {6} describes a simple first order kinetic process, similar to that described by Equation (5). This asymptotic limit is illustrated in Figure 6 by the dash lines.

On the basis of the model described above, we conclude that production of trapped electron centers (Q_{ot}) is required for subsequent production of interface states and charge (Q_{it}). We suggest that the most likely candidate for the mobile species transported from bulk to interface is neutral atomic hydrogen.

Acknowledgements

We wish to thank D. Dong, F. Pesavento, and H. Chew for technical assistance. This research was performed at the IBM Thomas J. Watson Research Center under partial support by the Defense Advanced Research Projects Agency (monitor: Deputy for Electronic Technology, RADC, Contract F19628-16-C-0249). F.J.Feigl was supported at Lehigh University by the Office of Naval Research, Electronics and Solid State Science Program.

References

{1} B.E. Deal, J.Electrochem.Soc. 127, 979 (1980).
{2} E.H.Nicollian, C.N. Berglund, P.F. Schmidt, and J.M.Andrews, J.Appl.Phys. 42, 5654 (1971).
{3} D.J.DiMaria, "Defects and Impurities in Thermal SiO$_2$," in The Physics of SiO$_2$ and Its Interfaces, edited by S.T. Pantelides (Pergamon Press, New York, 1978), pp. 160-178.
{4} D.R.Young, "Electron Trapping in SiO$_2$", in Insulating Films on Semiconductors 1979, edited by G.G.Roberts and M.J.Morant (The Institute of Physics, London, 1980), pp. 28-39.
{5} R.A. Gdula, J.Electrochem. Soc. 123, 42 (1976).
{6} D.R.Young, E.A. Irene, D.J.DiMaria, R.F.DeKeersmaecker, and H.Z. Massoud, J.Appl.Phys. 50, 6366 (1979).
{7} Y.Miura, K. Yamabe, Y.Komiya, and Y. Tarui, J.Electrochem. Soc. 127, 191 (1980)
{8} S.Lai and D.R. Young, This Conference
{9} F.J.Feigl, D.R. Young, D.J.DiMaria, and J.A. Calise,J.Appl. Phys. (to be published).

Trapping Characteristics in SiO$_2$

D.R. Wolters, and J.F. Verwey

Philips Research Laboratory
5600 MD Eindhoven, The Netherlands

Abstract

Avalanche injection is a valuable tool for use in investigating
trapping characteristics. The analysis of the data can in a
number of cases be improved by the use of the Zeldovitsj equation.
For a number of charging curves reproduced from published data
this yields an excellent fit and a large reduction in scatter
of the capture cross sections. A model is presented based on
trapping site generation.

Introduction

One of the extensively used techniques to study trapping phenomena in MOS capacitors is high frequency avalanche injection
(Nicollian, Goetzberger and Berglund {1}). The Si-substrate is
driven in deep depletion and by the large field in semiconductor
minority carriers are accelerated. They gain a high energy from
the field leading to impact ionisation and multiplication. Since
they acquire a relatively high energy a large number is injected
over the interface barrier into the oxide. High injection currents of 10^{-3} amps/cm^2 can be obtained. The attractive feature
is that there is a homogeneous emission of charge carriers and
charging of the dielectric takes place over the whole area.
Flat band voltage shifts or the injection field can be used to
monitor the charge density.

Traditionally the flat band voltage shifts are described in terms
of first order chemical reaction kinetics and the applicability
was more or less proven in the first papers of Nicollian et al.{1}.
In the later papers the introduction of more than one trap with
varying trapping efficiency (capture cross section) was needed
to explain the results (Ning and Yu {2}, Young {3}, DiMaria {4}).
In a number of recently published papers the large scatter in
the results for neutral traps with small capture cross section
became clear. Therefore doubts against the model used can be
raised {1} - {6}.

Presumably, the application of first order reaction kinetics is
not always allowed. We think that the first order chemical
reaction kinetics does not hold for a large number of charging
experiments.

Charging

In the papers on high frequency avalanche injection the basic assumptions are:
a. The trapping rate is proportional to the concentration of the free electrons.
b. The trapping rate is proportional to the number of empty traps.
c. A trap has a constant trapping probability expressed in a capture cross section.

Let J be the current density, n the density of filled traps, N-n the density of free traps, v_{th} the thermal v_d the drift velocity of the electron and σ_c the capture cross section of the traps. Then the charging rate is given by

$$dn/dt=(N-n)J\sigma_c v_{th}/qv_d . \qquad (1)$$

When J is constant integration yields

$$n=N\{1-exp\ (-J\sigma_c v_{th}t/qv_d)\}. \qquad (2)$$

Plotting ln(dn/dt) versus t should yield a straight line with slope $J\sigma_c v_{th}/qv_d$. When this is done for the charging curves it can be often seen that the curves are convex toward the t axis (see fig. 1). This occurs in a large number of charging curves. Ning and Yu {2}propose that more traps with varying σ_c's could account for this and all data are nowadays interpreted as multi-trap charging.

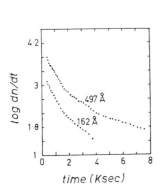

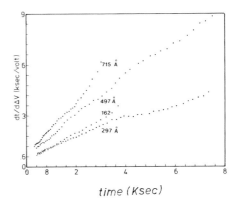

Fig.1. Replotted data from Young (Fig.6 Ref.{5}) to demonstrate that $\ln(d\Delta V_{fB}/dt)\neq t$

Fig.2. Same data as in Fig.1 but now replotted as $(d\Delta V_{fB}/dt)^{-1}$ vs t. The lines extrapolate to approximately the same value t_i-t_0 on the t axis

We propose a trapping probability P(n) which decreases exponentially with the number of trapped charges. Instead of Eq. (1) we propose

$$dn/dt=J\sigma_c v_{th}P(n)N/qv_d \qquad (3)$$

where

$$P(n)=exp(-n/N) \qquad (4)$$

112

The combination of Eqs. (3) and (4) yields
$$dn/dt = N\exp(-n/N)/t_0 \tag{5}$$
where
$$t_0 = qv_d/J\sigma_c v_{th} . \tag{6}$$
Eq. (5) is equivalent to the Zeldovitsj (or Elovich) (equation used extensively in chemisorption kinetics . Integration of Eq. (5) yields
$$\exp(n/N) - 1 = (t-t_i)/t_0 \tag{7}$$
where $n=0$ when $t=t_i$;
rearranging gives
$$n = N\ln((t-t_i+t_0)/t_0); \tag{8}$$
differentiating yields
$$dn/dt = N/(t-t_i+t_0) . \tag{9}$$
Plotting dn/dt vs. t yields the value of t_i-t_0 at $(dn/dt)^{-1}=0$. Incorporation of (t_i-t_0) in Eq.(8) makes it possible to plot n vs.$\ln(t-t_i+t_0)$. Extrapolating to $n=0$ yields the value of t_0 from which $\sigma_c v_{th}/v_d$ can be calculated.

<u>Data on charging</u>

The data on charging experiments have been extracted from literature. The most complete data sets were published by Young {5} giving plots of $\Delta V_{fB}(=nx_0\bar{x}q/\varepsilon)$ vs.t where x_0 is the oxide thickness and $\bar{x}$ is the centroid of the charge measured with respect to the gate, ε is the dielectric constant and q the electronic charge, n is the charge density. The data sets were reproduced by photographical magnification and digitalising the coordinates of arbitrarily chosen, approximately equidistant, points of the curves. A Philips microcomputer PM 4410 was used for replotting and algebraic conversions. Extrapolated values (t_0) and slopes (N) were calculated by the linear regression method. In fig. 2 Young's data (fig. 3 of ref. {5}) are replotted as $(d\Delta V_{fB}/dt)^{-1}$ versus time .

The line was extrapolated to $(d\Delta V_{fB}/dt)^{-1}=0$ which yields t_i-t_0. In fig. 3 ΔV_{fB} is plotted versus $\ln(t-t_i+t_0)$ and straight lines are obtained. The extrapolation to $\Delta V_{fB}=0$ gives the value of t_0. The same procedure was followed for figs. 4, 5 and the results are tabulated in table I.

In order to calculate N from $d\Delta V_{fB}/d\ln(t-t_i+t_0)$, i.e., the slopes of the plots in figs. 3, 4 and 5 it was assumed that $\bar{x}=1/2x_0$ and $N=(2\varepsilon/qx_0^2)d\Delta V_{fB}/d\ln(t-t_i+t_0)$. It must be mentioned that in fig.4 there is a large deviation at the last part of the charging curve. This is referred {5} to as the "turn around effect" and will be ignored here as it was also ignored in the treatment followed by Young {5}. See also {6}.

<u>Results</u>

In table I the values of σ_c and N for the three replotted data sets given in fig. 3, 4 and 5 are compiled. The first thing to notice is the excellent correspondence of the σ_c values. The scatter has been greatly reduced to less than half a decade while in one series (samples 5 to 8) the scatter is even within the limit of error to be expected from reproducing the data.

It is clear that for MOS samples having the same technological treatment σ_c as well as N have consistent values. They are in the range where we expect the neutral water related trap. The radius of the centre is calculated to be 1.6-2.4 Å which is the dimension of a water molecule or an O-H group. The trend in the series, having had different post oxidation treatments, is that the trap density increases when hydrogen is introduced in the oxide (cf. sample 10 and 11 vs. 9, 12, 13).

N_2 treatments reduce the trap density which is in accordance with our intuitive knowledge. To compare the traditional method and the one presented here, a plot was made of the data published by Young {5} (reference {5} table I) and the corresponding values acquired by the method presented here (sample 9-13 in table I here).

The values of samples 1-8 table I have been added for comparison. It can be seen that the trap density is almost one order of magnitude larger than found by Young. The values of 25-120 ppm seem, however, to be realistic. In the presentation of Young four different traps A, B, C, D are assumed to exist. It is, however, more likely that they all correspond to one and the same trap. The σ_c found by our analysis is very close to those reported by Ushirokawa {10}; his values 1.76, 1.85 and 2.57 ($\times 10^{-18} cm^2$) were acquired from avalanche injection experiments under constant bias.

Theory

In chemical kinetics logarithmic time laws are often met with (e.g. chemisorption). The relationship between rate and charge is referred to in literature as the Zeldovitsj Equation (or Elovich Equation) (Aharoni {8} Low {9}). The number of physical models having various concepts but leading to the same time laws (cf. Eqs. 8 and 9) is very abundant. The concept of an activation energy barrier for the trapping reaction cannot be used here since it predicts an increasing reaction rate in the case of charging, under increasing bias instead of a decreasing one (cf. Eq. 5). Ning has published an empirical relationship {7} between the capture cross section and the oxide field, which is given by

$$\sigma_c = \sigma_{co} \exp(-b\ E_{ox}) \tag{10}$$

where

$$\sigma_{co} = 1.60 \cdot 10^{-15} cm^2 \text{ and } b = 7.35 \cdot 10^{-7} cm/V.$$

Ning {7} explained the field dependence of capture by the positively charged centres by a combination of barrier lowering and electron heating but decided that for the neutral traps only electron heating effects may be expected. Substitution of σ_c in Eq. (1) by Eq. (10) directly leads to the Zeldovitsj equation Eq. (5) and would provide the base for the analysis of the data given in the previous section.

however, Ning {7} presented his relationship as an empirical relationship and provided no other explantation than the unknown electron distribution in SiO_2.

114

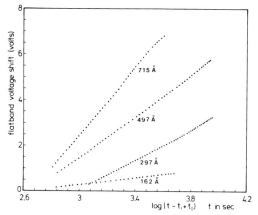

Fig.3 Same data as in Fig.1,2 but now re-plotted as ΔV_{fB} vs $\ln(t - t_i + t_0)$. The lines extrapolate to approximately the same value t_0 on the t axis. From t_0, $\sigma_c v_{th}/v_d$ can be calculated

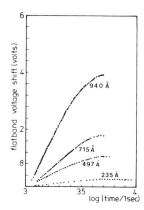

Fig. 4 Same plot as fig. 3 but now for data from reference {5} fig. 4.

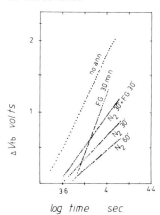

Fig.5 Same plot as Fig.3 but now for data from Ref.{5} Fig.13

Another argument is that the data used by Ning were analysed by the multiple trap model and it is not certain what is the exact value of the σ_c's. A very simple derivation of the exponential factor is the following {11}. Assume that an injected, subsequently trapped, electron annihilates a fixed volume h, where $h \ll V$, V is the total volume.

The free volume for trapping is decreased by a fraction $h' = h/V$ The probability of finding a non-annihilated part of the volume is $(1-h')$. When n electrons are trapped this probability is decreased to $(1-h')^n \approx \exp(-nh')$ the approximation is very good for n large and $h' \ll 1$. Given that n electrons have already been trapped the trapping rate will be given by

$$dn/dt = \exp(-nh')/t_0 h' \qquad (11)$$

By substitution of $N = (h')^{-1}$ we find a form equal to Eq. (5). The difficulty of providing a physical model for Eq.(11) is that the trapping continues also when $n > N$. This can be seen from

115

eq. (8) when $t-t_i>(e-1)t_0$, a situation which occurs in figs. 3, 4, and 5. It is therefore probable that trapping sites are generated during the process. The generation rate decreases with a reciprocal time law as can be seen from Eq. (9). There is a correspondence with wearout mechanisms, where failure rates decrease in time with (1/t) laws. See the review paper in this proceedings (12). If it is assumed that the mechanism for trapping site generation is the same as proposed for wearout, the trapping site generation starts at a void or a local disturbance of the dielectric constant. By the local acceleration electrons may gain >4 eV which is enough to break a Si-0 bond. This centre may function as an electron trap.

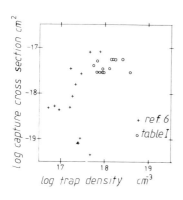

Fig. 6 Compilation of σ_c and N for data analysed by the traditional method (table I reference {5} corresponds to table I loc. cit sample 9-13) and all the data from table I.

Table I

sample	data from ref.[5] fig.	parameter (value)	σ_c [$\star 10^{18}$cm^2]	N [☆ 10^{18}cm^3]
		oxide thickn.		
1.		$x_0 = 16.2$ nm	5.72	2.7
2.		$x_0 = 29.7$ nm	2.90	3.6
3.	fig. 6	$x_0 = 49.7$ nm	5.76	1.8
4.	T = 77°K	$x_0 = 71.5$ nm	5.92	1.4
		oxide thickn.		
5.		$x_0 = 23.5$ nm	2.65	0.75
6.	fig. 4	$x_0 = 49.7$ nm	2.65	0.85
7.		$x_0 = 71.5$ nm	2.50	0.65
8.	T = 300°K	$x_0 = 94.0$ nm	2.61	0.90
		POA		
9.		No anneal	3.5	1.30
10.		F.G. 30'	5.9	1.66
11.	fig. 13	$N_2$30' + FG30'	3.3	0.86
12.		N_2 30'	5.2	0.64
13.	T = 293°K	N_2 60'	4.2	0.53

Trapping characteristics calculated with Eq.(8) from Young's data ref. [5] figs. 4, 6 and 13.
F.G. 30' stands for a **P**ost **O**xidation **A**nneal in Forming Gas for 30 minutes; $N_2$30' is the same treatment in N_2 etc.
See for assignments ref. [5].

When a void is frequently employed by passing electrons, all gaining energy, the contents may be ionised. The whole void acts as a conductor increasing the apparent polarization of the dielectric. The charge accumulation at both sides of the void would decrease the applied field in the local situation inhibiting further "trap generation". The reader is referred to {12} for an extensive description.

Since t_0 and N are derived from the extrapolation to $t \to 0$ we need not take the trap generating mechanism into account for the determination of σ_c and N.

Conclusions

1. The simple first order kinetic equation proposed to analyse charging by avalanche injection of SiO_2 is not adequate to describe the results in some cases.
2. The application of the Zeldovitsj equation used in chemisorption kinetics gives a perfect fit on data published recently in literature.
3. The acquired trapping characteristics for three sets of data are consistent.
4. A model based on trapping site generation is presented

The authors wish to thank T. Hoogestijn for the necessary software and H.J. de Wit, C. Crèvecoeur, H.C. de Graaff and F. Vollenbroek for valuable discussions. We want to acknowledge of H.J. de Wit and W. Rey the valuable contributions for the derivation of Eq. (11).

References

1. E.H. Nicollian, A. Goetzberger and C.N. Berglund, Appl.Physics Lett. 15(1969) 174. E.H. Nicollian and C.N. Berglund, J. Appl. Phys. 42 5654 (1971)
2. T.H. Ning and H.N. Yu, J. Appl. Phys. 45,5373 (1974).
3. D.R. Young, E.A. Irene, D.J.DiMaria and R.F. De Keersmaecker, J.Appl.Phys. 50(1979) 6366
4. D.J.DiMaria (1978) Proceedings of the International Topical Conf. on the Physics of SiO_2 and its interfaces and S.T. Pantelides N.Y. Pergamon Press p. 160.
5. D.R.Young, Ins. of Phys. Conf. Series 50(1980) 28.
6. R.A. Gdula, J. Electrochem. Soc. 123 42 (1976)
7. T.H. Ning, J. Appl. Phys. 49 5997 (1978)
8. C.A. Aharoni and F.C. Tompkins, Advances in Catalysis and related subjects 21 (1970).
9. M.J.D. Low, Chem.Rev. 60(1960) 267.
10. A. Ushirokawa,   E. Suzuki and M. Warashina, Jap.J.Appl. Phys. 12 398 (1973).
11. The essential form of Eq. (11) was contributed by H.J. deWit and W. Rey.
12. "Dielectric breakdown and wearout". D. Wolters. This proceedings.

Interface Effects in Avalanche Injection of Electrons into Silicon Doixide

S.K. Lai and D.R. Young

I.B.M. Thomas Watson Research Center
Yorktown Heights, NY 10598, USA

In the study of the dependence of electron trapping on processing, a turn-around effect of the flatband voltage shift has been observed [1,2]. Instead of increasing monotonically with increasing injection of electrons, the flatband voltage shift reaches a maximum and then decreases with further injection. This effect was observed after $10^{16}/cm^2$ to $10^{17}/cm^2$ electrons were injected into the oxide. Similar interface effects had been observed when electrons were injected into silicon dioxide by different techniques [2,3,4] and they could be reduced significantly by dry oxidation and annealing [1,2,5]. It is believed that this effect is related to the trapping of electrons in water related centers in the bulk of silicon dioxide [2,6]. Some details of the process have been studied [6]. In the present paper, the electronic properties of the turn-around effect will be presented.

The observed turn-around effect is due to the generation of fast and slow interface states at the silicon-silicon dioxide interface. The process can best be understood by studying Fig. 1. After the trapping of electrons,

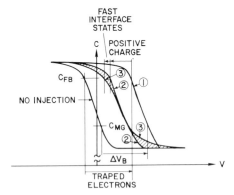

Fig. 1 Schematic capacitance curves showing the shifts due to different charge components in the oxide and at the interface after injection of electrons by avalanche in silicon

there is the bulk negative charge which will shift the capacitance curve in the direction of positive voltage (curve 1). The shift is parallel since there is no charge exchange with silicon during the voltage sweep. Then there are the slow states which are donor states. They are partially empty of electrons to give a positive charge. The effect of the positive charge is to shift the capacitance curve in the negative voltage direction (curve 2). The shift is again parallel as these slow states do not exchange charge with silicon during the course of a voltage sweep. Finally, there are the fast interface states which do exchange charge with silicon to give a distortion in the capacitance curve (curve 3). The sign of charge in fast interface states had been studied in oxides damaged by radiation [7]. Because of the special shape of the interface state spectrum, it was possible to conclude that the fast interface states are acceptors above midgap towards the conduction band and donors below midgap towards the valance band. There is no charge due to interface states at midgap. In the present experiments, similar distortion in the interface state spectrum is observed and the above signs for fast states are also consistent with present results. Therefore, the distortion in curve 3 is pivoted around the midgap point. The turn-around effect is caused by the sum of positive charge in the slow and fast states. Most of the positive charge are from the slow states with the fast states giving only a small contribution. Experimentally, the density of trapped electrons in the bulk of the oxide is determined by the photo I-V technique which measures only the net charge in the bulk of the oxide and is not sensitive to charge within 3 nm of the interface [2,8]. From the difference between the positive photo I-V shift and the shift in midgap voltage, the density of positive charge is determined. The density of fast states is measured by quasistatic capacitance technique [9].

For a 50 nm oxide, the capacitance curves before and after injection of $10^{18}/cm^2$ electrons are shown in Fig. 2. The generation of interface states

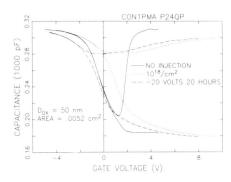

Fig. 2 High frequency and quasistatic capacitance curves in a 50 nm oxide before and after injection of $10^{18}/cm^2$ electrons and after subsequent bias stress at room temperature

can be seen from the distortion of the high frequency capacitance curve and the quasistatic curve. Also shown are the capacitance curves after a room temperature bias stress of -20 volts for 20 hours. The curves are shifted in the negative voltage direction showing an increase in positive charge. However, there is little change in the quasistatic curve showing that there is no increase in interface state density. The positive charge comes from emptying of electrons from the slow states. Applying a positive bias stress has the opposite effect. The process is, however, slow and not very reproducible.

The charging and discharging of the slow states is enhanced at elevated temperature. Figure 3 shows the midgap voltage as a function of tempera-

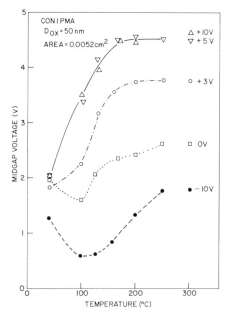

Fig. 3 Midgap voltages as a function of temperature for different capacitors in a 50 nm oxide after annealing under different biases

ture for different bias stress. Biases of +5 volts or above give the same effect, showing that the process is not enhanced by electric field. For these biases, the shifts in midgap voltages are the same as positive photo I-V shifts showing that all the positive charge is neutralized. For biases less than 5 volts, the slow states are partially empty after the temperature stress, giving some positive charge. For negative biases, there is an increase in positive charge at lower temperature due to the emptying of electrons. However, at higher temperatures, there is some annealing of the slow states and the density of positive charge is decreased even though it is under a negative bias. The difference between +5 volts to -10 volts gives a lower limit to the slow state density of $10^{12}/cm^2$. Another effect of the temperature is to make it more difficult to charge and discharge the slow states. After a positive or negative heat bias stress, reversing the stress field for the same time and temperature only give small changes in the flatband voltages. However, it was observed that when the wafer was heated repeatedly during measurement of other capacitors over a period of months, the flatband voltage was shifted back to that characteristic of zero bias. The slow states are still present but the time constants are increased. The fast states are also annealed by the heat bias stress, and the annealing is much more efficient under a positive bias than under a zero or negative bias. It is possible that the slow states communicate with silicon through the fast states by some thermally activated process. This will explain the temperature dependence and the observation that the time constants for slow states are increased after annealing.

In conclusion, both fast and slow interface states are generated in the process of avalanche injection of electrons into silicon dioxide and their subsequent trapping in the bulk of silicon dioxide. These interface states cause instability in the flatband and inversion voltages in MOS systems. In the study of electron trapping in the bulk of the oxide, it is necessary to carry out avalanche injection at elevated temperature. This will minimize the generation of fast interface states and with the average positive DC bias during avalanche, the slow states are filled by electrons. The interface effects are thus insignificant. The best way to minimize both the electron trapping and the interface states is to use a ultra-dry oxide [5].

Acknowledgement

The authors would like to thank D.J. DiMaria of Watson Research Center and F.J. Feigl of Lehigh University for many very valuable discussions on electron trapping and interface effects. The technical support and sample preparation by F.L. Pesavento and J.A. Calise are greatly appreciated. They would also like to thank D.J. DiMaria and M.H. Brodsky for a critical reading of the manuscript. The work was supported by the Defence Advance Research Projects Agency, Washington D.C. and monitored by the Rome Air Development Center, Deputy for Electronic Technology, Solid State Science Division, Hanscom AFB, MA 01731 under contract F19628-78-C-0225.

References

1. R.A.Gdula, *J. Electrochem. Soc.* **123**, 42, 1976.
2. D.R. Young, E.A. Irene, D.J. DiMaria, R.F. DeKeersmaecker, H.Z. Massoud, *J. Appl. Phys.* **50**, 6366, 1979.
3. S.Pang, S.A. Lyon, W.C. Johnson, *The Physics of MOS Insulators Proceedings of the International Topical Conference* Ed G. Lucovsky, S.T. Pantelides, F.L. Galeener (New York: Pergamon) pp285-289
4. D.J. DiMaria, R. Ghez, D.W. Dong, *J. Appl. Phys.* **51**, 4830, 1980.
5. S.K. Lai, D.R. Young, J.A. Calise, F.J. Feigl, *submitted to J. Appl. Phys.*
6. F.J. Feigl, D.R. Young, D.J. DiMaria, S.K. Lai, J.A. Calise, *accepted by J. Appl. Phys.*
7. G.A. Scoggan, T.P. Ma, *J. Appl. Phys.* **48**, 294, 1977.
8. D.J. DiMaria, *J. Appl. Phys.* **43**, 4073, 1976.
9. M. Kuhn, *Solid State Electronics* **13**, 873, 1970.

Interface State and Charge Generation by Electron Tunneling into Thin Layers of SiO$_2$

K.R. Hofmann and G. Dorda

Siemens Research Laboratories, Otto-Han-Ring 6
D-8000 München 83, Fed. Rep. of Germany

Introduction

The generation of interface states and the accumulation of positive charge at (or near) the Si-SiO$_2$ interface of MOS-structures have been observed by several authors under various physical conditions [1-9]. This paper presents the first quantitative study of the interface state and charge generation in connection with the injection of electrons into the SiO$_2$ layer by Fowler-Nordheim tunneling. The dependence of these effects on the number of injected electrons, the polarity and strength of the applied electric field, and on the oxide thickness have been investigated.

Sample preparation and experimental techniques

Our samples were poly-Si/SiO$_2$/Si MOS capacitors fabricated on p-type (100) Si substrates of 20 Ωcm resistivity. The oxide layers were thermally grown in dry O$_2$ with 2 % HCl at 900 °C. Their thickness t_{ox} ranged from 8.0 to 42.5 nm. The poly-Si electrodes were prepared by low-pressure CVD, then n$^+$-doped to N_D = 3 x 10^{20} cm^{-3} with phosphorus. After an electron-gun metallization with Al/Si/Ti and photolithography the devices were annealed at 475 °C for 30 minutes. The mobile ion content of the capacitors as determined by temperature bias stress was less then 5 x 10^{10} cm^{-2}. The current injection by Fowler-Nordheim (FN) tunneling of electrons through the interface barrier into the SiO$_2$ was achieved either from the Si-substrate or the poly-Si electrode by applying positive or negative bias voltages (V_{inj}) to the field plate of the capacitors. Electric fields between 6 and 9 MV/cm were used, corresponding to current densities between 1.3 x 10^{-9} and 2 x 10^{-5} Acm^{-2}. The current-field characteristics of the samples very accurately followed a FN-law. During the injection experiment the current density usually was held constant at j = 5 x 10^{-6}Acm^{-2} for injected electron densities $N_{inj} \leq 10^{16}$ cm^{-2} and at j = 2 x 10^{-5}Acm^{-2} for N_{inj} above this value. At chosen time intervals the injection was interrupted, and the high frequency capacitance-voltage curve and the conductance-voltage curve were measured. From these the flat-band shifts ΔV_{FB} and the fast interface state density N_{ss} near midgap were determined as a function of the injected electron density N_{inj}. In order to gain additional information about the distribution of the charge in the oxide also the low current FN-characteristic (for injection from the Si-substrate) was measured during the injection intervals [6]. Thus the parallel voltage shifts of the FN-characteristic, the "FN-shift" ΔV_{FN} was obtained as a function of N_{inj}.

Results and discussion

The results of an injection experiment with positive applied voltage (electron injection from the Si substrate) are shown in Fig.1. A negative flat-band shift ΔV_{FB} is found, which increases sublinearly with the injected electron density N_{inj} until reaching a saturation around $N_{inj} \approx 10^{17} cm^{-2}$ (curve a). At the same time with growing N_{inj} a strong increase of the density of fast interface states N_{SS} occurs (curve b) running almost parallel to curve a. That this behaviour indeed depends only on N_{inj}, the number of electrons per unit area passed through the oxide, was checked by using different current densities for the injection experiment. An injection current 25 times smaller than usual (and correspondingly longer time intervals) gave the same result as that in Fig.1. Even long-time experiments using a 3 orders of magnitude smaller current density ($1.3 \times 10^{-9} Acm^{-2}$) within a factor of 2 led to the same ΔV_{FB} at $N_{inj} = 10^{16} cm^{-2}$.

The experiments with negative applied voltage (injection from the poly-Si electrode) yielded a very similar behaviour (Fig.2). The flat-band shift is also negative, it increases sublinearly with N_{inj} and saturates above $N_{inj} = 10^{17} cm^{-2}$. Its value is approximately 3 times higher than in the case of positive V_{inj}.

In Fig.3 the measured increase of the fast interface state density near midgap, ΔN_{SS} is plotted against ΔN_{FB}, which represents the effective density of positive elementary charges at the interface corresponding to ΔV_{FB} observed at the same stage of injection. For both polarities of injection the increase in fast interface state density ΔN_{SS} is approximately proportional to ΔN_{FB} which points to their close interrelation. At a given ΔN_{FB} (ΔV_{FB}) a considerably larger generation of interface states is observed for the positive polarity than for the negative.

In experiments performed on a series of capacitors with different oxide thicknesses t_{ox} similar $\Delta V_{FB}/N_{inj}$ and N_{SS}/N_{inj} curves were obtained. But it was found that for a given N_{inj} the flat-band shift ΔV_{FB} of the capacitors is approximately proportional to the square of the oxide thickness t_{ox}. This is demonstrated by Fig.4, where the values of $-\Delta V_{FB}/t_{ox}$ at $N_{inj} = 10^{16} cm^{-2}$ are plotted against the corresponding oxide thicknesses t_{ox}. This $\Delta V_{FB} \sim t_{ox}^2$ dependence suggests that the positive charge responsible for the negative flat-band shift is homogeneously distributed in the bulk of the oxide layer.

The obvious conclusion of a positive bulk charge is, however, not confirmed by the FN-shifts measured (for injection from the Si-substrate) at the injection experiments. If the negative flat-band shift ΔV_{FB} was caused by a positive bulk charge, a negative Fowler-Nordheim shift ΔV_{FN} of approximately the same magnitude should result. By way of contrast always a positive ΔV_{FN} is found which is absolutely smaller than ΔV_{FB} and increases with N_{inj} (Fig.1 and 2, curve c). This indicates a net negative charge in the oxide under the conditions of FN-injection from the Si-substrate.

This discrepancy between a positive bulk charge following from the quadratic dependence of the negative ΔV_{FB} on t_{ox} and a net negative charge following from the observed FN-shifts can be explained as follows: 1. the positive charge appearing in the flat-band shift derived from the C-V measurement actually represents an interface charge at the SiO_2/Si-substrate

123

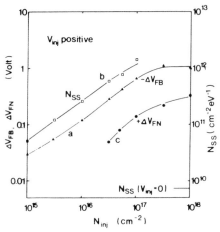

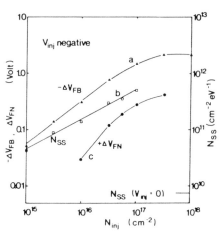

Fig. 1 Injection from the Si-substrate. (a) flat-band shift ΔV_{FB} (b) fast interface state density N_{SS} (c) Fowler-Nordheim shift ΔV_{FN} plotted against the injected electron density N_{inj}. t_{ox} = 21.6 nm

Fig. 2 Injection from the poly-Si electrode. (a) flat-band shift ΔV_{FB} (b) fast interface state density N_{SS} (c) Fowler-Nordheim shift ΔV_{FN} plotted against the injected electron density N_{inj}. t_{ox} = 21.6 nm

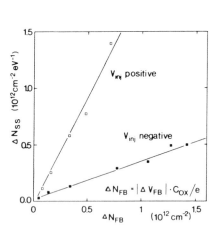

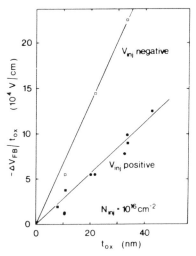

Fig. 3 Increase of fast interface state density ΔN_{SS} versus the effective density of elementary charges ΔN_{FB} corresponding to ΔV_{FB}. t_{ox} = 21.6 nm

Fig. 4 Flat-band shift over t_{ox} plotted against t_{ox} at N_{inj} = 10^{16} cm^{-2} for experiments with different oxide thicknesses. The straight lines prove the $\Delta V_{FB} \sim t_{ox}^2$ relationship.

boundary. It has only little effect on the FN-characteristic, either because it is located very close to the SiO_2/Si-substrate interface [6], or
because it consists largely of donor-type interface states which are occupied by electrons under the condition of a high applied positive voltage
[5] prevailing during the measurement of the FN-characteristic. 2. The
effective negative charge responsible for the positive FN-shift ΔV_{FN} is
due to trapped electrons in the oxide. A residue of this negative trapped
charge also shows up in a small positive ΔV_{FB} found in injected samples
after annealing in Argon for 20 minutes at 400 °C. 3. The quadratic dependence of the flat-band shift on the oxide thickness is not the consequence of a homogeneous positive charge distribution in the oxide. It
rather indicates an effect occuring in the bulk of the oxide by which surface states and interface charge are generated in proportion to the volume
and thus to the thickness of the oxide layer.

As this effect is essentially independent of the polarity of injection,
simple explanations like impact ionization followed by trapping of holes
at the interface [2,8] seem to be excluded, at least for the case of $V_{inj} < 0$.

It can therefore be assumed that under the condition of electron injection by Fowler-Nordheim tunneling a neutral agent is produced in the
bulk of the SiO_2, which diffuses to the SiO_2/Si-substrate interface and
causes there the observed generation of interface states and positive
charge. A possible agent for this process could, for example, be excitons,
as has been proposed by WEINBERG and RUBLOFF [8].

References

1 D.J. DiMaria, Z.A. Weinberg, J.M. Aitken, J. Appl. Phys. 48 898 (1977)
2 M. Shatzkes, M. Av-Ron, J. Appl. Phys. 47, 3192 (1976)
3 K.O. Jeppson, C.M. Svenson, J. Appl. Phys. 48, 2004 (1977)
4 L. Risch, this conference
5 D.R. Young et al, J. Appl. Phys. 50, 6366 (1979)
6 P. Solomon, J. Appl. Phys. 48, 3843 (1977)
7 P.M. Solomon, J.M. Aitken, Appl. Phys. Lett. 31, 215 (1977)
8 Z.A. Weinberg, G.W. Rubloff, Appl. Phys. Lett. 32, 184 (1978)
9 Genda Hu, W.C. Johnson, Appl. Phys. Lett. 36, 590 (1980)

Modelling of Flat-Band Voltage Shift During Avalanche Injection on MOS Capacitors

Massimo V. Fischetti

Physics Group, SGS/ATES Componenti Elettronici S.p.A.
20019 Settimo Milanese, Milano, Italy

1. Introduction

It has been suggested {1} that the creation of donor type electronic states at the Si-SiO$_2$ interface during electron avalanche
injection is the mechanism responsible for the "turn around"
of the flat band voltage shift ΔV_{fb} observed in Al-gate MOS
capacitors at room temperature {2}. In the present work we take
(and modify) this idea as a starting point to develop a
phenomenological model which explains-on empirical grounds-the
observed behaviour of ΔV_{fb} vs. injected charge over a wide range
of temperatures (from 77°K to 400°K). We shall consider features
that a simple buildup of positive charge does not account for,
such as the absence of the "turn around" at temperatures different
from room temperature. Hints to construct the simple model are
taken from results of observations performed in our laboratories
{3} and from results reported in the literature.

2. The Model

The model is based on the following considerations:

1. The growth of surface states density N_S(cm^{-2}) created-directly
or indirectly-by the hot electrons can be described by an
exponential approach to a saturation value with time constant τ_f.
They are annihilated at temperatures as low at 250°C with annealing
time constant τ_{an}.

2. We shall assume that the kinetics of formation and annealing
of the (initially) positive oxide defects is "similar" to the
kinetics of the surface states. This does not imply that positive
charges and surfaces states are necessarily the same objects,
but that they are at least created and annihilated by similar
mechanisms.

3. The positive defects created by hot electrons probably act
as electron traps themselves. This is suggested by the characteristics of the instability of ΔV_{fb} observed after the injection
has been turned off {2} which may be viewed as a charging and
discharging of the positive traps by some phonon assisted electron
tunneling (and/or hopping) {4}. Furthermore the electronic
occupancy of the positive (neutral) defects does not affect their
annealing behaviour (which is likely to be an ionic process).

Thus the formation, annealing and neutralization of the positive charges can be described by a first order kinetics. For the concentration N_d (cm^{-2}) of the defects we get

$$\frac{dN_d(t)}{dt} = \frac{1}{\tau_f}\left[N_T - N_d(t)\right] - \frac{1}{\tau_{an}} N_d(t) \qquad (1)$$

where N_T (cm^{-2}) is the total concentration of "sites" at which defects may be formed.
The fraction N_n of those defects which are occupied by electrons will be given by

$$\frac{dN_n(t)}{dt} = \frac{1}{\tau_c}\left[N_d(t) - N_n(t)\right] - \frac{1}{\tau_{an}} N_n(t) \qquad (2)$$

where $\tau_c^{-1} = J\sigma_c/q$ is the electron capture rate of the defects. σ_c is their "equivalent" capture cross section and J the injected current density.

The overall flat-band voltage shift can be obtained by solving (1) and (2) with $N_d(0)=N_n(0)=0$ and by subtracting the net positive charge $N_d(t) - N_n(t)$ from the negative charge trapped in the oxide. Thus

$$\Delta V_{fb}(t) = \frac{qtox}{\varepsilon_{ox}}\left\{ \sum_{i=1}^{n} \bar{x}(i) N_{ox}^{(i)} (1-e^{-t/\tau_{ox}^{(i)}}) - \right.$$

$$-\frac{N_T}{\tau_f}\tau_{eq}\left[\left(\frac{\tau_L}{\tau_C}\right)\frac{\tau_L}{\tau_L-\tau_{eq}} (e^{-t/\tau_L}-e^{-t/\tau_{eq}})+ \right. \qquad (3)$$

$$\left.\left. + (1-\frac{\tau_L}{\tau_C})(1-e^{-t/\tau_{eq}})\right]\right\}$$

where $\tau_L = (\tau_{an}^{-1} + \tau_c^{-1})^{-1}$ and $\tau_{eq}=(\tau_{an}^{-1}+\tau_f^{-1})^{-1}$. We have assumed that n types of traps exist in SiO$_2$, each with concentration $N_{ox}^{(i)}$, capture rate $\tau_{ox}^{(i)-1}$ and centroid $\bar{x}(i)$.

3. Summary of the Experimental Results and Discussion

Measurements have been performed on samples having 1090Å thick wet oxides on 0.3Ω-cm, <100> p-type substrates with 1.26×10^{-2} cm^2 Al-gate and sealed in TO-5 in N$_2$. A 330 kHz, 33% duty cycle square wave was automatically adjusted in amplitude to yield a constant current of 3.02×10^{-7} Amperes.

In fig. 1 and table 1 we show the results obtained by least square fitting the experimental curves with (3). Figure 1(a) shows that the positive oxide defects are quickly annihilated for temperatures higher than 350°K and quickly created for T≤250°K. Their neutralization by electrons is strongly enhanced at low

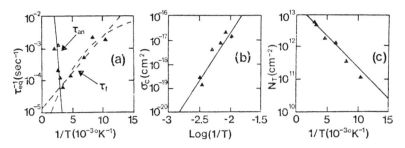

Fig.1 Temperature dependence of the parameters τ_{an}, τ_f, σ_c and N_T. We have $\tau_{an}=8.5\times10^{-8}\exp(0.79eV/kT)$sec; $\tau_f=1.0\times10^{-4}T^{3.3}$sec(or $9.2\times10^4 x \exp(-0.045eV/kT)$sec); $\sigma_c=3.5\times10^{-9}T^{-4.11}$ cm²; $N_T=3.1\times10^{13}\exp(-0.047eV/kT)$cm²

Table 1   Parameters relative to pre-existing oxide traps

$N_{OX}\{cm^{-3}\}$	$\sigma_{OX}\{cm^2\}$	$\bar{x}$	Energy of thermal detrapping{eV}
$(6\pm2)\times10^{17}$	$(2.7\pm0.7)\times10^{-18}$	0.5 tox (ref.2)	1.3 ± 0.1
$(1\pm0.5)\times10^{17}$	$(4.5\pm1.5)\times10^{-19}$	0.5 tox (ref.2)	1.8 ± 0.2
$(1\pm0.5)\times10^{17}$	$(2\pm1)\times10^{-16}$	tox (ref.5)	0.3 ± 0.05(ref.5)

temperatures (fig.1(b)), while the total number of sites available for formation of a defect is increasing with T exponentially
(fig. 1(c))

We see in fig. 2 the consequences of this picture: for T≤250°K
not much positive charge. can be formed and, once created, it is
quickly neutralized. For T≥350°K, annealing proceeds very fast.
Thus in neither case is the turn around observed. Only for
temperatures close to 300°K, being both annealing and neutralization of the positive charge slow processes, we do observe
the "turn-around".

To conclude, it is evident that additional experimental work is
needed to discriminate among the various mechanisms which may be
invoked to support on a physical ground the phenomenological picture emerging from the model we presented. However we would like
to stress that our work is consistent with the idea that trivalent

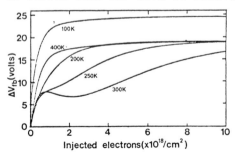

Fig.2 Flat-band voltage shift
as a function of the injected
charge at several temperatures
simulated by (3) and with the
results shown in fig.1.

Si defects {6} and Si-H bonds may play a relevant role. Indeed theoretically computed energies of the $\equiv Si_S^{\cdot}$ and $\equiv Si_0^{\cdot}$ defects {7} agree with our observations: a strong peak of N_{SS} is found at 0.35eV above the silicon valence band ($Si_S^{\cdot}$) and the symmetry with respect to opposite
biases exhibited by the instability of ΔV_{fb} after avalanche suggests that the positive charge is energetically close to the Si conduction band (Si_0^+).

Moreover our observations about the time dependence of the buildup of surface states do not agree with the behaviour expected in the case of diffusion limited processes ($\propto t^{1/4}$ {8} or $\propto t^{1/8}$ {9}). Since interstitial hydrogen is a fast diffuser, the breaking of Si_0-H and Si_S-H bonds is a plausible origin of the defects. We are not able to say whether hydrogen {10}, excitons {11} or something else could be the causes of this bond breaking.

As a possible cause of the low temperature annealing, the diffusion of hydrogen originated to the $Al-SiO_2$ interface may be invoked.

While the temperature dependence of the parameters τ_f, τ_{an} and N_T may be shown to be at least not inconsistent with this picture, we have no explanation for the T^{-4} behaviour of σ_c. Indeed, as noticed by NING {12}, high field and electron heating effects should kill the temperature dependence of the coulombic electron capture cross section via acoustical phonon emission.

References

1. Y.Miura, K. Yamabe, Y. Komiya and Y.Tarui, J.Electrochem. Soc. 127, 191 (1980)
2. D.R.Young, E.A.Irene, D.J.DiMaria and R.F.De Keersmaecker, J.Appl.Phys. 50, 6366 (1979)
3. M.V. Fischetti, R.Gastaldi, F.Maggioni and A.Modelli, unpublished results
4. D.J.Breed, Solid St.Electr. 17, 1229 (1974)
5. T.H.Ning, J.Appl.Phys. 49, 5997 (1978)
6. C.M.Svensson, in Proceedings of the International Conference on the Physics of SiO_2 and Its Interfaces (Pergamon, New York 1978) p. 328
7. R.B. Laughlin, J.D.Joannopoulos and D.J.Chadi, in Proceedings of the International Conference on the Physics of SiO_2 and Its Interfaces (Pergamon, New York, 1978) p. 321
8. K.O.Jeppson and C.M. Svensson, J.Appl.Phys.48,2004 (1977)
9. R.B.Fair and R.C.Sun, IEEE Trans. Electron Device ED 28, 83(1981)
10. Z.A. Weinberg, D.R.Young, D.J.DiMaria and G.W. Rubloff, J.Appl.Phys. 50, 5757(1979)
11. Z.A.Weinberg and G.W.Rubloff, Appl.Phys.Lett. 32, 184(1978)
12. T.H.Ning, J.Appl.Phys. 47, 3203(1976)

Influence of Electron-Phonon Scattering on Photoinjection into SiO_2

J.v. Borzeszkowski, M. Schmidt

Zentralinstitut für Elektronenphysik der AdW Berlin
Berlin, DDR

In this note it is shown that the energy loss of photoinjected electrons by polar optical scattering influences essentially the photoinjection current in SiO_2. To describe this effect, a simple model is proposed which takes into account the multiple scattering of electrons and its energy dependence. The results obtained by numerical calculations are supported by experimental investigations of photoinjection.

Numerical Calculations and Experimental Results

The photoinjection current is determined by the excitation conditions and by the transport of excited charge carriers in the emitter, and, on the other hand, by scattering processes within the insulator in the region of the image force potential (see Fig.1). Until now, however, the influence of scattering has been accounted for only under the following assumptions [1]:

1) Only electrons not scattered along its way to x_0 contributes to the current. The photoinjection current then is given by

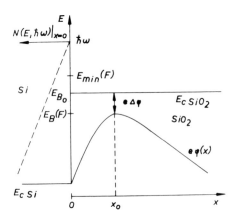

Fig. 1. Simplified representation of an Si/SiO_2 heterojunction for photoinjection from Si. $N(E_0,\hbar\omega)$ — supply function, E_B — zero field barrier energy, $E_B(F) = E_B - e\Delta\varphi$ — field dependent barrier energy, $e\Delta\varphi \sim \sqrt{F}$ — barrier lowering, $x_0 \sim 1/\sqrt{F}$ — position of barrier maximum, $E(x) = E_0 - E_{pot}(x)$, $E(x)$ — kinetic energy, E_0 — injection energy, $E_{pot}(x) = e\varphi(x)$ — potential energy determined by image force, internal and external electric fields

$$I(\hbar\omega,F) = c \int_{E_B(F)}^{\hbar\omega} N(E_0,\hbar\omega) \left[\int_0^{\theta_{max}} P(F,\cos\theta,E_0)d(\cos\theta) \right] dE_0 \quad , \tag{1}$$

where $N(E_0,\hbar\omega) = (E_0-\hbar\omega)$ is the energetic distribution of the electrons injected at the insulator interface (supply function)[2]. $P(F,\cos\theta,E_0)$ is the probability that an electron injected at an angle θ, with respect to the normal direction at the interface, will reach x_0 without scattering,

$$P(F,\cos\theta,E_0) = \exp\left\{ -\frac{1}{\cos\theta} \int_0^{x_0} \frac{dx}{\ell[E(x)]} \right\} \quad . \tag{2}$$

Here $\ell(E) = v(E)\tau(E)$ is the mean free path of an electron with the velocity $v(E)$ and the mean free flight time $\tau(E)$ between two interactions.

2) The mean free path of the electrons is assumed to be constant, independent of the excitation conditions and the electric field strength in the insulator.

Under assumptions 1 and 2, comparing the photoinjection current $I(\hbar\omega,F)$ to the photoemission current I_0 into the vacuum, $I(\hbar\omega,F)$ is only modified by an attenuation term, $I(\hbar\omega,F) = I_0 \exp(-x_0/\ell)$. However, it must be mentioned that these assumptions simplify the electron transport in polar insulators in a manner which cannot be justified by physical arguments. As one knows, in such insulators the polar scattering of electrons on optical phonons predominates.

In [3] it is shown that the predominating interaction of electrons with polar optical phonons does not lead to a constant attenuation length and that it is not possible to interpret it microscopically as the mean free path between two electron-phonon interactions.

As long as one adheres to assumption 1, even the consideration of the energy dependence of the electron-phonon interaction does not make expression (1) to a suitable description of the photoinjection current.

The bad correspondence between theory and experiment in Fig.2 is due to the strong anisotropy of the polar-optical scattering. This anisotropy leads, for electrons with an energy being large in comparison with the phonon energy, to a small angle scattering (forward scattering); by this scattering, the momentum direction of the electrons is not changed significantly such that one expect that electrons injected with a sufficiently high energy can overcome the barrier maximum in the insulator, also after multiple interaction with phonons.

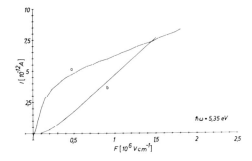

Fig. 2. Experimental photoinjection current (a) in comparison with photoinjection current calculated via eq.(2) (d_{0X} = 114 nm, $V_{FB} \simeq \Phi_{ms}$). The photon energy is given by $\hbar\omega$ = 5.35 eV. The numerical calculations were carried out by consideration of two modes of optical photons $\hbar\omega_{q1}$ = 0.06 eV, $\hbar\omega_{q2}$ = 0.15 eV

To summarize this point, $I(\hbar\omega,F) = I_0 \exp(-x_0/\ell)$ describes the photoinjection current only phenomenologically. The microscopic interpretation is also not possible on the basis of (2).

Our concept is to consider the multiple scattering of electrons within a simple model. For this end, we assume that the electrons exchange energy with the electric field F at a rate given by e F v and by small-angle scattering events at a rate given by (3), ($W^{\pm}$—electron-phonon scattering rates [4])

$$\left(\frac{\partial E}{\partial t}\right)_{SC} = \sum_i \hbar\omega_{q_i} (W_i^- - W_i^+) \qquad (i=1,2) \quad . \tag{3}$$

The electrons then undergo the average energy loss

$$\Delta E = \int_0^{x_0} \frac{1}{v} \left(\frac{\partial E}{\partial t}\right)_{SC} dx \tag{4}$$

between the emitter surface and the barrier maximum in the insulator. Only electrons with a minimal injection energy

$$E_{min}(F) = E_B(F) + \Delta E(F) \tag{5}$$

may overcome the potential barrier. The photoinjection current then simply amounts to

$$I(\hbar\omega,F) = c \int_{E_{min}}^{\hbar\omega} N(E_x,\hbar\omega)dE_x \quad . \tag{6}$$

$N(E_x,\hbar\omega)$ results from the projection of the isotropic supply function $N(E_0,\hbar\omega)$ on the emitter surface. E_{min} increases more rapidly than the Schottky barrier (see Fig.3). This reflects the strong influence of the multiple electron-phonon scattering for field strength smaller than 10^6 Vcm^{-1}.

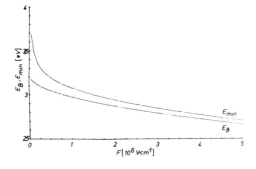

Fig. 3. Comparison of the field-dependent Schottky barrier energy $E_B(F)$ with the minimal injection energy $E_{min}(F)$

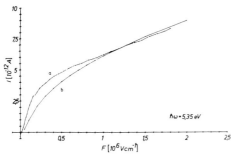

Fig. 4. Experimental photoinjection current (a) in comparison with the calculated photoinjection current (b) via eq.(6) (d_{0x} = 114 nm, $V_{FB} \simeq \Phi_{ms}$). The photon energy is given by $\hbar\omega$ = 5.35 eV

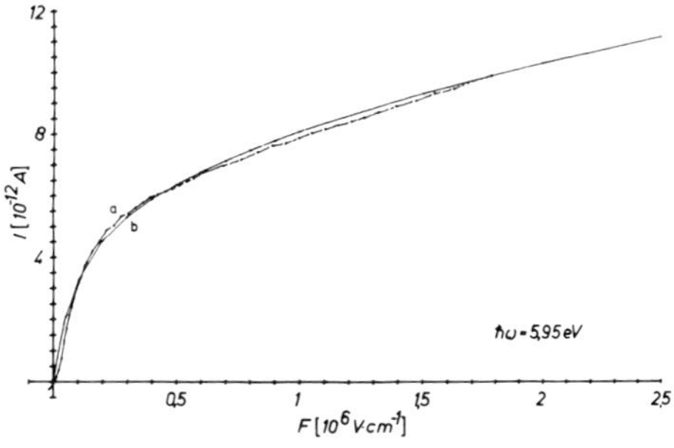

Fig. 5. Experimental photoinjection current (a) in comparison with the calculated photoinjection current (b) via eq.(6) (d_{0x} = 114 nm, $V_{FB} \simeq \Phi_{ms}$). The photon energy is given by $\hbar\omega$ = 5.95 eV

To test the quality of equation (6) photoinjection measurements were performed at 300 K and at 77 K. The samples used were MIS capacitors fabricated on n-type silicon wafers [10 Ωcm, (111)] with an optical semitransparent Al gate electrode. Figures 4 and 5 show that even the simple incorporation of multiple scattering of electrons described by (6) leads to a satisfying correspondence between theory and experiment. The discrepancies yet existing for small exciting energies and small field strength (see Fig.4) are caused by those electrons which have the energy $E \simeq \hbar\omega_q$ near x_0; contrary to our assumption, those electrons are not scattered by small angles. The removal of this shortcoming of our model and also the contribution of photon assisted tunneling will be discussed in a forthcoming paper [5]. The experimental observed independence of the photoinjection on temperature indicates that the spontaneous emission of phonons is the dominant mechanism of energy loss.

Conclusions

The model proposed here demonstrates that one must take into account the multiple scattering of electrons and its energy dependence in order to find a microscopic interpretation of the photoinjection current. Consequently, this point of view requires rediscussing the methods of determining the barrier energy, the insulator charge, etc., by photoinjection measurements.

References

1. C.N. Berglund, R.J. Powell: J. Appl. Phys. *42*, 573 (1971)
2. G.W. Gobeli, F.G. Allen: Phys. Rev. *127*, 141 (1962)
3. J. v. Borzeszkowski, M. Schmidt: Proc. "Physik der Halbleiteroberfläche", AdW (ZIE) 1980
4. E.M. Conwell: Solid State Phys., Suppl.9 (Academic, New York 1967)
5. J. v. Borzeszkowski, M. Schmidt: to be published in Phys. Status Solidi b

Part IV

Multilayer Structures

Charge Loss in MANOS Memory Structures

H. Teves, N. Klein*, and P. Balk

Institute of Semiconductor Electronics/SFB 56 "Festkörperelektronik"
Technical University Aachen, Templergraben 55
D-5100 Aachen, Fed. Rep. of Germany

Introduction

The metal-Al_2O_3-Si_3N_4-SiO_2-Si (MANOS) system is an attractive gate system for MIS non-volatile memory devices since it permits storage of negative charge in a well defined area (the Si_3N_4 layer) close to the substrate /1/. It exhibits relatively low write and erase voltages, but the rate of loss of negative charge is comparable to that of MNOS devices /2/. In the latter case the leakage at short storage times has been interpreted as being caused by direct tunneling of electrons from the nitride traps into the silicon; at longer times they first have to drift through the nitride before reaching tunneling distance /3-10/. The present study shows that the same basic model also fits the charge drift in and loss from the layered dielectric in the MANOS structure. By increasing the thickness of the SiO_2 film the tunneling rate decreases and the charge redistribution in the Si_3N_4 becomes apparent.

Samples and test methods

MANOS capacitors were produced with thermally grown SiO_2 layers ranging from 2 to 10 nm, 10 nm thick pyrolytic Si_3N_4 and 60 nm thick pyrolytic Al_2O_3 films, all prepared at 900°C on 1 Ωcm n or p substrates. The thinner SiO_2 films were grown in 10 % O_2/N_2, these of 5 nm and thicker in O_2. Part of the samples were subsequently annealed in O_2 since this treatment is known to reduce the current through the Al_2O_3 at a given field /11/. Al dots (0.8 mm diam.) were prepared by e-gun evaporation through a mask and annealed for 15 min. in N_2. The charge condition of the samples was determined from standard high frequency C-V measurements, supplemented by current-time measurements during charge loss under short-circuit conditions. The flat band voltages V_{FBO} of the samples as prepared were a few tenths of a volt and positive.

Experimental results

Samples were charged with a positive gate pulse to obtain an initial flat band voltage of +10 V, measured approximately 10 sec after charging. Typical results of V_{FB} vs t measurements at room temperature are presented in Fig.1 for samples with SiO_2 layer thickness w_s of 2 and 10 nm. The charge decays roughly logarithmically with time but in the O_2 annealed samples with $w_s \gtrless 4$ nm only after an initial increase. This effect could be enhanced by application of a small negative voltage and diminished with a small positive voltage applied to the gate. The rate of decay decreased drastically for the sample with thicker SiO_2 film. Samples with n and p type substrates behave identically. Fig.2 shows the relative rate of charge loss per decade of time, $r_a = \left| \left[(V_{FBO}\bar{V}_{FB})/\log t \right]/V_{FBO} \right|$. The main feature of the data is

---

*On leave from Technion, Haifa, Israel

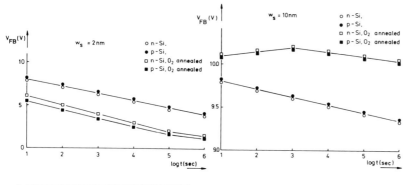

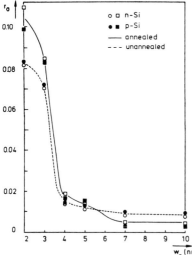

Fig.1 V_{FB} vs log t for short-circui-
ted MANOS samples with SiO_2 thicknesses 2 and 10 nm

Fig.2 Relative rate of charge loss
per.decade of time vs SiO_2 thickness

the rapid change in r_G between w_s=3 and 4 nm. At smaller w_s values the change
is rather modest; for larger values r_G appears to become nearly constant.
Measurement of the temperature dependence of r_G for thick oxide samples between 50 and 250°C yielded only small activation energies ($\leqslant$ 0.3 eV).

The short circuit current during discharge, measured between 10^{-3} and 10^3
sec after discharging, exhibited a roughly inverse dependence on time in all
cases. The current was larger for thinner samples but flowed always in a
direction opposite to that of the charging current.

Discussion

We will discuss the data using the concepts of tunneling and drift mentioned
in the introduction. As shown in Fig.3 we make the simplifying assumption
that the stored charge is confined to the 1 eV deep Si_3N_4 potential well /2/,
which is certainly not quite correct for the unannealed samples /12/, and
that the charge is evenly distributed over this region immediately after
charging. After /3/ charge is lost by direct tunneling from the nitride

traps into the silicon. Whereas initially most charges emerge from traps near the SiO_2-Si_3N_4 interface, the depletion front gradually moves inward, leading to a logarithmic time dependence for V_{FB}. Assuming that during the discharge process the field in the SiO_2 has the same constant value in all cases the slope of the V_{FB} vs log t plot should be independent of w_s. The plots will only move to larger values of t with increasing w_s and, in the case of MNOS structures, the decrease of V_{FB} with log t would first start at $t=10^6$ sec (the end point of our measurement) for w_s values of approximately 3.5 nm. Roughly the same r_α values were indeed observed for MANOS samples with w_s=2 and 3 nm, together with a considerable shift of the decreasing V_{FB} vs log t branch with w_s.

At the same time a drift process will take place in the nitride, leading by itself to an increased charge concentration near the SiO_2-Si_3N_4 interface where the field is largest and thus giving larger V_{FB} values. This implies, that in agreement with our observations V_{FB} cannot yet have reached a saturation value and that a fraction of the traps in the nitride are still unoccupied. The drift towards the Al_2O_3 will be much smaller due to the considerably smaller field in the part of the nitride adjoining the Al_2O_3. When for capacitors with $w_s > 3$ nm the linear decrease of V_{FB} with log t moves towards the end of the observation window the competing process of charge drift should make itself increasingly felt. This should lead to a smaller decrease or even to an initial increase of V_{FB}, as indeed exhibited by the data of Figs.1 and 2. It should also be mentioned, that the data points for O_2 annealed samples which show an initial increase for $w_s > 4$ nm change slope at the same value of t (10^3 sec). Refering to Fig.3 it may be argued, that the value of the field in the SiO_2 (3 MVcm^{-1}) means for example that for $w_s > 4$ nm all electrons tunneling to the silicon from states in the nitride more than 1.2 eV below the top of the Si_3N_4-SiO_2 barrier see the same triangular barrier. This statement is a fair approximation since it implies that the field in the SiO_2 remains the same, i.e. both charge loss and w_s dependence of the field at a given charge are negligeably small. In the unannealed samples the charge will be smeared out over a larger depth region and extend into the Al_2O_3. Consequently, it may be argued that their r_α values for thin SiO_2 capacitors should be somewhat smaller (lower electron concentration near the SiO_2-Si_3N_4 interface), their r_α values for thick oxide capacitors larger (slower redistribution) than thos measured on O_2 annealed samples.

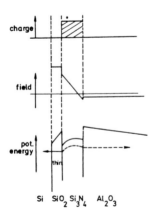

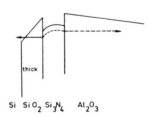

Fig.3 Stored charge, electric field and band diagrams showing potential energy distribution in MANOS structures with thin and thick SiO_2 films.

138

The effects of positive and negative bias during the tunnel and drift processes fit the above qualitative model and so does the sign of the "discharge" current. Appreciable charge drift towards the gate electrode is not to be expected due to the small field near and in the Al_2O_3 layer and because the presence of the Si_3N_4-Al_2O_3 barrier. The activation energy for the charge loss process for samples with thick SiO_2 films is difficult to interpret. However, its small value suggests a dominant role of the tunneling process through the SiO_2 barrier in this region of the V_{FB} vs log t plot.

By the use of a relatively large range of SiO_2 thicknesses this study was able to qualitatively separate the effects of internal drift and tunneling. The influence of the field (i.e. the amount of stored charge) is presently being investigated.

Acknowledgments

The authors are indebted to the generous help of H. Kliem with the short-circuit current measurments. They wish to thank D.R. Young for useful discussions.

References

1. P. Balk, Proc. 8th Int. Vacuum Congress (Ed. F. Abélès and M. Croset), Soc. Franc. du Vide, Paris, Vol. I, 525, 1980
2. F. Stephany, M. Schumacher and P. Balk, Rev. Phys. Appl. 13, 829 (1978).
3. L. Lundkvist, I. Lundström and C. Svensson, Solid-State Elecr. 16, 811 (1973)
4. H.E. Maes and R.J. van Overstraeten, J.Appl. Phys. 47, 667 (1976)
5. L. Lundkvist, C. Svensson and B. Hansson, Solid-State Electr. 19,221 (1976)
6. C.A. Neugebauer and J.F. Burgess, J. Appl. Phys. 47, 3182 (1976)
7. A.V. Ferris-Prabhu, IEEE Trans. Electron Dev., ED-24, 524 (1977)
8. K. Lehovec and A. Fedotowsky, Appl. Phys. Lett. 32, 335 (1978)
9. H.E. Maes and G.L. Heyns, J. Appl. Phys. 51, 2706 (1980)
10..H. Yamamoto, H. Iwasawa and A. Sasaki, IEEE Electron Dev. Lett., EDL-2, 21 (1981)
11. H. Kampshoff, F. Stephany and P. Balk, J. Electrochem. Soc., 124,1761 (1977)
12. A. El-Dessouky, PhD Thesis, Aachen, 1979

Surface-State Density Evaluation Problems in MNOS Structures

P. Tüttő, J. Balázs, Zs.J. Horváth

Research Institute for Technical Physics of the Hungarian, Academy of Sciences
1325 Budapest, Ujpest 1, P.O. Box 87, Hungary

1. Introduction

The capacitance-voltage investigations are of peculiar importance in the study of interface properties in MIS structures.
In our research work on MNOS memory structures we investigated
the quasi-static and high-frequency C-V characteristics of MNOS
capacitors and found that the usual evaluation methods give
contradictory results.

2. Experimental

The samples were prepared on 1 ohmcm n-Si substrate. The slices
were chemically oxidized to a thickness of about 1.7nm. The CVD
Si_3N_4 films had a thickness of about 40-80nm. The deposition
conditions were chosen to obtain optimal memory properties. On
the slices Al dots with an area of 1.4×10^{-4} cm^2 were formed.

To depress the influence of electrical instabilities the quasi-
static capacitance C_{lf} and the high-frequency capacitance C_{hf}
were recorded simultaneously (see Fig.1.a). C_{lf} and the bias V_m
were recorded as a function of C_{hf} to simplify the evaluation.
Fig. 1.b shows a typical measured C_{lf}-C_{hf}-V_m characteristic.

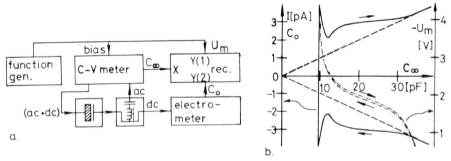

Fig. 1   a: The scheme of the measurement,
        b: Experimental C_{lf} and V_m vs. C_{hf} plot measured on MNOS
        memory structure (d_N=62 nm).

The different evaluation methods gave contradictory results:
-High frequency C-V characteristics were almost ideally abrupt,
$N_{ss}(V)$ evaluation from their slope {1} gave moderate values

$(10^{11} cm^{-2} eV^{-1})$.
- Quasi-static curves measured alone and also together with the high-frequency ones {2, 3} resulted in rather high N_{ss} values ($\sim 10^{12} cm^{-2} eV^{-1}$), moreover the surface potential evaluation by integrating the quasi-static curve {3} proved to be not reliable, the change of the surface potential was too small.

From the contradiction it is clear that one or more of the generally accepted assumptions is not valid for these MNOS structures:
- Strong inhomogeneity in the dopant concentration near to the Si interface can not explain this behaviour for measurements on other structures with thicker oxide layers on the same slices showed negligible inhomogeneity.
- The presence of deep traps with a density in the order of that of the dopants can also be excluded because of the fairly good minority carrier lifetime (1-100μs at 310°K).
- Quasi-static interaction of traps within the insulator layer with silicon could result in similar C-V behaviour. The quantitative evaluation however showed that these traps should be 10-50 nm away from the Si interface, which is impossible.

Thus we concluded that the insulator has an internal polarization, characterised by a polarization potential V_p, which follows the bias at low electric fields, while it cannot follow the high-frequency signal. Thus the voltage on the metal electrode:

$$V_{metal} = V_{surface} + \frac{Q_{space-charge} + Q_{surface-states}}{C_{insulator}} - V_p(V_{insulator}).$$

(1)

C_{lf} and C_{hf} can be received in modified form {4}:

$$C_{lf} = \left\{ \frac{1}{C_{sc}+C_{ss}} + \frac{1/C_i}{1+(dV_p/dV_i)} \right\}^{-1} ; \quad C_{hf} = \left\{ \frac{1}{C_{sc}} + \frac{1}{C_i} \right\}^{-1}. \quad (2)$$

The insulator polarization results in a virtual decrease of the insulator thickness in the quasi-static capacitance, while it changes the high-frequency C-V curves through affecting the potential distribution (1).

From (1) and (2) we can derive the surface-state density distribution and the differential polarization:

$$N_{ss} = \frac{C_i}{q} \left\{ \frac{C_{lf}}{C_i} \left(\frac{dV_m}{dV_s} \right)_{lf} - \frac{1}{(C_i/C_{hf}-1)} \right\} ; \quad \frac{dV_p}{dV_i} = \frac{C_{lf}}{C_i} \frac{1}{1-(dV_s/dV_m)_{lf}} - 1.$$

(3)

Because of the insulator polarization we have to calculate the surface potential from the high-frequency capacitance {5}. This calculation is limited to a narrower band, from about flat-band to inversion. Fig. 2 shows typical surface-state distribution (a) and differential polarization plots (b) evaluated according to (3).

As can be seen in Fig. 2a, this type of insulator polarization can strongly affect the evaluation of the surface-state densities. Even a very little polarization effect in the insulator seriously affects the evaluation in strong accumulation and inversion; the small contribution to the quasi-static current results in a steep increase of the virtual surface-state density distribution towards the edges of the band gap.

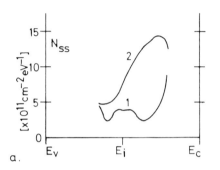 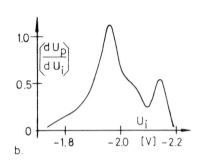

a. b.

Fig.2.a: Evaluated surface-state distribution at the Si-chemical SiO$_2$ interface. Curve 1. with taking into account of the insulator polarization; Curve 2. supposing that there is no polarization.
b: Evaluated differential polarization of the insulator.

The nature of the polarization in MNOS structures is not clear yet, however it can be explained if we suppose the presence of "free" charge carriers which move in a rather inhomogeneous built-in electric field found earlier {6}.

In accumulation the evaluation gave impossible negative values for dV_p/dV_i, indicating the validity limit of the surface potential evaluation, caused by the presence of surface-states which follow the 1 MHz high-frequency signal.

We have gathered more information about this fast-state effect in the measurement of the insulator capacitance C_i . One can obtain the insulator capacitance simply recording $d(1/C_{hf})/dt$ as a function of $1/C_{hf}$ in strong accumulation using linear ramp as bias.

$$\frac{d}{dt}\left(\frac{1}{C_{hf}}\right) = \left(\frac{dV_m}{dt}\right)C_i\left\{\frac{1}{C_s}\frac{d}{dV_s}\left(\frac{1}{C_s}\right)\right\} \qquad (4)$$

In accumulation when the surface-state density is negligible, $C_s = C_{sc}$, this plot is a parabola, the zero point of which yields the insulator capacitance.

Measuring MNOS capacitors the "parabola" had a characteristic symmetrical distortion (Fig.3.a) which was independent of the actual flat-band voltage values. This distortion appeared at biases less than the memory threshold voltage. The measured

$d(1/C_{hf})/dt$ plots clearly demonstrated the effect of a discrete surface-state level fast enough to follow the high-frequency signal. The degree of the distortion depended on the sample preparation. Samples with normal thermal oxide (>10nm e.g.) showed almost an ideal parabolic plot.

Calculated surface-state effect can be seen in Fig. 3.b for various densities of states having an energy level of E_C-150meV. It can be seen that this type of fast discrete states increases or decreases the slope of the high-frequency C-V curves depending on the surface potential. It should be noted that the distortion of the C-V curves appears at rather high semiconductor surface capacitance values, thus in normal C-V plots it can hardly be seen.

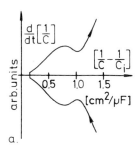 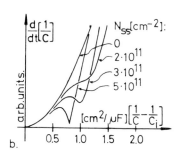

Fig. 3.a: Experimental $d(1/C_{hf})/dt$ vs. $1/C_{hf}$ plot measured on
MNOS memory structures (N_{SS}=3 x 10^{11}cm^{-2}; E_{SS}=E_C-150meV)
b: Calculated $d(1/C_{hf})/dt$ vs. $1/C_{hf}$ plots in accumulation
approximation.

3. Summary

Quasi-static and high-frequency C-V characteristics of MNOS memory structures were measured. From the analysis of the curves it follows that in the case of contradictory lf and hf C-V characteristics, one has to take into account the polarization of the insulator layer. From the two curves both the surface-state density distribution and the insulator polarization can be evaluated without further measurements. An important restriction is that the surface potential can not be received from the quasi-static curves.

In our MNOS structures a discrete surface-state level was found at about E_C-150meV, the density of which was technology-dependent. From the analysis of the effect of this type of states, which follows the high-frequency signal, it can be concluded that the joint presence and response of slow and very fast states can give rise to similar anomalous C_{lf}-C_{hf} behaviour. Both the insulator polarization and surface-state spectrum with distributed response time seems to be general in MIS structures made on compound semiconductors.

It is important that if we know the dopant distribution in the semiconductor the $C_{lf}-V$ and $C_{hf}-V$ curves have certain redundancy: if the surface states are slow "normal" states one can obtain information on the insulator polarizability, or in the case of a stable insulator we can distinguish between slow and fast states.

References
{1}   L.M.Terman, Sol.-St.Electronics 5, 285 (1962)
{2}   C.N. Berglund, IEEE Trans. on El. Devices ED-13, 701 (1966)
{3}   M.Kuhn, Sol.-St.Electronics 13, 873 (1970)
{4}   P. Tüttő, et al., Sol-St.Electronics to be published
{5}   A. Goetzberger, et al., CRC Crit.Rev. in Sol.-St.Sci. 6,1(1976)
{6}   G. Stubnya, et al., paper at ESSDERC 1980, York

Dye-Sensitized Photodischarge of Metal-Dye-Oxide-Silicon (MDOS) and Metal-Dye-Nitride-Oxide-Silicon (MDNOS) Capacitors

I.A.M. Wilson* and A.E. Owen

Department of Electrical Engineering, University of Edinburgh, King's Buildings Edinburgh EH9 3JL, Scotland, United Kingdom

1. Introduction

When a semiconductor or semi-insulator is coated with a suitable organic dye, or has the dye dispersed within it, a photo-current is often observed to flow through the sample if it is illuminated with light of energy less than the band-gap of the substrate. This is known as *dye-sensitisation*. The phenomenon is well-known in the fields of photography and electrophotography but few studies have been reported on wide band-gap solids such as silicon dioxide (SiO_2) and silicon nitride (Si_3N_4). For substrates of this sort even the basic question of whether sensitization is due to an electron transfer or energy transfer mechanism remains to be resolved. The work reported in this paper is an extension of that by FLYNN et al. [1,2] and HOLWILL et al. [3]; a type of xerographic discharge experiment is used to complement the previous results which were obtained mainly under conditions of constant applied bias. Voltage decay rates and photovoltages have been measured as a function of initial bias (magnitude and polarity), illumination (intensity and wavelength), and dye thickness, for various MDOS and MDNOS samples.

2. Experimental Details

Capacitor-like structures were prepared on 2 inch diameter silicon wafers, moderately doped either n- or p-type. The silicon was first thermally oxidised to produce SiO_2 layers with thicknesses ranging from 50 to 900 nm. For the dual-dielectric (MDNOS) samples the oxide layer was 100 to 120 nm thick and the Si_3N_4 layer was deposited on top of the SiO_2 by a chemical-vapour-decomposition technique; the thickness of the Si_3N_4 layer was between 120 to 160 nm on different samples. The SiO_2 was removed from the reverse side of the silicon wafer and an aluminium base contact deposited by thermal evaporation. Sample preparation was continued by thermal evaporation in vacuum of metal-free phthalocyanine (H_2Pc) onto the top insulating layer (SiO_2 or Si_3N_4) using an out-of-contact mask; the thickness of the H_2Pc layers varied from 50 to 400 nm. Finally, semi-transparent electrodes of gold or aluminium were evaporated onto the dye. For control purposes, some samples were also made without the dye-layer, i.e. the semi-transparent top electrodes were deposited directly onto the upper insulating layer. The top metal electrodes were in the form of fingers, with an effective area of 3.5 mm^2, but extending to larger aluminium contact pads.

The basic procedure of the experiment was first to charge a sample capacitor *in the dark*, with the bottom electrode grounded. A Cary 401 vibrating-reed electrometer was connected across the sample to monitor the decay in voltage on disconnection of the supply, and the analogue output of the electrometer was fed to an X/Y-time recorder with a voltage-offset facility. A typical trace from an un-illuminated dye-sensitised sample shows a slow (dark) decay. On illumination with light in the visible spectrum, there is a sudden increase *or* decrease in voltage. This is a *photovoltage*. The photovoltaic effect is followed immediately by an increased rate of voltage decay so long

---

*Present address: IRD, Fossway, Newcastle-upon-Tyne, England.

as the sample is illuminated. When the illumination is turned off, the photovoltage pulse is again observed but in the reverse direction and the slower *dark* decay rate resumes at a lower voltage than if the sample had not been illuminated. A control sample, without the H_2Pc layer, shows a similar dark decay and photovoltage, but no photo-induced discharge. Discharge rates were calculated from the slopes of the voltage versus time traces. The samples were rigidly mounted in a screened, evacuated and temperature-controlled enclosure which was light-proof except when a perspex light pipe, used to direct the illumination, was uncapped.

3. Results and Discussion

A complicating factor in experiments of the kind reported here is the presence and redistribution of space-charge in the oxide and nitride layers, *and* in the semi-conducting H_2Pc layer. This factor is very likely to be involved, either directly or indirectly, in the mechanism of dye-sensitization but since the space-charge density is altered by every application of voltage and illumination, and because relaxation times can be very long, experimental results are not precisely repeatable within a working timescale. Reproducible patterns can be seen, however, in the results of measurements of photo-discharge rates within the first few seconds of illumination.

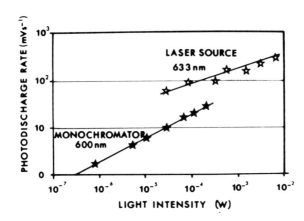

Fig. 1 Photodischarge rate as a function of light intensity (log-log plot) for an MDOS sample - dye contact biased positively.

The importance of the properties of the H_2Pc is illustrated by three experiments with the simpler MDOS structure. Firstly, the spectral dependence of the initial photo-discharge rate, for a given initial bias, closely matches the absorption spectrum of the dye and also its photoconductivity spectrum [3]. Secondly, the intensity dependence of the photo-discharge rate (V_{ph}) is close to what would be expected for the dye alone (i.e. $V_{ph} \propto I^{1/3}$ [4]) and is almost the same for both negative and positive bias. The intensity dependence is shown in Fig. 1, for an initial positive bias at a field of about 10^5 V cm^{-1}, and over an intensity range for red light of almost five orders of magnitude. (The two sets of data do not coincide exactly, partly because of the small difference in wavelength of the two sources and partly because they were obtained on adjacent samples which had different electrical histories.) The absence of any indication of saturation in the photo-discharge rate at high intensities (up to 10^{16} photons/sec. on the electrode area) suggests that the quantum efficiency of the process is low since the highest rate is equivalent to a current of about 5×10^8 charge carriers/sec. These results do not help decide between resonant energy transfer and direct electron transfer but the results of Fig. 2 suggest that energy transfer is unlikely. Fig. 2 plots the photo-discharge rate versus initial voltage (positive and negative bias), with the dye thickness as a parameter. Note that for positive bias there is a clear dependence on dye thickness whereas for negative bias, while there is some scatter in the results, V_{ph} seems to be independent of dye thickness. These results indicate that resonant energy transfer is very unlikely, certainly for positive bias and, to a lesser extent, also for negative bias. This follows because it is known that the range of

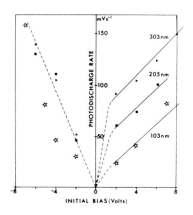

Fig. 2 Dependence of the photo-discharge rate on biasing voltage for an MDOS sample, with dye thickness as a parameter.

excitons in H_2Pc is much less than that of the dye thicknesses used in these experiments [5]. If excitons were to take part in energy transfer at the dye-insulator interface, they would have to be generated close to that inter-face. The probability of that happening decreases as the dye thickness increases and hence the discharge rate should *decrease* rapidly with thick-ness rather than increase, as is the case for positive bias, or remain essentially constant, as is the case for negative bias (Fig. 2). The above results on dye thickness and light intensity also apply to MDNOS and MDNS structures, since a broadly similar behaviour has been observed during photo-discharge experiments with varied initial biasing conditions.

The influence of the silicon substrate on the efficiency of the dye-sensitization process is emphasized by the results shown in Figs. 3 and 4. Both diagrams refer to an experiment in which an MDNOS capacitor on an n-type silicon substrate of resistivity 1 ohm-cm, with a top semi-transparent contact of gold, was charged to various initial voltages and then illuminated with *white* light using a microscope lamp. Fig. 3 shows that there is a sharp minimum in both dark and photo-discharge rates at initial negative biases between -5 and -7 volts. Otherwise the graphs are almost linear (similar to the MDOS samples of Fig. 2, under negative bias). In Fig. 4 the photovoltage is plotted against initial bias for two intensities of illumination, the stray room light being at least three orders of magnitude lower in intensity than the incident light from the microscope lamp.

With positive and small negative biases the photovoltage generated by illumination of relatively high intensity (the microscope lamp) is negative (i.e. the top contact is negative). As the bias becomes more negative, the "high intensity" photovoltage approaches zero and becomes positive when the bias is about -6 volts. At approximately -4 volts a photovoltage is observed under room light and this increases to about half the "full" light value at

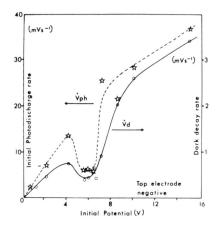

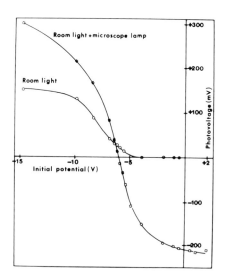

Fig. 3 Light and dark decay rates versus initial biasing potential for an MDNOS structure.

Fig. 4 Illustrating the effect of the level of illumination and initial potential on the photo-voltage of an MDNOS capacitor.

at about -10 volts bias; both curves are apparently tending to a saturation value of the photovoltage for biases greater than -10 volts. A similar behaviour was observed for an adjacent MNOS capacitor (i.e. without the dye), indicating that these photovoltages are not associated with the H_2Pc layer. It seems that two photovoltages are generated, one at the $Si-SiO_2$ interface, the other of opposite polarity most likely at the $SiO_2-Si_3N_4$ interface. At low illumination levels insufficient light reaches the silicon so the $SiO_2-Si_3N_4$ photovoltage predominates. For higher intensity illumination the two photovoltages balance at about -6 volt bias, which is also the bias at which an internal barrier is formed to charge carriers in the dark (Fig. 3). Further work is required to investigate the variation of the crossover point with different sample parameters in order to obtain more information on barrier heights at the several interfaces.

4. References

1. B.W. Flynn, A.E. Owen and J. Mavor. J. Phys. C.: Solid-St. Phys. 10, 4051 (1977).

2. B.W. Flynn, J. Mavor and A.E. Owen. Solid-State and Electron Devices (IEE), 2, 94 (1978).

3. R. Holwill, A.E. Owen and J. Mavor. J. Non-Cryst. Sol. 40, 49 (1980).

4. G.H. Heilmeier and S.E. Harrison. J. Appl. Phys. 34, 2732 (1963).

5. R.C. Nelson. J. Opt. Soc. Am. 51, 1186 (1961).

Part V

Interface Characterization Techniques

EPR on MOS Interface States

Edward H. Poindexter and Philip J. Caplan

US Army Electronics Technology and Devices Laboratory
Fort Monmouth, NJ 07703, USA

1. Introduction

The detailed structure and physical chemistry of the Si/SiO_2 interface assume even greater importance as integrated circuit technology advances into the submicrometer regime. Beyond the continued quest for knowledge of the origins of interface states, atomic features take on added interest because of defect carrier scattering in near-ballistic transport devices. Among common analytical techniques, electron paramagnetic resonance (EPR) has shown substantial ability to reveal the atomic nature and disposition of defects in the Si/SiO_2 system [1]. It is also useful in certain cases, such as very thin oxides, where conventional methods may not be applicable.

A number of different MOS defects have been observed by EPR. The most easily seen EPR signal is that at g=2.0055 [2] produced by simple physical damage -- grinding, nuclear radiation, etc. Reduced elemental iron (Fe^0) is surprisingly common in device-grade silicon, and gives rise to a signal at g=2.07, observed by NISHI [3], and designated P_c. A second common signal, P_a [3,4] is found near the interface at g=1.999 in phosphorus-doped wafers, and presumably arises from donor electrons with an undetermined degree of independence from the P atom. A variety of somewhat ill-defined recombination centers has been observed by both regular "dark" EPR and with light [5-8]. All of the above represent impurities or damage extrinsic to the MOS system. Even the best quality Si/SiO_2 interface, however, shows an inherent characteristic EPR signal, designated P_b [3], which has been later assigned [1] to unbonded orbitals on trivalent silicon (Si^{III}) at the unavoidably lattice-mismatched crystal boundary. It is this particular defect center which has the greatest interest for MOS physics, and is the subject of the studies discussed here.

We will be concerned with the atomic structure, crystallographic situation, and chemical behavior of this P_b center. The long-suspected connection of dangling silicon orbitals with interface states will be examined. The energy levels corresponding to different charge states of the P_b center, and response of the center to external stimuli such as electric fields or optical radiation are also of interest in establishing the role of this defect with respect to MOS device electrical features.

2. Experimental Details

Sample preparation procedures for these EPR studies have been presented elsewhere [1]. In this work, additional experiments involved visible or ultraviolet light. For the former, an ordinary slide projector was focused onto the wafer sample via the EPR cavity window. For the latter, a mercury lamp not near the spectrometer was used, since it was not necessary to irradiate samples while in the cavity. Wafer orientations were established by the Laue method, and silicon resistivity $\geq 50\Omega$cm was used throughout. Unless otherwise noted, EPR observations were taken at 295°K. Other details will be added in the following sections, as appropriate.

3. Results and Discussion

3.1 Crystallography of P_b Defects

Representative EPR signals from oxidized (110) silicon wafers are shown in Fig. 1. On (111) wafers, the signal-noise ratio is generally adequate (up to 40:1) to allow direct real-time display. On (100) the inherent defect concentration is several times lower, and the signal is split into components; the (100) signal was usually recorded after several hundred accumulations in a signal averager. Wafers sliced with (110) face are intermediate in difficulty, but are usually recorded directly (as shown).

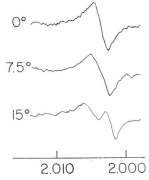

Fig. 1 Typical EPR spectra of P_b centers on oxidized (110) silicon wafers.

The P_b signals are all anisotropic; i.e., g-values vary with sample position in the spectrometer magnetic field; and multi-component signals change their appearance quite drastically. The diverse appearance of P_b signals on wafers of different orientation is clear proof that the resonance is crystallographically constrained, and the anisotropy is not an interface artifact. The P_b (111) center was earlier found to have perfect (within observational precision) rotational symmetry, with the minor axis of the g-tensor ellipsoid normal to the interface [1]. The principal values of the g-tensors for all three wafer orientations are included in Table I.

Table 1 Principal values of g-tensor for P_b signals on oxidized (111), (100) and (110) silicon wafers. Values for (110) are approximate.

| | (111) | (100) | | (110) |
	P_b	P_{b0}	P_{b1}	P_b
g_1	2.0013	2.0015	2.0012	2.0016
g_2	2.0086	2.0080	2.0076	2.0075
g_3	2.0086	2.0087	2.0052	2.0081

The principal values of the (111) g-tensor are very similar to those assigned to various forms of silicon vacancy structures in radiation-damaged bulk silicon [9]; P_b (111) does not at all resemble any other common defect in either bulk Si or SiO_2 [10]. The respective values for P_b (111) and P_{b0} (100) are nearly equal, indicating the equivalence of these centers. The dangling orbital of the $\cdot Si \equiv Si_3$ center on (100) is oriented in strict accord with the symmetry distinguishable Si-Si bond directions. The situation of the P_{b0} center on (100) precludes rotational symmetry, and so g_1 and g_2 are not quite equal.

The second center observed on (100), P_{b1}, has principal g-values which are unlike any common center ever observed in Si or SiO_2, and it is thus unique to the interface itself. No principal g-value is aligned with a possible dangling orbital direction, but g_1 and g_2 lie in the plane defined by a possible SiO bond and a dangling orbital on a singly-oxidized silicon atom. Thus P_{b1} is crystallographically consistent with a $\cdot Si \equiv Si_2O$ structure, and it is tentatively so assigned.

The other major wafer orientation is (110), and its anisotropy map (Fig. 2) is intermediate in complexity. The g-tensor is again not quite rotationally symmetric. Unlike (100), the anisotropy map of P_b (110) is not completely defined by a single sample rotation, in the (111) plane, since the presence of the interface eliminates possible dangling bond directions. The curves connecting the points in Fig. 2 are not a best fit to the (110) data points. Rather, the lines are calculated from the principal g-values observed on (111), with the theoretical tensor ellipsoid oriented to fit bond directions appropriate on the (110) face. Even though the (110) anisotropy study is not yet completed, the near-perfect fit indicates that the P_b (110) center is the same species as P_b (111) and P_{b0} (100). Again, it is oriented so that the dangling orbital is exactly along possible Si-Si bond directions as they intercept the (110) face. Only canted dangling orbitals are present on (110); no dangling Si orbitals are found in the plane of the (110) face itself.

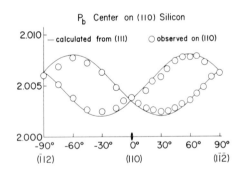

P_b Center on (110) Silicon

Fig. 2 Anisotropy of g-tensor for P_b centers on (110) silicon. Rotation plane $(1\overline{1}1)$

The observed P_b centers are portrayed schematically in Fig. 3. Since $P_b(111) \approx P_{b0}(100) \approx P_b(110)$, these centers are all labeled P_{b0}. Unobserved broken bond possibilities are not included. It may be stated as a generality that at the Si/SiO_2 interface, broken bond structures occur only in dispositions which may easily be imagined and portrayed at the advancing oxide front. There are no broken Si-Si bonds in the various possible directions beneath the interface, as might be expected if there were strain or distortion of the Si lattice. There is no evidence of oxygen penetration beneath the first row of silicon atoms, which would give rise to dangling orbitals in the other possible directions, not constrained by the interface plane. In summary, EPR portrays only the simplest possible interface defect structure.

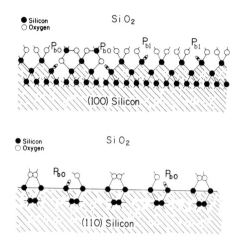

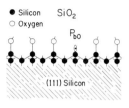

Fig 3. Trivalent silicon defects at the Si/SiO_2 interface revealed by EPR

3.2 Defects not Observed

It has been emphasized above that Si^{III} structures revealed in P_b signals occur only in certain allowed orientations depending on the interface crystallographic plane. Dangling orbitals in other directions are not observed, either for $\cdot Si \equiv Si_3$ or other (tentative) centers such as $\cdot Si \equiv Si_2 O$. Also not observed near the interface of thermally oxidized wafers is the well-known E' center of damaged SiO_2, $O_3 \equiv Si^+ --- \cdot Si \equiv O_3$ [10]. The E' center is a fairly stable, easily-produced oxygen-deficiency center in crystalline and amorphous quartz, and glass. It is a tempting candidate for a source of fixed oxide charge N_f (nee Q_{ss}). Also not observed to date is the non-bridging oxygen hole center [10], also often invoked in models of interface states and traps. Its unobservability at this stage should not be regarded too seriously, since its EPR line is so much broader than E' or P_b, or even damage signals, that special adjustments would be needed. However, since it is an oxygen-excess center, it seems chemically improbable near the strongly reducing atmosphere of the silicon front. Experiments are underway to seek a definitive answer.

On the silicon side of the interface, the common triplet silicon defects formed by bridging orbitals over a vacancy (Si-A or B_1 center) are also absent [9].

3.3 Physico-Chemical Aspects of P_b Centers and Interface States

Since the major interest in this study is the possible connection of $\cdot Si \equiv Si_3$ with interface states, a number of pertinent chemical and physical situations were tested. These have been described in detail elsewhere [1, 11] and will be summarized but briefly here:

153

(1) P_b concentration and midgap interface state density D_{it} are quanti-tively proportional over a wide variety of sample conditions; approximately, D_{it} x 1.1eV = (1 to 2) x P_b . For wafers oxidized in dry O_2 at 1000°C, with fast pull, P_b = 1 x 10^{12}cm^{-2}

(2) P_b and D_{it} are strong on (111), weak on (100), intermediate on (110).

(3) P_b and D_{it} are both drastically reduced by H_2 anneal or "alneal." They are both restored by subsequent moderate-temperature N_2 anneal. They are both much reduced in the case of steam oxide growth.

(4) P_b and D_{it} follow the Deal oxidation triangle.

(5) The numerical correlation is very good for both O_2 and Ar treatments, but less clearcut for high temperature N_2.

(6) P_b and N_f are correlated only as N_f and D_{it} are correlated.

3.4 Electrical Behavior of P_b Centers

The strong physico-chemical correlation between P_b and D_{it} suggests an at-tempt to vary the occupancy of the P_b center by application of an electric field while observing the EPR signal. If the energy of either the $(+\leftrightarrow 0)$ or the $(0\leftrightarrow -)$ levels is in the band gap of silicon, a surface potential should vary the occupancy as the P_b level traverses the Fermi level.

Several samples with an MOS structure of 1 cm^2 area were prepared, as in Fig. 4. Application of voltage did indeed cause a significant change in EPR signal amplitude, Fig. 5 [12]. The change was most rapid as the sample was swept through the flat band voltage, where the disposition of the surface carriers undergoes a change. Further testing, however, revealed that most

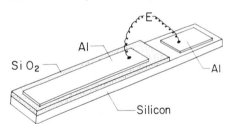

Fig. 4 MOS sample for voltage-controlled EPR study of P_b defects

of this change is due to the electron relaxation time T_{1e}. A variation of T_{1e} affects the signal amplitude when adjusted for maximum signal. Signal S varies with microwave field strength H_{1e}, $S \sim A H_{1e}\, T_{2e}(1 + \gamma_e^2\, T_{1e}\, T_{2e}\, H_{1e}^2)^{-1}$ where T_{2e} is proportional to the inverse of the observed EPR linewidth, and γ_e is the electron gyromagnetic ratio.

Fig. 5 Voltage-controlled P_b signal

It is thus necessary to determine S at several values of H_{1e} to deconvolve both the intrinsic signal strength and the relaxation time T_{1e}. The results are plotted in Fig. 6, for both T_{1e} and net EPR polarization A. The relaxation change of 3:1 completely overshadows the polarization change, rendering the latter very inaccurate. At best, a weak downward drift under positive voltage is observed, suggesting accretion of the second electron in the P_b center as the bands are bent downward. This would require the (0↔-) level of the P_b center to be somewhere between midgap and the conduction band. A similar downward drift with negative voltage would place the (+↔0) level between the valence band and midgap, but the experiments are not clear on this possibility. An analogous effect with high positive or negative corona discharge appears to substantiate this view [13].

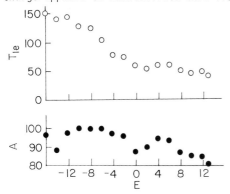

Fig. 6 Electron relaxation time T_{1e} and net polarization A of P_b centers as a function of applied voltage

The decisive relaxation effect is, nevertheless, a significant result. The postulated mechanism is relaxation via conduction electrons attracted to the interface by positive applied voltage. The mechanism for such electron spin relaxation might be either dipole-dipole or exchange. A calculation of the possible dipole-dipole effect by the well-known approach [14] follows,

$$T_{1e}^{-1} = N\gamma_e^4 \hbar^2 \int_0^\infty (r^2/10D)(2\pi r)(r^2 + d^2)^{-3} \, dr \, . \tag{1}$$

Here, T_{1e} is P_b electron relaxation time, N is carrier concentration, D is thermal diffusion constant for the conduction electrons, and d is the effective distance between P_b electrons and the carrier sea. We obtain

$$T_{1e}^{-1} = \pi N \gamma_e^4 \hbar^2 (10Dd^2)^{-1} \, . \tag{2}$$

To explain the observed T_{1e} (10^{-5} sec) requires d=0.1A, which is unacceptably small. Thus, the dipole-dipole mechanism is entirely too weak, by several orders of magnitude. This confirms the results of other studies in defect-center spin relaxation [15]. It has been deduced that exchange of defect electrons with the rapidly-relaxing conduction electrons is a common mechanism for defect relaxation [16]. The relaxation time of conduction electrons at 295°K is about 10^{-10} sec, and even a very infrequent spin exchange would produce the observed T_{1e}, 10^{-5} sec. It will be seen later that exchange is pertinent in developing working hypotheses for the optically stimulated effect.

The voltage controlled -P_b study is marred by several problems. First, difficulties have occurred on making large-area MOS samples with desired high resistivity oxides. Leakage not only reduces the surface potential, but also introduces non-uniformity near current pathways. Second, the high concentration of P_b defects may pin the Fermi level, obscuring the interpretation. Third, oxide breakdown has prevented use of desirably high negative gate voltage. Hopefully these problems will soon be overcome.

3.5 Behavior of P_b Centers Under Optical Irradiation

Optical irradiation is a natural suggestion for determination of the energy states of $Si\equiv Si_3$ centers. Wafers were suspended in a double-wall dewar, containing liquid nitrogen or not, for study at 77^OK or 295^OK. Wafer positions both parallel and perpendicular to the magnetic field H_o were tested, with resultant normal or grazing incidence of the light. The observed resonance was identified with the usual dark P_b signal by its g-anisotropy. Other resonances were not observed.

An example of the optical enhancement of P_b signals is shown in Fig. 7 for (111) wafers at 77^OK. The gain-corrected enhancement is nearly +10X and is more or less independent of wafer angular position in the cavity. Much weaker effects could be observed at 295° with some samples, varying considerably between samples. Nevertheless, the positive sign of the enhancement immediately distinguishes this effect from the thin-oxide photoconductive resonance

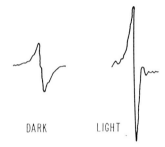

DARK   LIGHT   Fig. 7 Optical enhancement of P_b signals
                         from oxidized (111) silicon wafers

(PCR) phenomenon, which is always negative [7]. The onset of signal distortion at higher light intensity is shown in Fig. 8; this may arise from the Dyson effect in the high local interface conductivity, due to light-injected carriers.

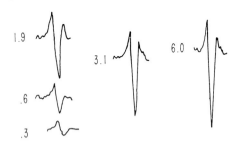

Fig. 8 EPR signal enhancement vs. light intensity

Both light-stimulated and dark EPR signals saturate with microwave power more or less normally. This should be contrasted with the thin-oxide PCR signal, which at low power varies as $(Power)^{3/2}$ [7] and does not decline in the saturation limit. Thus, in view of both the positive sign and saturation of the enhanced (111) P_b signal, the observed effect on P_b appears to arise from a greater net spin polarization, rather than from a PCR spin-dependent recombination mechanism.

The source of the increased EPR signal amplitude is not clear at present. Conceivably, it might arise from increased abundance of P_b centers in the paramagnetic charge state ($\cdot Si \equiv Si_3$) under optically induced surface-potential changes, but this seems unlikely in view of the order-of-magnitude signal enhancement. Such a change would imply an interface defect concentration of 10^{13}, unacceptably high in view of initial D_{it} values.

Another possibility is direct filling and emptying of P_b charge states by the light, with a steady-state polarization determined by relative transition rates [17]. This would imply centers which are either empty or doubly occupied at dark equilibrium. Paramagnetic (neutral) P_b-like centers are readily observed in bulk silicon in several ramifications, mainly in multiple vacancies or in the presence of some nearby impurity or dopant which produces a slight perturbation of the defect. (As a result, perfect rotational symmetry of the EPR g-tensor is not generally observed; the (111) P_b center is nearly unique in this respect). The single $\cdot Si \equiv Si_3$ center of bulk Si in pristine local surroundings, however, is much harder to observe, requiring low temperatures and much care. Recently, it has been proposed and substantiated by electrical evidence that the pristine silicon vacancy defect is a negative-U center [18]. As a result, it is not commonly observed in the single-electron state. In the case of P_b centers, however, the overlying oxide would provide the necessary perturbation to stabilize the single-electron state, and the negative-U model seems much less likely to apply. Furthermore, the negative-U mechanism would again require defect concentration of 10^{13} to explain the observed enhancement; and thus, on this basis alone, it is as unlikely as the surface-potential argument above.

A final postulated mechanism is dynamic electron polarization (DEP). The phenomenon of spin-dependent recombination at certain defect centers has been well explored [5 - 8]. It is reasonable that electron-hole generation at these centers would also be spin-dependent, and might thereby produce a non-equilibrium spin polarization. In the previous section, it was shown that the spin relaxation time of the P_b center is controlled by exchange with interface conduction electrons; thus, any non-equilibrium polarization of the latter would readily be transferred to the P_b electrons. Conceivably, the P_b center itself could act as the generation site. The electron occupancy of the center, of course, need not change in these DEP arguments.

It should be emphasized that the mechanism of light-enhanced EPR from P_b centers is by no means settled at this time. A direct recombination mechanism like PCR, however, seems to be precluded.

3.6 Ultraviolet Bleaching of P_b Centers

The transitory effect of optical irradiation leads one to examine the influence of higher-energy ultraviolet (UV) radiation. It was quickly determined that UV from a Hg lamp produces a long-lasting bleaching of the P_b EPR signal (Fig. 9). This bleaching is produced in a few minutes under the lamp, and persists for hours afterward, thus obviating the need for in-situ cavity illumination. The recovery is also shown in Fig. 9. A trapping phenomenon seems indicated.

The recovery of the signal, although requiring several hours at room temperature, was found to be imperceptibly slow at 77°K. Immersion in water,

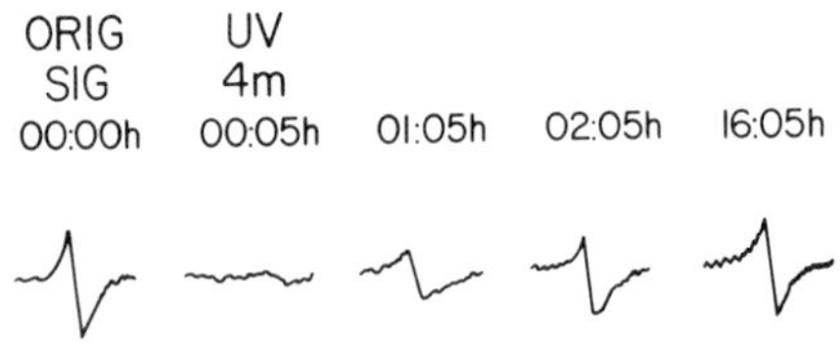

ORIG   UV
SIG   4m
00:00h 00:05h   01:05h   02:05h   16:05h

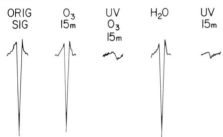

Fig. 9 Ultraviolet bleaching of
the P_b resonance and recovery

however, restored the P_b signal in a few minutes. In order to establish
the energy required for the bleaching, filters to isolate the 3660 A and
2537 A lines were employed. Only the short wave radiation was effective,
indicating an energy ≥ 3.3eV. Since a UV lamp produces quantities of ozone,
it was necessary to show whether ozone itself had a role in the bleaching.
Exposure to a strong, dark source of ozone did not produce any perceptible
bleaching. With concurrent UV irradiation, bleaching was observed as above.
In continued tests of the gaseous ambient, it was found the UV bleaching did
not occur in vacuum or in nitrogen; air or oxygen was necessary. The ozone
tests and the effects of water are shown in Fig. 10.

ORIG O_3 UV H_2O UV
SIG   15m O_3 15m
                15m

Fig. 10 Absence of ozone effect
on UV bleaching; restoration of
EPR by water

The UV bleaching was observed initially only in native oxides, and so a
series of wafers with different oxide thickness was tested. The effect declines rapidly with thickness beyond 40 A, suggesting either tunneling to
oxide surface traps or an electric field effect due to outer oxide charging.

A contrasting role for UV was discovered in extension of the measurement
to the initial stages of oxide formation. When an oxidized silicon wafer was
stripped of all oxide in HF, no P_b signal could be detected initially. After
exposure to air for an hour or more, a P_b signal appeared, which could be
UV-bleached in the usual way. However, if exposed to UV while in air immediately after HF stripping, the P_b signal grew very rapidly. Finally, after
reaching a limiting amplitude, further application of UV bleached the signal,
as usual. These results reinforce the oxide location of the traps.

An interpretation and model for these observations must be tentative at
this time. First, it appears that the initial application of UV to a bare
silicon surface significantly enhances the growth rate of the first few layers of oxide (19), which are reaction-rate-limited, rather than diffusion-limited. The bleaching phenomenon itself, however, is also dependent on the
presence of oxygen and it is very tempting to assign it to oxygen adsorbed
on the outer oxide surface. Oxygen photoadsorption on semiconductor surfaces
is very well known (20); and indeed, oxygen is the only common gas to so
adsorb, due to its large electron affinity (with an energy which may well be

situated in or even below the silicon bandgap). UV radiation with certainty generates carriers, and further is able to excite both conduction and valence electrons above the conduction band of silicon. The requirement of 2537 A light suggests they are raised to the conduction band of SiO_2. Once there, they travel with ease to a surface oxygen molecule (or atom), rapidly building up a strong surface negative charge. A steady state is reached when the SiO_2 bands are bent well upward at the oxide surface, so that tunneling back to the silicon balances the outward travel of the heated electrons. At this condition, the silicon bands are also bent upward sufficiently to bring the $(+ \leftrightarrow 0)$ level of $Si \equiv Si_3$ above the Fermi level, and the dangling orbital is thus unoccupied.

Of course, it is also conceivable that the P_b center is directly discharged by the UV, but such an effect would not be distinguishable without delicate experiments. The UV cross-section of localized P_b orbital electrons may be much less than silicon valence or conduction electrons. In any event, the final steady state result would be the same, whether or not they are directly involved in the excitation process. The magnitude of the UV energy required for the bleaching, if involved, indicates that the $(+ \leftrightarrow 0)$ level of the $Si \equiv Si_3$ center may be lower than expected for a dangling orbital, perhaps even in the valence band of Si.

The role of water in the recovery process is most likely a simple short-circuiting of the charge separation across the oxide via surface conduction or permeation through oxide pores. It is highly unlikely to be a chemical neutralization at the interface. Any such role for H_2O would almost surely be a reaction and bonding to the $\cdot Si \equiv Si_3$ orbital, as in hydrogen or steam annealing, with resultant elimination of the P_b EPR signal.

4. Concluding Remarks

The presence of oriented $\cdot Si \equiv Si_3$ defects at the Si/SiO_2 interface now seems well-established by the very consistent and convincing crystallographic behavior of the P_b signal in differently oriented wafers. The correlation and identical physico-chemical responses of P_b and D_{it} strongly suggest a close and possibly causal connection. Electrical variation of P_b by applied voltage in an MOS structure supports a connection between P_b and D_{it}, but does not make clear whether it is direct or indirect. Intimate exchange with conduction-band electrons is indicated by a strong relaxation effect on the P_b resonance. The latter is partially confirmed by optical enhancement of the P_b signal, which requires exchange with spin-polarized carriers in one explanation. Finally, successful bleaching of the P_b resonance by UV-enhanced trapping establishes the physical dischargeability of the $\cdot Si \equiv Si_3$ center, which is to be contrasted with the ease of chemical neutralization by hydrogen. Extended studies will, it is hoped, enable deduction of the energy levels of charge states of the $Si \equiv Si_3$ center, development of a useful atomic and crystallographic model for the center, and establishment of its roles in initial and annealed interface traps in MOS structures.

5. Acknowledgments

Initial phases of this work were done in cooperation with B. E. Deal and R. R. Razouk. Voltage-controlled P_b studies were conducted with N. M. Johnson, D. K. Biegelsen, and M. D. Moyer. C. Svensson kindly revealed his corona charging experiments before publication. Very fruitful discussions with S. R. Morrison in regard to adsorbed oxygen traps are deeply appreciated.

159

6. References

1. P. J. Caplan, E. H. Poindexter, B. E. Deal and R. R. Razouk, J. Appl. Phys. 50, 5847 (1979).

2. R. C. Fletcher, W. A. Yager, G. L. Pearson, A. N. Holden, W. T. Read and F. R. Merritt, Phys Rev 94, 1392 (1954).

3. Y. Nishi, Japan J. Appl. Phys. 10, 52 (1971).

4. A. G. Revesz and B. Goldstein, Surface Sci. 14, 361 (1969).

5. I. Solomon, Solid State Commun. 20, 215 (1976).

6. D. Kaplan and M. Pepper, Solid State Commun. 34, 803 (1980).

7. G. Mendz, D. J. Miller, and D. Haneman, Phys Rev B 20, 5246 (1979).

8. I. Shiota, N. Miyamoto and J. -I. Nishizawa, Surf. Sci. 36, 414 (1973).

9. Y. -H. Lee and J. W. Corbett, Phys. Rev. B 8, 2810 (1973).

10. D. L. Griscom in the Physics of SiO2 and its Interfaces, S. T. Pantelides, Ed. (Pergamon, New York, 1978) 232.

11. P. J. Caplan, E. H. Poindexter, B. E. Deal and R. R. Razouk in The Physics of MOS Insulators, G. Lucovsky, S. T. Pantelides, and F. L. Galeener, Eds. (Pergamon, New York, 1980) 306.

12. E. H. Poindexter, P. J. Caplan, J. J. Finnegan, N. M. Johnson, D. K. Biegelsen and M. D. Moyer in The Physics of MOS Insulators, G. Lucovsky, S. T. Pantelides, and F. L. Galeener, Eds. (Pergamon, New York 1980) 326.

13. C. Brunström and C. Svensson, Solid State Commun. 37, 399 (1981).

14. A. Abragam, The Principles of Nuclear Magnetism, (Clarendon, Oxford, 1961) 300.

15. E. Abrahams, Phys. Rev. 107, 491 (1957).

16. D. J. Lepine, Phys Rev B 2, 2429 (1970).

17. R. A. Street and D. K. Biegelsen, Solid State Commun. 33, 1159 (1980).

18. G. D. Watkins and J. R. Troxell, Phys. Rev. Letters 44, 593 (1980).

19. R. Oren and S. K. Ghandhi, J. Appl. Phys. 42, 752 (1971).

20. S. R. Morrison, The Chemical Physics of Surfaces (Plenum, New York, 1977) 308.

The Non-Equilibrium Linear Voltage Ramp Technique as a Diagnostic Tool for the MOS Structure

L. Faraone

RCA Laboratories, Princeton, NJ 08540, USA

J.G. Simmons

University of Bradford, Bradford, BD7 1DP, United Kingdom

A.K. Agarwal

Lehigh University, Betlehem, PA 18015, USA

1. Introduction

The quasi-static linear voltage ramp technique of KUHN [1] has been widely used to determine the interface properties of MOS devices. Following the theoretical work of BOARD and SIMMONS [2] on the non-equilibrium ramp-response, several experimental studies were undertaken which studied the effect of temperature and sweep rate [3,4]. In this paper, an analytical and experimental study is presented of the non-equilibrium response of MOS devices subjected to a linear voltage ramp. Various areas of the resulting I-V plots are identified with physical quantities such as gate charge (and hence electric field at the Si-SiO$_2$ interface), and surface potential, a knowledge of which is very useful in studying non-equilibrium effects. To circumvent the difficulty of tedious graphical integration involved in the process of extracting these quantities from the experimental I-V plots, an experimental technique is presented which facilitates their direct measurement. The resulting I-V, ϕ_s-V and Q_g-V plots are related to each other in a manner which gives interesting insight into the physical mechanisms occurring within the device. Furthermore, agreement with existing theory is found to be extremely good.

2. Basic Equations

Using standard notation, the equations describing the Q_g-V ramp-response for an n-type MOS capacitor are [5];

$$Q_g(V) = Q_{FB} - \frac{1}{\alpha} \int_{V_{FB}}^{V} I\,(V)\,dV \quad ; \quad V = V_o - \alpha t, \tag{1}$$

$$Q_g(V_B) = Q_{FB} - (A_1 + A_2)/\alpha \quad ; \quad V = V_B, \tag{2}$$

$$Q_g(V) = Q_g(V_B) + \frac{1}{\alpha} \int_{V_B}^{V} I\,(V)\,dV \quad ; \quad V = V_B + \alpha\,(t - t_B), \tag{3}$$

where α is the voltage ramp-rate, V_B is the gate voltage at the point of sweep reversal, areas A_1 and A_2 are shown in Fig.1(a), and Q_{FB} is the gate charge at flat-band given by $Q_{FB} = (V_{FB} - \phi_{ms})\,C_{ox}$, where ϕ_{ms} is the metal-semiconductor work function difference. Eq.(1) applies for the forward sweep, (2) at the point of sweep reversal, and (3) for the reverse sweep. The integral appearing in the above equations is indicated in Fig.1(a); thus, a graphical integration of the experimental I-V curve allows $Q_g(V)$ to be determined.

161

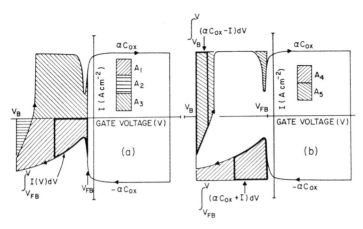

Fig.1 Graphical integration of the I-V curve to obtain (a) $Q_g(V)$, or (b) $\phi_s(V)$

Furthermore, this is a direct measurement of electric field E_s at the Si-SiO$_2$ interface, since, by Gauss' law, $E_s = (Q_g - Q_{FB})/\varepsilon_s$ where ε_s is the absolute permittivity of the semiconductor.

Using BERGLUND's integral [6], the following equations can be derived for the ϕ_s-V ramp-response [5];

$$\phi_s(V) = \frac{1}{\alpha C_{ox}} \int_{V_{FB}}^{V} (\alpha C_{ox} + I)\, dV \qquad ; V = V_o - \alpha t, \qquad (4)$$

$$\phi_s(V_B) = -A_4/\alpha C_{ox} \qquad ; V = V_B, \qquad (5)$$

$$\phi_s(V) = \phi_s(V_B) + \frac{1}{\alpha C_{ox}} \int_{V_B}^{V} (\alpha C_{ox} - I)\, dV ; V = V_B + \alpha (t - t_B), \qquad (6)$$

where area A_4 is shown in Fig.1(b). Eq.(4) applies for the forward sweep, (5) at the point of sweep reversal and (6) for the reverse sweep. The above integrals are indicated in Fig.1(b); thus, $\phi_s(V)$ can be determined by graphical integration of the I-V curve. An interesting observation to be made, is that since $Q_g(V_{FB})$ is the same for the forward and reverse sweeps, then from (2) and (3) we can write $A_1 + 2A_2 = A_3$, where these areas are shown in Fig.1(a). Similarly, since $\phi_s(V_{FB}) = 0$, then from (5) and (6) we obtain $A_4 = A_5$, as indicated in Fig.1(b).

3. Experimental Procedure

The MOS capacitors used in this study were fabricated from n-type silicon wafers of <100> orientation and 10 Ω-cm resistivity. An 1100 Å gate oxide was used and the aluminum field plate was 1 mm in diameter. The V_{FB} was measured to be -1.2V, which gives a value of $Q_{FB} = -2.7 \times 10^{-8}$ coul. cm^{-2}. Figure 2 shows the experimental setup used to obtain I-V, Q_g-V and ϕ_s-V

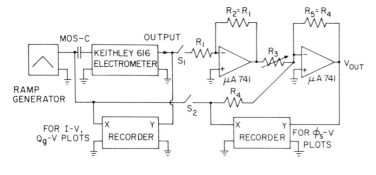

MOS-C   OUTPUT   R₁   R₂=R₁   R₅=R₄

Fig. 2 Experimental setup to measure and plot the I-V, Q_g-V, and ϕ_s-V ramp-response of a MOS capacitor

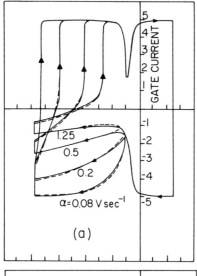

(a)

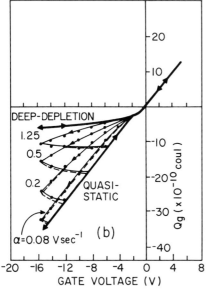

(b)

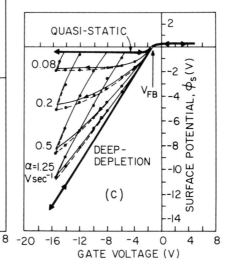

(c)

Fig.3 Non-equilibrium I-V, Q_g-V and ϕ_S-V curves; ——, experimental; and ---, theoretical fit. The points indicate calculations based on the graphical integration of experimental I-V curves.
For $\alpha = 0.08$, I x $(0.4 \times 10^{-11}A)$;
$\alpha = 0.2$, I x $(1.0 \times 10^{-11}A)$;
$\alpha = 0.5$, I x $(2.5 \times 10^{-11}A)$;
$\alpha = 1.25$, I x $(6.25 \times 10^{-11}A)$

plots, and is based on the techniques described in [7,8]. It will suffice to state that with switches S_1 and S_2 open, the I-V or Q_g-V plots can be obtained by having the electrometer in the current mode or coulomb mode, respectively. With switches S_1 and S_2 closed, and the electrometer in the coulomb mode, the output voltage V_{out} is directly proportional to ϕ_s if the potentiometer R_3 is appropriately set [5].

4. Results and Discussion

The experimental I-V, Q_g-V and ϕ_s-V curves at 300^OK for various ramp-rate α are shown by the full lines in Fig.3(a), (b) and (c), respectively. Calcu-lated curves based on the non-equilibrium theory [2,4,5], using a generation rate of 1.2×10^{14} cm^{-3} sec^{-1}, are indicated by the dotted lines. The points represented by the full circles in Fig.3(b) and (c), were obtained by graph-ical integration of the experimental I-V curves as described in Section 2. As evident from Fig.3, excellent agreement is obtained between experiment, theory and the results of graphical integration. A study of Fig.3(b) and (c) allows some general trends to be determined. As ramp-rate α is increased, at a given gate voltage during the forward sweep the magnitude of Q_g is reduced, and the magnitude of ϕ_s is increased. The above observations are consistent with the fact that a higher ramp-rate drives the device into deeper depletion. The quasi-static and deep-depletion Q_g-V and ϕ_s-V plots are also shown in Fig.3(b) and (c). It is interesting to note that these two cases represent the two limiting situations. All other non-equilibrium ϕ_s-V pnd Q_g-V curves at a given temperature necessarily lie between these two extremes regardless of ramp-rate.

5. Acknowledgments

This work was supported by the National Science Foundation (under grant ECS-7908364) and the Sherman Fairchild Foundation.

6. References

1. M.Kuhn: Solid-State Electron. 13, 873 (1970)
2. K.Board, J.G.Simmons: Solid-State Electron. 20, 859 (1977)
3. K.Board, J.G.Simmons, P.G.C.Allman: Solid-State Electron. 21, 1157 (1978)
4. A.G.Nassibian, L.Faraone, J.G.Simmons: J. Appl. Phys. 50, 1439 (1979)
5. L.Faraone, J.G.Simmons, A.K.Agarwal, P.D.Tonner: Solid-State Electron. (in press)
6. C.N.Berglund: IEEE Trans. ED-13, 701 (1966)
7. H.C.G.Ligtenberg, J.Snijder: Electron. Lett. 16, 523 (1979)
8. P.D.Tonner, J.G.Simmons: Rev. Sci. Instrum. 51, 1378 (1980)

MOS Characterization by Phase Shift Impedance Technique

J. Boucher and M. Lescure

Laboratoire de Semiconducteurs et d'Optoélectronique ENSEEIHT
2, rue Charles Camichel
F-31071 Toulouse Cedex, France

M. Mikhail

Military Technical College, Kobry el Coppa, Cairo, Egypt

J. Simonne

LAAS-CNRS 7, avenue du Colonel Roche
F-31400 Toulouse, Cedex, France

The electrical characterization of MOS structures is most often based on a global impedance or admittance measurement. When a low level modulation signal superimposed on the bias applied to the device is used, the method has a good sensitivity as long as the real and imaginary parts of the impedance are of the same order of magnitude. However, the range of frequency of the measurement may not be sufficient to extract the internal parameter that we wish to evaluate. The transient method is another standard way, unfortunately yielding poor accuracy due to the wide-band electronic system which introduces the noise of the MOS structure into the measurement.

MOS characterization through a phase shift measurement performed over a wide range of frequencies is definitely a more accurate technique allowing the definition of a multi-parameter model, for the following reasons:
- a 0.6° phase shift appears two decades before the cut-off frequency of a first order system,
- amplitude fluctuations have no effect on the measurement
- the information given by this method accounts for delay time due to propagation, and expresses most completely the inertial behaviour of the structure.

With this procedure, the phase shift selectivity is equivalent to a quality coefficient higher than 10^5, for a frequency band of the experiment ranging from 1 Hz up to 100 MHz. With an absolute accuracy of 0.1° on phase, 100MHz corresponds to a time resolution of 3×10^{-12}s.

Basic principles of the technique
The measurement technique block diagram is given in Fig. 1.

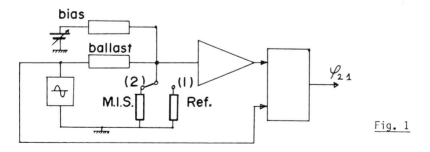

Fig. 1

The input signal, which is supplied by a frequency generator, serves also as a reference signal for the phase shift measurement and can be switched either across the MOS structure or a passive element used as a reference. It is a capacitive element. The MOS structure and the capacitance are fed through the impedance Z_1 of a ballast allowing the adjustment of the sensitivity in the frequency range of operation.

The method consists of a measurement of the phase shift difference φ_{21} between voltage across the MOS and across the passive element, as a function of the signal frequency ω. By taking the difference we cancel unknown phase shifts which could take place in connecting cables and measuring devices. φ_{21} depends on ω and Z_1, on the input characteristics of the amplifier A, on the reference and on the MOS structure.

The search for a good fitting between the response of a given model of the MOS and the experiments, results in a numerical evaluation of the parameters of the model. The high accuracy of the phase shift measurement, the wide range of frequencies covered, the ability to diversify the experimental conditions using different values of Z_1, allow one to check the chosen model with a high number of significant and precise parameters.

Interpretation of the results

The models given in Fig. 2 have been successfully used.

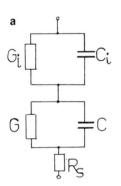
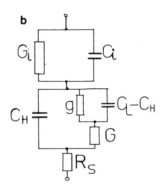

Fig. 2a,b

C_i, G_i account for the insulator capacitance and the dielectric loss, R_S is the bulk resistance. The model of Fig. 2(a) is the standard representation through C and G of the generation term related to the depletion zone and is valid in inversion conditions. The model of Fig. 2(b), derived from the Pierret and Sah model is more complex and valid in depletion conditions. A conductance g has been added on this representation and can be evaluated through experiment.

166

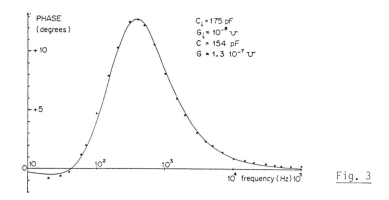

Fig. 3

In Fig. 3 are represented the theoretical and actual response of an inverted Si-SiO$_2$ MOS structure. An excellent fitting is obtained, with the parameters as given.
The validity of the model, including the numerical values of the parameters, should be checked over a wide range of frequencies. However, systematic experiments have shown that only a few measurements performed on a restricted range of frequencies allow a good evaluation of the parameters.

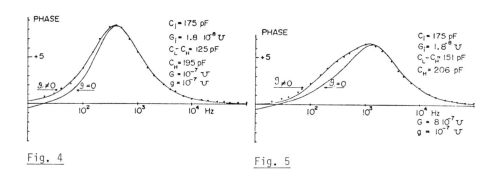

Fig. 4

Fig. 5

Figs. 4 and 5 represent the behaviour of the same MOS device biased into depletion both in the dark (Fig.4) or under illumination (Fig. 5).

These curves confirm the necessity to include the conductance g to validate the model towards the low frequencies. Furthermore, the dependence of the parameters on the light is evident.

Similar studies have been carried on GaAs and InP MIS structures. As an example, Fig. 6 and 7 exhibit results obtained with two anodic oxide- GaAs MIS structures.

167

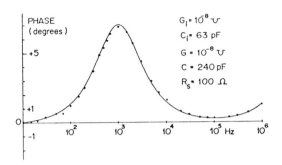

Fig. 6

In the figure: PHASE (degrees), +5, +1, 0, -1 on vertical axis; 10^2, 10^3, 10^4, 10^5 Hz, 10^6 on horizontal axis.

$G_l = 10^{-8} \; \mho$
$C_i = 63 \; pF$
$G = 10^{-8} \; \mho$
$C = 240 \; pF$
$R_s = 100 \; \Omega$

Fig. 6 - is similar to the case of Fig. 2(a), with the neglig-
able conductance value characteristic of the depletion mode.
The latter is noticed at all bias values.

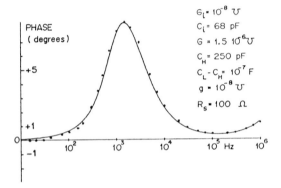

Fig. 7

In the figure: PHASE (degrees), +5, +1, 0, -1 on vertical axis; 10^2, 10^3, 10^4, 10^5 Hz, 10^6 on horizontal axis.

$G_i = 10^{-8} \; \mho$
$C_i = 68 \; pF$
$G = 1.5 \; 10^{-6} \mho$
$C_H = 250 \; pF$
$C_L - C_H = 10^{-7} \; F$
$g = 10^{-8} \; \mho$
$R_s = 100 \; \Omega$

Fig. 7 - corresponds to the model of Fig. 2(b), indicates the
existence of a high interface state density. The result is
consistent with the C(V) characteristics.

Si/SiO₂ Properties Investigated by the CC-DLTS Method

E. Klausmann

Fraunhofer-Institut für Angewandte Festkörperphysik
D-7800 Freiburg, Fed. Rep. of Germany

In recent years the DLTS (= deep level transient spectroscopy) method has also been used for characterising MOS interfaces. There are several variants [1] - [8]. Some of them have found interest because they are very sensitive and a full automatisation [9] of the measurements may be accomplished.

As shown in a previous paper [7], measurement data have to be treated mathematically in different ways depending on whether the traps under investigation are discrete like the bulk traps or are continuously distributed in energy like the interface states. With this new understanding the evaluation of the CC- (= constant capacitance)-DLTS method has been improved. However, a drawback has shown up. Samples with small interface state densities or samples with even rather small instabilities could not be measured reliably because of a strong error propagation. A new CC-DLTS variant is therefore suggested. It corresponds to the regular DLTS method by TREDWELL and VISWANATHAN [8].

In the papers mentioned [6], [7], the basic ideas and the instrumentation of the CC-DLTS method have been described. We will directly start with the pertinent equations. In the following formulae the usual properties of the interface states are stipulated. The traps are of the Shockley-Read type with capture cross sections that may depend on the energy E in the forbidden gap but which are independent of the temperature T. Because of the fixed oxide charges, random fluctuations of the surface potential with the variance σ_g are assumed. If minority carrier effects are excluded as usual, the correlation signal for a p-type sample is given by

$$\Delta V_G(t, T, \overline{E}_s) = \frac{A}{kT \cdot C_{ox}} \int_{-\infty}^{+\infty} \frac{\exp(-((E_s - \overline{E}_s)/kT)^2 / 2 \cdot \sigma_g^2)}{\sigma_g \cdot \sqrt{2\pi}} \cdot$$

$$\int_{E_v}^{E_c} \frac{N_{ss}(E) \exp(-t/\tau(E)) \cdot (1 - \exp(-t/\tau(E)))}{1 + \exp((E - E_s)/kT)} \, dE \, dE_s \qquad (1)$$

with the time constant

$$\tau(E)^{-1} = \sigma(E) \cdot v_{th}(T) \cdot N_V(T) \cdot$$

$$\cdot (\exp(-(E - E_V)/kT) + \exp(-(E_s - E_V)/kT)) \quad . \tag{2}$$

The DLTS delay time is t. E_s is the energy of the Fermi level at the inter-face. All energies refer to the majority carrier band edge $E_V = 0$ at the interface. The other quantities have the usual meaning (cf. Fig. 1 in [7]). The bias applied to the MOS capacitor determines E_s, and vice versa the energy E_s can be calculated with the Poisson equation from the measured MOS hf capacitance C and the known semiconductor data.
In the papers cited above, the formulae (1) and (2) have been interpreted more intelligibly: The DLTS technique is a spectroscopy of the emission time constant τ of the traps. Only traps within an interval of some kT around an energy E_0 contribute to the correlation signal substantially. We have to distinguish three cases:

(i) In the case $E_0 \ll E_s$ (or $E_0 \gg E_s$ for n-type samples) all traps within this interval fully contribute to ΔV_G.
The time constant τ is selected with the DLTS set-up by an appropriate choice of the delay times t and 2t. The relation between τ and t is given by

$$\tau = \alpha t . \tag{3}$$

In the case of discrete traps the constant α is equal $1/\ln2 = 1.44$ [10], and in the case of traps distributed continuously $\alpha = 2.52$ [7]. More-over the correlation signal is independent of σ_g. Eq. (1) can be simplified to

$$\Delta V_G = \frac{A}{C_{ox}} \cdot qN_{ss}(E_0) \cdot \frac{kT \ln2}{1 + kT \cdot \dfrac{d \ln \sigma}{dE_s}} \quad . \tag{4}$$

(ii) In the case $E_0 \simeq E_s$ the correlation signal decreases and in general it depends strongly on σ_g. In the particular case $E_0 = E_s$, the correlation signals are virtually independent of σ_g, because all possible curves with different σ_g intersect in almost one point [11], [8]. It follows from numerical calculations that this point is nearly half (approx. 0.48) the value of ΔV_G given by (4). Instead of $\alpha = 2.52$ the value $\alpha = 2.06$ is a better approximation.

(iii) In the case $E_0 \gg E_s$ (or $E_0 \ll E_s$ for n-type samples) the correlation signals vanish gradually.

In order to benefit from the properties (i) and (ii) the CC-DLTS feedback circuit must permit monitoring ΔV_G as a function of the MOS capacitance C (which is equivalent to E_s). This was not possible with the feedback system described in [6]. A new electronically controlled circuit has been built. This circuit will be published elsewhere.

The concept of the method suggested here will be shown in separate steps in the following. The procedure is illustrated by a p-type MOS capacitor as example. Its oxide was grown to 120 nm in an O_2 + HCl atmosphere at 1000 °C. The sample was then annealed at the same temperature in N_2 for 30 min. Dots of Al were evaporated onto the oxide. Then a last anneal took place at 450 °C in N_2 + H_2 for 30 min. The oxide capacitance C_{ox} was measured to be 110 pF and the doping density $3.2 \cdot 10^{15}$ cm^{-3}.

First, in order to survey the magnitude of the correlation signal and the possible pecularities, the temperature dependence of ΔV_G was plotted. The DLTS delay times were 10 and 20 ms. The MOS capacitance was adjusted to 52 pF. This is equivalent to $E_s - E_v = 0.62$ eV. Such a measurement in the deep depletion region was possible owing to the small minority carrier generation rate, especially below 250 K. The plot showed the correlation signal typically decreasing toward low temperatures, except for a conspicuous peak at 177 K.

Then starting from 60 K in steps of about 25 K the correlation signals were recorded as a function of the MOS capacitance C. A typical example is given in Fig. 1a. The decrease of ΔV_G from a constant value to zero can be seen clearly. The energy of the interface states contributing to ΔV_G was calculated from the capacitance C at the half value point: $E_o = 0.41$ eV. Then using (2) and (3) the capture cross section was found to be $\sigma = 3.4 \cdot 10^{-16}$ cm^2. Once several capture cross sections $\sigma(E)$ were determined from plots of ΔV_G vs. C at different temperatures (Fig. 2a), they were fitted to a curve whose logarithmic derivative could easily be obtained. Then (4) could be applied to the constant branch of the $\Delta V_G(C)$ curves and $N_{ss}(E)$ was determined. (Fig. 2b) With this set of $N_{ss}(E)$ and $\sigma(E)$ correlation curves $\Delta V_G(C)$ were calculated numerically with (1) varying σ_g (Fig. 1a). The variance σ_g then resulted from a comparison of these theoretical curves with the measured ones.

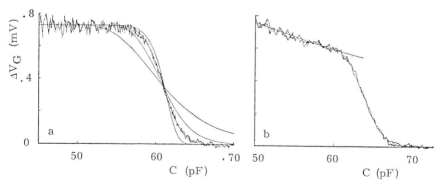

Fig. 1   Correlation signals vs. MOS capacitance at 236 K (a) and 177 K (b). The solid curves are calculated for σ_g = 0, 1, 2 and 4 (a) and for bulk traps (b).

$N_{ss}(E)$ and $\sigma(E)$ obtained in this way were checked with the conductance method [12] . As it can be seen from Fig. 2a/b the results of the two methods agree well. One may note that E_s was calculated on the assumption of a constant doping profile. As this holds only true approximately, a systematic error for both methods of about an order of magnitude has to be expected.

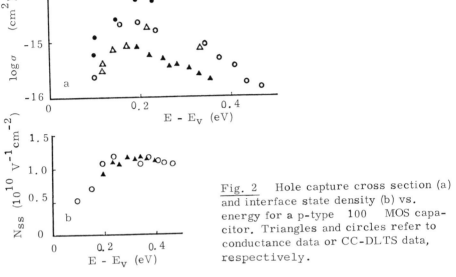

Fig. 2   Hole capture cross section (a) and interface state density (b) vs. energy for a p-type 100   MOS capacitor. Triangles and circles refer to conductance data or CC-DLTS data, respectively.

The foregoing procedure is not sufficient to explain the decreasing correlation signal $\Delta V_G(C)$ at 177 K (Fig. 1b) and the peak in the ΔV_G vs. T-curve at the same temperature. These effects can be attributed to bulk states. The measured ΔV_G is a superposition of a correlation signal ΔV_{Gss} due to the interface states and a correlation signal due to bulk states. The emission from the bulk states depends strongly on the band bending and therefore also on the capacitance C. If the basic ideas about this effect [4] , [3] are carried over to the CC-DLTS method and if at the same time a constant bulk trap concentration N_T is assumed, which is much smaller than the doping density N_0, then a   quantitative relation is found,

$$\Delta V_G = \frac{1}{4} \cdot q N_T \cdot \frac{A}{C_{ox}} \cdot (\epsilon \epsilon_0 \cdot \frac{A}{C_{ox}} \cdot (\frac{C_{ox}}{C} - 1) - \sqrt{\frac{2 \epsilon \epsilon_0}{q N_0} \Phi_T}) +$$

$$+ \Delta V_{Gss} .$$

(5)

The difference between the energy of the bulk traps and the Fermi level (in the neutral region outside the space charge   region) is denoted by $q \Phi_T$.

The concentration of the bulk states can be calculated from the slope of the ΔV_G vs. $1/C$-curve. N_T is found to be $9 \cdot 10^{12}$ cm^{-3}. The best fit according to (5) is drawn as solid curve in Fig. 1b. As the contributions of the bulk states and the interface states vanish at the same MOS capacitance C, it can be concluded that the bulk states lie at the same energy level as the interface states ($E_0 - E_v = 0.30$ eV) and they possess almost the same capture cross sections.

Summarizing the results obtained here, we state that the newly suggested CC-DLTS variant compares in sensitivity with the conductance method and is superior at low temperatures and in the presence of large surface potential fluctuations. The method is sensitive for detecting bulk states.

I thank Dr. K. Eisele for many helpful comments, and I am grateful to Dr. G. Sixt with AEG-Telefunken, who supplied the MOS capacitor.

1   K.L. Wang, A.O. Evvaraye: J. Appl. Phys. 47, 4574 (1976)
2   M. Schulz, N.M. Johnson: Appl. Phys. Lett. 31, 622 (1977)
3   N.M. Johnson, D.J. Bartelink, M. Schulz: in Physics of SiO$_2$ (ed. S.T. Pantelides), 421 (1978)
4   K. Yamasaki, M. Yoshida, T. Sugano: Jap. J. Appl. Phys. 18, 113 (1978)
5   N.M. Johnson: Appl. Phys. Lett. 34, 802 (1979)
6   M. Schulz, E. Klausmann: Appl. Phys. Lett. 18, 169 (1979)
7   E. Klausmann: in Ins. Films on Semic. (Inst. Phys. Conf. Ser. No. 50) 97 (1979)
8   T.J. Tredwell, C.R. Viswanathan: Solid State Electr. 23, 1171 (1980)
9   M.D. Jack, R.C. Pack, J. Henriksen: IEEE Trans. ED - 27, 2226 (1980)
10 D.V. Lang: J. Appl. Phys. 45, 3014 and 3023 (1974)
11 E. Klausmann, A. Goetzberger: US-Army ERO Report DAJA 37-79-C-0547 (March 1980)
12 E.H. Nicollian, A. Goetzberger: Bell Syst. Techn. J. 46, 1055 (1967)
13 G.H. Glover: IEEE Trans. ED-19, 138 (1972)

Study of Ellipsometry:The Computation of Ellipsometric Parameters in a Nonuniform Film on Solid Substrate

Luo Jinshen, Chen Mingqi

Department of Electronic Engineering, Xi'an Jiaotong University
Xi'an Jiaotong, P.R. China

1. Introduction

In this paper, we worked out the method and program for computing the ellipsometric parameters (ψ and Δ) in a nonuniform
film (absorbing or nonabsorbing) on solid substrates by means of
a many-layer model. As example of an application we determined
the complex refractive index profiles of ion-implanted layers
of various doses from the ψ and Δ profiles measured by ellipsometric methods combined with layer stripping, and studied the
damage profiles in these layers {1}. This method can also be
used to study other nonuniform films.

2. Theory of Computation

We imagine that the nonuniform film consists of many thin uniform layers as shown in Fig. 1, and represent them by 1, 2, ...
1-1, respectively. Let n_1^* , n_2^*, ...$n*_{1-1}$ be the complex refractive indices of these layers, respectively, n_0 and n_1^* being
those of air and substrate, respectively {2}.

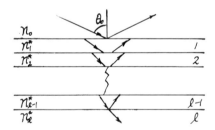

Fig. 1 The many-layer model

Let a monochromatic light beam be incident on the surface at
an angle Θ_0. The complex reflection coefficient of this many-
layer system for S polarized waves is given by

$$r_s = \frac{(m'_{11} + m'_{12} p_1) p_0 - (m'_{21} + m'_{22} p_1)}{(m^*_{11} + m'_{12} p_1) p_0 + (m'_{21} + m'_{22} p_1)} . \qquad (1)$$

Here $p_0 = n_0 \cos \Theta_0$, $p_1 = n_1^* \cos \Theta_0$, and the matrix

174

$$\begin{pmatrix} m'_{11} & m'_{12} \\ m'_{21} & m'_{22} \end{pmatrix} = \overset{1-1}{\underset{j=1}{\prod}} \begin{vmatrix} \cos(\delta_j) & \frac{i}{p_j}\sin(\delta_j) \\ ip_j\sin(\delta_j) & \cos(\delta_j) \end{vmatrix} \qquad (2)$$

where $p_j = n_j^* \cos \Theta_j$, $\delta_j = k_0 n_j^* \delta Z_j \cos\Theta_j$, $k_0 = 2\pi/\lambda_0$,
λ_0 is the light wave length in vacuum, and δZ_j is the thickness
of the jth layer. The equations for computing the complex refraction coefficient for P polarized waves can be obtained by
replacing p_j by $q_j = \cos \Theta_j / n_j$ in (1) and (2). According to
SNELL's law, we have

$$n_0\sin \Theta_0 = n_1^* \sin \Theta_1 = \ldots\ldots = n_{1-1}^* \sin\Theta_{1-1} = n_1^*\sin\Theta_1. \quad (3)$$

Finally the ellipsometric parameters satisfy the following
equation

$$e^{i\Delta} \, \text{tg}\psi = \frac{\Gamma_p}{\Gamma_s} \, . \qquad (4)$$

If replacing j=1 by j=i in (2), we can evaluate the ellipsometric parameters ψ and Δ for the case when the upper i-1
layers are removed. Let i take the integral numbers from 1 to 1-1
in turn, we can compute the ψ and Δ profiles in the nonuniform layer. As example of an application we used this computing method to evaluate the complex refractive index profiles
of ion-implanted layers from the values of ψ and Δ measured
by the ellipsometric method combined with layer stripping. For
each sample we chose the optimum $n^* = n-ik$ profile such that the
computed ψ and Δ profiles best fitted the experimentally measured data.

3. Experiments and Comuted Results

Samples of Si were implanted with P^+ at ion energy 100keV to
the doses of $2.10^{14}/cm^2$, $5.10^{14}/cm^2$, $1.10^{15}/cm^2$ and $1.10^{16}/cm^2$.
A He-Ne-Laser ($\lambda = 6328\text{Å}$) was used for all ellipsometric
measurements.
Both ellipsometric and backscattering measurements were carried
out for samples of each implanted dose. The results of our
ellipsometric measurement and computation are shown in Figs.2
to 5. In these figures, the symbols ● and ▲ represent the
measured values of ψ and Δ , respectively, and the solid and
dashed lines in each lower figure show the refractive index n
and extinction coefficient k profiles which were chosen so
that the computed ψ and Δ profiles best fitted the measured
data. The solid lines in each upper figure show the computed ψ
and Δ profiles. In the case of high doses, e.g. the
doses of $1.10^{16}/cm^2$, $1.10^{15}/cm^2$, and $5.10^{14}/cm^2$, it was observed
that the ψ and Δ profiles oscillate with increasing depth of
the stripped layer and the evaluated complex refractive index
profiles have a constant portion each as shown in Figs. 2 to 4.
It is obvious that this constant portion should correspond to
an amorphous layer caused by the ion implantation of high dose.

In this way we could determine the thickness and location
of the amorphous layer from the ellipsometric measurement. In
the case of low doses, e.g. the dose of $2.10^{14}/cm^2$, it was
proved that the complex refractive index profile has a peak
which corresponds to the damage peak of the implanted layer.
The damage profiles determined by the backscattering method and
our ellipsometric method are compared in Table 1. A good agreement is obtained. The results measured by the backscattering
method are shown in Fig. 6.

Table 1 Damage profiles determined by both methods

Sample No.	dose (1/cm²)		The results measured by ellipsometric method	The results measured by backscattering method
1	1.10^{16}	The thickness and location of amorphous layer	2000Å, from the surface to the depth 2000 Å	1980 Å, from the surface to the depth 1980 Å
2	1.10^{15}		1650Å, from the surface to the depth 1650 Å	1524 Å, from the surface to the depth 1524 Å
3	5.10^{14}		970 Å, from the depth 480 Å to 1450 Å	From the depth about 400 Å to 1334 Å
4	2.10^{14}	The position of damage peak	At a depth of 1000 Å	At a depth of 992 Å

4. Conclusion

The method and program for computing the ellipsometric parameters of a nonuniform film were worked out. The computing
method can be used to determine damage profiles of ion-implanted
layers and to study other nonuniform films. We obtained the
following conditions for determining the damage profiles of ion-
implanted layers by the ellipsometric method.
1) If the Δ and ψ profiles oscillate with increasing depth
   of the stripped layer as observed in Figs. 2a and 3a, a
   uniform amorphous layer is formed which extends from the
   surface to a limited depth as indicated by the decay of the
   curves in Figs. 2b and 3b.

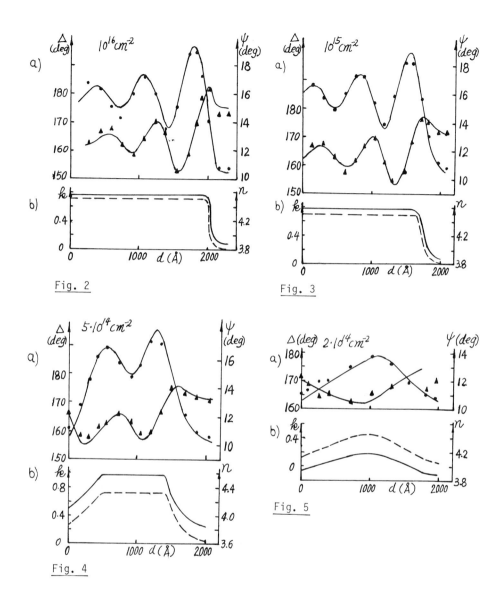

Fig. 2

Fig. 3

Fig. 4

Fig. 5

Figs. 2-5
a) Ellipsometric parameters Δ and ψ as a function of sectioning depth d(Å) for various implanted phosphorous doses as shown in the insert. Circles and triangles represent measurement data of Δ and ψ, respectively. Solid lines show the best fit of computations.
b) Profiles of the refractive index n (solid line–right scale) and extinction coefficient k (dashed line-left scale) as obtained by the fit of the computed curves in Figs. 2a-5a.

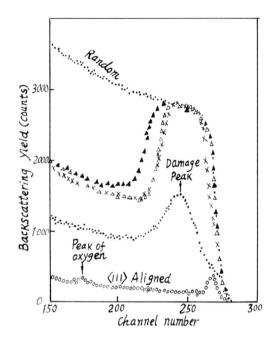

Figure text within image area:

E$_{He}$+ = 1 MeV

Channel Width = 2.12KeV/Channel

▲ 1.10^{16}/cm²
△ 1.10^{15}/cm²
× 5.10^{14}/cm²
● 2.10^{14}/cm²

Fig. 6 Backscattering measurements

2) If the Δ and ψ profiles have a shape as shown in Fig. 4a, the amorphous layer is only formed in the center of the damage region.

3) If the ψ profile shows a peak and the Δ profile has a valley in the central region (see Fig. 5a), no amorphous layer is formed and the maximum of the damage is located between the Δ valley and ψ peak.

Acknowledgement

The assistance of Mr. CHEN HOANSHEN, who performed the back-scattering measurements, is gratefully acknowledged.

References:

1) Adams, J.R., Surface Science, 56, 307 (1976)
2) Born, M. and Wolf, E., Principles of Optics (Pergamon, New York 1978) Chapter I

Part VI

Breakdown and Instability of the SiO$_2$-Si System

Breakdown and Wearout Phenomena in SiO$_2$

D.R. Wolters

Philips Research Laboratories
5600 MD EIndhoven, The Netherlands

ABSTRACT

Technological improvements have largely cancelled the effects of contaminations but the mechanisms of shorting at high and medium fields have not been understood so well. New experimental evidence is incompatible with existing breakdown theories. Novel electrical testing techniques have given evidence that charge incorporation or charge flux through the dielectric plays an essential role in the deterioration of the insulator. In the absence of a sound theory a model will be discussed assuming gaseous discharges.

INTRODUCTION

The trend to scale down the lateral dimensions of integrated circuits implies application of ultra-thin layers of SiO$_2$. Applied field strengths are now in the order of 2 MV/cm and may increase in the future. It is therefore of importance to understand the mechanisms and parameters which play a role in breakdown and life time. The topics of dielectric strength and life time failures have been covered recently by a number of excellent reviews, cf. Budenstein [1], Klein [2], Solomon [3], Osburn [4] and Kern [5].

This paper is a critical discussion on this problem. It is focussed on common electrical testing techniques applying electrical fields to capacitors. The fields are ramped or stepped or are kept constant over a long period of time. In some cases constant current sources are employed [6].

Breakdowns are detected by sudden voltage drops, current instabilities or simply by detecting whether the current exceeds a preset level. In most cases the capacitor is destroyed. When the destruction is in the high field region (> 8 MV/cm) the term "intrinsic breakdown" is often used. At lower fields the term "defect related breakdown" is used while when the breakdown is only achieved after a prolonged electrical stress the breakdown is denoted by "wearout". It is preferred here to denote the "intrinsic" by "high field" breakdown and the "defect related" by "medium field" breakdown.

In the section on **Statistics** the significance of the applied distribution laws are shown by referring to their specific properties. Area, current density and time dependence of the distributions will be discussed. These dependences must be anticipated when choosing the proper conditions for tests of capacitors with different thicknesses. The results of these tests will be discussed in the section on **High field breakdown.** Time and mechanical stress dependence will be investigated in the section on **Medium field breakdown.** In the last section **Theories on high field breakdown** models based on charge carrier multiplication and recombination are shortly discussed. A model based on gas discharges and proposed for other dielectrics will be related to SiO$_2$.

180

STATISTICS

Since the probability of finding pinholes, defect- or contamination-related weak spots is substantial for large areas it is difficult to distinguish intrinsic and extrinsic effects. Hence, there is a preference to investigate high field or "intrinsic" effects on small area capacitors ($< 10^{-4} cm^2$) [6].

In this section a discussion will be given about statistical distributions which are usually applied. It will be argued that among these, extreme value distributions turn out to be the most appropriate. Extreme value statistics describe the distribution of largest or smallest values of other unlimited distributions [7,8].

The significance of extreme value statistics

De Wit et al. [9] and later Solomon et al. [10] derived extreme value breakdown probabilities based on Poisson distributions over area and time (i.e. random processes). Kristiansen [11] proposed to use Weibull distributions, which are one form of the extreme value distributions, but unfortunately the wrong variable was used; $Ln(E)$ instead of E where E is the field strength. As discussed below the significance of extreme value statisitics can be checked by the so-called "stability postulate" [12].

The "stability postulate" states that smallest (largest) values of groups of observations are distributed in the same way as the original distribution **if, and only if** the original distribution itself consists of smallest (largest) values. Consider a distribution of m observations of smallest (largest) values of a distributed variable, divide it arbitrarily in groups of n samples ($n < m$); then the smallest (largest) values of each n samples are distributed in the same way. They are only shifted to smaller (larger) values of the variable. The shift is calculable when n, m are known. On extremal probability paper, where the cumulative fraction of failures F is plotted as $\ln\ln(1 - F)^{-1}$ versus the value of the variable, the distribution shifts parallel towards higher probabilities by an amount

$$\ln\ln(1 - F)^{-1} = 1/\sigma_n \cdot \ln(m/n) \tag{1}$$

where σ_n is a theoretical value only dependent on n.

$$\sigma_n \rightarrow \pi/\sqrt{6} \text{ for } n \rightarrow \infty \quad .$$

When the slope on probability paper is given by $\tan\alpha$ then the shift of the curve towards smaller values of the variable (e.g. the field strength E) is

$$\Delta E = (\sigma_n \tan\alpha)^{-1} \ln(m/n) \quad . \tag{1a}$$

If Eq. (1a) is obeyed the distributions are 'stable' (cf. [8] sct. 6).

The property of a parallel horizontal shift is unique for extreme value statisitics and by checking this property other distribution laws can be precluded [12].

The most common extreme (smallest) value distribution is given by

$$F(E) = 1 - \exp(-\exp(y)); \qquad f(E) = (\sigma_n/S) \cdot \exp(-y) f(E) \tag{2}$$

where F(E) is the cumulative probability, f(E) is the probability density and the quantity y is given by

$$y = (\sigma_n/S)(E - \tilde{E}) \qquad (3)$$

where σ_n is the theoretical quantity mentioned above, S is the standard deviation and $\tilde{E}$ the modal value of the variable E.

Essentially the same equation was derived by de Wit et al. [9] and Solomon et al. [10] where E is then a time dependent function of the applied field strength. Rearranging Eq. (2) yields

$$\ln\ln(1 - F)^{-1} = y = (\sigma_n/S)(E - \tilde{E}) \quad . \qquad (4)$$

When plotting $\ln\ln(1 - F)^{-1}$ versus the breakdown field strength E a straight line must be achieved with y = 0 when E = $\tilde{E}$ (i.e. F = 1 − 1/e = 0.63).

The modal value of E = $\tilde{E}$ can be calculated from the arithmetic mean $\bar{E}$ by

$$\tilde{E} = \bar{E} + (S/\sigma_n)y_n \qquad (5)$$

Where y_n is another theoretical quantity ([8] table 6.2.3). It approaches $\gamma = 0.577$ .... for $n \to \infty$.

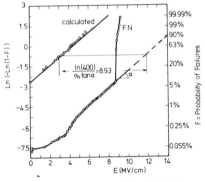

Fig. 1. Demonstration of the "stability postulate" of extremal distributions. The dispersion (slope) does not change in the distribution of minima of grouped data. The bottom-curve is the initial distribution (12000 cap.). The upper curve 30 minima of groups of 400 capacitors each. In the vertical part no breakdown occurs.

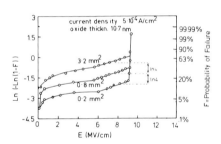

Fig. 2. The shift of the failure distribution is proportional to the ratio of areas (ln 4).

The mentioned check of the stable extremal properties, the invariance of the distribution for grouping of observations, is demonstrated by our own data in fig. 1. A large number of MOS capacitors (m = 12000) was tested for breakdown by ramping the field strength in steps of 0.1 MV/cm in 5 msec. Gate oxide thickness was 40.2 nm, area 0.02 mm². A capacitor was considered to have failed when the current density through the gate oxide exceeded 5×10^{-4} A/cm². The gate oxide was thermal oxide grown in pure oxygen at 950°C in standard processing equipment. Gate material was phosphorous doped polysilicon. The substrate was 2 - 5 Ω cm Si.

The cumulative plot on extremal probability paper was made by plotting $\ln\ln(1-F)^{-1}$ vs E. The measurements were arranged so that 30 sequential runs of 400 capacitors each were tested (n = 30, m = 12000). The total cumulative plot of the 12000 capacitors is given by the lower curve in fig. 1. A second test immediately after each breakdown revealed that in the vertical part of the curve capacitors were not irreversibly destroyed. This was verified by microscopic inspection. In this high field strength region Fowler-Nordheim tunneling is expected. Current level and applied fields are in agreement with the data of Lenzlinger and Snow [14]. It was also found that for various thicknesses the current density was the same for the same field strength. The lower part of the curve consists of real breakdowns. Approximately a straight line is obtained from 5 to 8 MV/cm.

When of the 30 runs of 400 tests each the smallest values of each run is plotted on the same probability paper a parallel curve appears. A line can be calculated by Eq. (4) and Eq. (5) with $\sigma_{30} = 1.11238$ and $y_{30} = 0.53662$ and the mean value $\bar{E}_{30} = 3.57$ MV/cm and standard deviation $S_{30} = 1.68$ MV/cm, following from the observations. It fits excellently to the data points. The slopes of the calculated line and of the lower part of the initial curve are equal. The vertical distance between the two lines is $\ln(400)/\sigma_{30} = 5.39$. The horizontal distance is $\ln(400)/a/\sigma_{30} = 8.95$ MV/cm where a is the slope (a = 0.603 $(MV/cm)^{-1}$).

If the 400 capacitors of one run had been interconnected in a device of 8 mm^2 the latter would have failed at the smallest value of E. The distribution of 30 of such devices will be located on the same line. The distribution of an unlimited number of devices could have been calculated by a vertical shift of $\ln(400) = 5.99$.

In general, if an unlimited distribution of small area capacitors is known, the unlimited distribution of large capacitors can be found by a translation in the vertical direction of $\ln(A_1/A_2)$, where A_1, A_2 are the areas, provided that the distributions are extremal. This same area dependence of the yield $(1-F)^{-1}$ has been proposed based on random distribution of defects. The Poisson (random) distributions and extreme value distributions are related to processes with small probabilities. The Poisson distribution describes the **number** while the extreme value statistics describe the **size** of rare events ([8] p.2).

To demonstrate this for geometrical different areas three series of capacitors with areas 0.2, 0.8 and 3.2 mm^2 respectively produced on a single wafer were tested. The current density level was 5.10^{-4} A/cm^2. The cumulative plots are given in fig. 2. The shifts in vertical direction are approximately a factor $\ln(4)$. The parts of the distributions where tunneling occurs coincide.

Area dependence of mean breakdown voltage

A logarithmic area dependence of the mean voltage to breakdown was recently demonstrated by Anolick and Nelson [15].

It is shown in fig. 3. The average breakdown voltage for the various capacitors is plotted versus the area A. The linear curve corresponds to Eq. (1a) when $\ln(m/n)$ is replaced by $\ln A$. From the slope of the line a it is possible to reconstruct the dispersion of the distributions. It follows from Eq. (1a) that

$$d\Delta E/d\ln A = 1/(a\sigma_n) = S\sqrt{6}/\pi \text{ for } n \longrightarrow \infty \; . \tag{6}$$

The fact that in fig. 3 a straight line is obtained over two orders of magnitude is consistent with a random distribution of breakdown spots and with the application of extremal distribution laws.

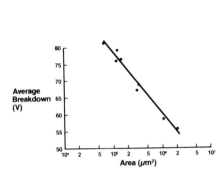

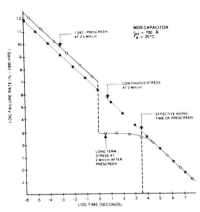

Fig. 3. The mean voltage for breakdown shifts proportional to the areas as predicted by extreme value statistics (fig. reproduced from [15]).

Fig. 4. The failure rate for wearout is given by $h = (k/t)^{\alpha}$ with $k = 10^{15}$ sec and $\alpha = 0.93$. It shows that log-normal statistics must be precluded (fig. reproduced from [6]).

An alternative area dependence of the cumulative failure probality F is used by a number of workers [16,17]. It is given in a form equal to

$$(1 - F)^{-1} = 1 + A \cdot D \tag{7}$$

where D is the latent defect density and A is the area. (D is time dependent.) The relationship is based on Bose-Einstein statistics and it is assumed that the random distributed effects are indistinguishable. The argumentation for this was commented on by Solomon [3]. Eq. (7) is used often in combination with the log-normal statistics employed for wearout [15 - 18].

The theoretical justification for the log-normal distribution was given by Metzler [18]. It is based on the following assumptions.
1. The probability of failure per unit time is proportional to the current density.
2. Oxide currents are due to tunneling (F.N.).
3. Electron trapping induces a time (and current) dependent field distortion.
4. The distortion is distributed at random over the lateral dimensions of the capacitor.
The latter assumption was necessary in order to fit the experimental and theoretical distributions. Metzler introduces an extra factor in the log-normal distribution law called a "field enhancement factor".

Wearout, a non-Gaussian process
Crook [16] has produced plots of the so-called hazard rate or failure rate, of the wearout distributions. One of the plots is reproduced in fig. 4 ([16], fig. 10). One should note that the failure rate h is given over 13 orders of magnitude; which is an extremely large range. An important feature is that the slope on the double logarithmic plot is approximately -1. From this it follows that $h \overset{\sim}{\propto} 1/t$. It is the failure rate of the Pareto type distribution which is principally non-Gaussian ([7], p. 245). The log-normal distribution should have $h \propto (1/t) \cdot \ln(t - \bar{t})$. Hence one can conclude that the log-normal distribution does not apply.

Alternatively, it should be possible to derive the distribution laws by simple integration of the failure rate. The failure rate is defined by

184

$$h = d\ln(1 - F)^{-1}/dt \tag{8}$$

and was seen from fig. 4 to be given by

$$d\ln(1 - F)^{-1}/dt = (k/t)^{\alpha} \tag{9}$$

where $\alpha = 0.94$ and $k = 10^{15}$ s for the 2MV/cm stress.
Integration of Eq. (9) yields

$$\ln (1 - F)^{-1} = (1 - \alpha)^{-1} t^{1-\alpha} k^{\alpha} \tag{10}$$

where $F = 0$ when $t = 0$.
Taking logarithms on both sides it follows that

$$\ln\ln(1 - F)^{-1} = (1 - \alpha)\ln t + \ln(k^{\alpha}/1 - \alpha) \quad . \tag{11}$$

From Eq. (11) it can be seen that there is a linear relation between $\ln\ln(1 - F)^{-1}$ and $\ln(t)$. This proves that extreme value distributions apply.

Crook's data plotted on log-normal probability paper are given in fig. 5. The range on the cumulative failure scale is so small that replotting on extremal probability paper would not have revealed any difference. It is known that log-normal distributions can closely "mimic" extreme value distributions ([7], table I, p. 279).

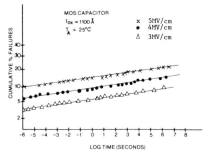

Fig. 5. The cumulative plot on log-normal paper closely mimics extreme value distributions (see however fig. 4). From the fact that the slopes are small $((1 - \alpha) \rightarrow 0)$ it can be deduced that wearout is almost absent (fig. reproduced from [6]).

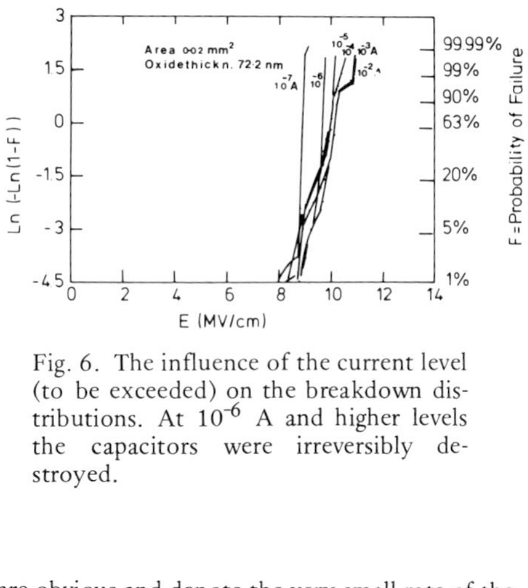

Fig. 6. The influence of the current level (to be exceeded) on the breakdown distributions. At 10^{-6} A and higher levels the capacitors were irreversibly destroyed.

In fig. 5 the very small slopes (i.e. $(1 - \alpha) \approx 0$) are obvious and denote the very small rate of the wearout. The distance between the curves is proportional to the difference in stressing fields. Notice that the slopes are equal. This means that the stressing field itself does **not** accelerate the wearout, e.g. in the case that $\alpha = 1$, the wearout would be absent. The same follows from the equal slopes in fig. 4 for the fields at 2 and 2.5 MV/cm. The failure rate adapts itself even to the milder stressing condition at 2 MV/cm after 1 sec prescreen of 2.5 MV/cm and h decreases further with the same $(k/t)^{\alpha}$ law. The conclusion must be that when the failure rate decreases fast enough wearout will not give additional breakdown (e.g. $h = k/t$).

185

Wearout must be caused by a mechanism like charge trapping in or conduction through the oxide or both. Both have been reported to decrease in time by 1/t laws [3,6,19,20]. However, if the 1/t law would be strictly obeyed then from the above discussion it would be concluded that wearout should be absent; it is the deviation of the 1/t laws for instance by a de-trapping or non-trapping of injected charge carriers which causes the long run deterioration. We shall return to this subject below.

HIGH FIELD BREAKDOWN

Current density dependence

From the shift of the cumulative breakdown distributions towards lower probabilities, while decreasing testing areas, (fig. 1), it can be understood that somewhere in the high field strength region the distribution must become truncated, i.e. somewhere at a field strength the probability of breakdown will be one, and a very steep dispersionless part must show up. The onset of Fowler Nordheim tunneling can mask this part (see previous section). In this case larger current densities must be chosen to get actual breakdown.

In the following it will be indicated that this current density is also area dependent. Harari [21] realised that the area of the investigated capacitors must be decreased in order to investigate the "intrinsic" breakdown. The smallest area used in fig. 1 were 0.02 mm^2, but in the steep part, which is 80% of the curve, capacitors did not breakdown. It was the tunnel current which exceeded the 5.10^{-4} A/cm^2 current density level. The field strength of 9.2 MV/cm is comparable to that found by Lenzlinger and Snow [14]. Harari reported breakdown current densities as high as 50 A/cm^2 for capacitors with an active area of 2.3×10^{-4} mm^2. We have therefore tested capacitors similar to those of fig. 1 at several current density levels and the breakdown distributions are given in fig. 6. Current densities were chosen in intervals of a factor 10 up to 50 A/cm^2. Microscopic inspection revealed that above 5.10^{-3} A/cm^2 all capacitors had experienced a breakdown. It was also noted that light emission occurred at these higher current densities.

Area dependence

Harari's data also indicate a trend towards increasing breakdown current densities at decreasing area. In the case of fig. 6 a current density of 5×10^{-3} A/cm^2 is sufficient to obtain a breakdown. Harari reports that current densities as high as 50 A/cm^2 were needed to breakdown capacitors of 2.3×10^{-4} mm^2 area.

The area dependence of breakdown current density can be explained by the following. In order to cause destruction of the dielectric a certain local current must occur. This area-independent current consists of the current from the external source plus the discharge current of the capacitor. We may write

$$I_{destr} = AJ_{ext} + d(CV)/dt \tag{13}$$

where I_{destr} is the destruction current, J_{ext} the external current density, A the area, C the capacitance, V the voltage prior to breakdown and t the time.

By substitution of $C = \epsilon A/d$ and $V/d = E$ where ϵ is the dielectric constant, d thickness of the dielectric and E the field strength, we find:

$$J_{ext} = I_{destr}/A - \epsilon dE/dt \quad . \tag{14}$$

Hence, J_{ext} is an inverse function of the area. The few observations made, seem to confirm this but there are insufficient data available to check this relation.

186

Time dependence

As has been shown [9] the thickness dependence of the maximum field strength for breakdown should be investigated by using equal ramping rates for the field strength instead of using voltage ramps. It standardizes time dependent effects. De Wit et al. [9] show that under these precautions the maximum breakdown strength was independent of oxide thicknesses. Fig. 7 show their results.

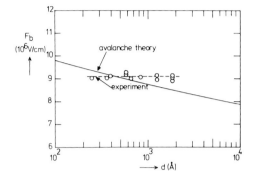

Fig. 7. For large area breakdown with low ramp rates the maximum voltage is independent of oxide thickness. Here, demonstrated for Al_2O_3 (fig. reproduced from [9]).

▼ Fig. 8. The time for breakdown in the "intrinsic" region is correlated to the voltage shift ΔV_{BD} needed to keep a high current constant until breakdown (fig. reproduced from [6]).

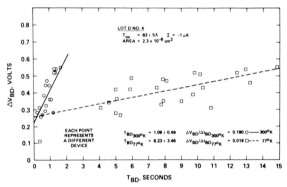

We explored the same ramping field strength measurements for SiO_2 and found that the distributions were indeed invariant for oxide thickness provided that the current densities are small. This means that distributions of large area capacitors have a thickness independent maximum which is close to the field where substantial tunnel currents are found. However, decreasing the area increases the current and field for breakdown above the tunneling region, as has been shown in the preceding section. Measuring techniques such as used by Harari [6] forcing a preset current through the oxide and recording voltage and time to breakdown are suitable to examine the time dependence of the high field breakdown. A clear demonstration of this time dependence is given in fig. 8 [6].

Samples from the same lot were tested either at 77°K or at 300°K and display a significant difference in time to breakdown. There is a correlation between the maximum breakdown voltage and the time to breakdown. From fig. 8 it will be clear that the time rather than the voltage is temperature dependent which follows also from other experiments [6]. The measurements revealed the relation $\Delta V_{BD} = \beta t_{BD}$ where t_{BD} is the time to breakdown and ΔV_{BD} the voltage difference between the voltage to breakdown V_{BD} and the initial voltage V_{in} needed to force

187

the current through the sample. β is a constant which was found to be independent of oxide thickness. β is temperature dependent but not with a simple activation energy. ΔV_{BD} and t_{BD} were found to vary practically independent of V_{in} for the same lot.

Harari [6] found a reciprocal dependence of t_{BD} on the preset current density level which is demonstrated in fig. 9. It suggests the relation $I_{BD} \cdot t_{BD} = \text{const}$.

Probably at this high fields a constant amount of charge (1 - 2 $Coul/cm^2$ which is $1.6 - 3.5 \cdot 10^{19}$ charges/cm^2) must be injected in order to induce breakdown.

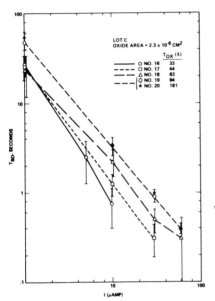

Fig. 9. The time to breakdown is proportional to the reciprocal value of the preset current. This suggests that a certain charge has to be injected prior to breakdown (fig. reproduced from [6]).

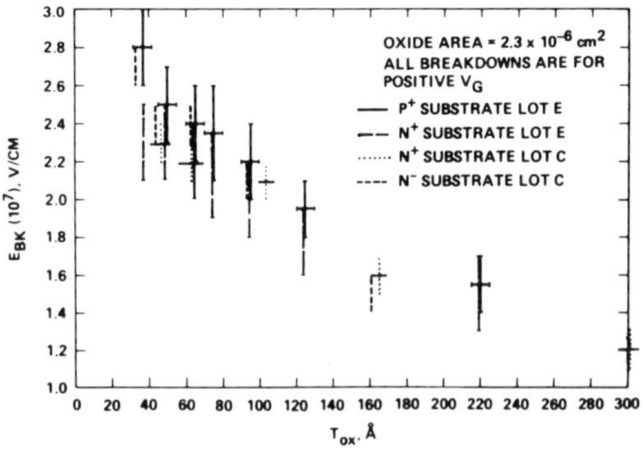

Fig. 10. The maximum field strength for breakdown is dependent on oxide thickness for small area capacitors reaching very high values of almost 30 MV/cm.

Thickness dependence

As was mentioned in the previous section small areas and high current densities must be used when exploring the "intrinsic" region. With his constant current method Harari found a thickness dependence in the 3 - 30 nm region [6] see fig. 10. While Harari gains a lot of information on current and time dependence the method introduces a larger scatter of the observations by the fact that breakdown time is rather large and wearout effects play a role. We have applied the constant field strength ramps with high ramprate and found a thickness dependence quite comparable to that of Harari [6,21] and Osburn [22]. The results are given in figs. 11 and 12 for samples prepared in dry oxygen and dry oxygen with trichlorethane respectively. It can be seen from the observations (each point represents the mean of > 800 capacitors) that there is a definite thickness dependence up to 80 nm. The distribution becomes more dispersed for decreasing oxide thickness; this is indicated by the vertical bars (see for the specifications fig. 1).

The influence of other parameters as temperature, electrode material, substrate doping was not significant [6]. Harari found a small decrease (5 - 10%) of breakdown voltage on negative gate polarities.

MEDIUM FIELD BREAKDOWN

Time dependence

It can be seen in fig. 1 that for a number of capacitors breakdown occurs below the region of field strength where tunneling currents are found. The effect of increasing the area of the capacitor will be that the cumulative distribution exhibits a parallel shift to higher probabilities, i.e. a larger fraction of the capacitors will show low field breakdown. The same will occur when the ramping rate of the field strength is smaller [9]. When the ramp rate is zero, as it is in wearout tests, a large fraction, if not all, capacitors will breakdown in the long run.

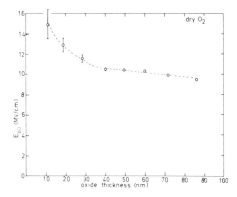

Fig. 11. The thickness dependence of breakdown measured with constant ramp rate of the field is significant up to 80 nm. The ramp rate was 20 MV/cm/sec.

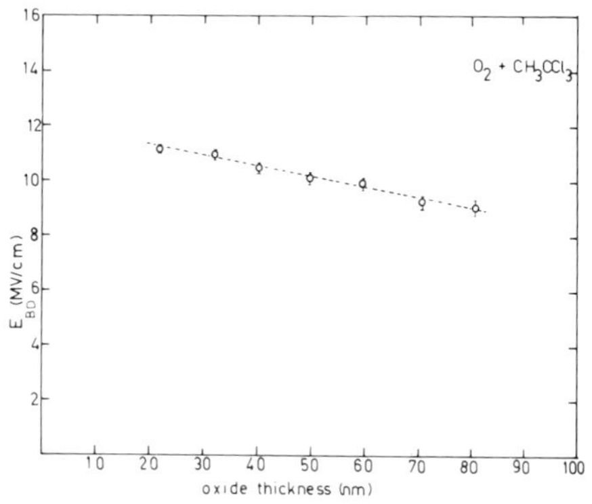

Fig. 12. The same as fig. 11 for oxides grown in CH_3CCl_3.

The fact that ramping rates are so important for the whole breakdown distribution and that even the highest breakdown field strength are time dependent suggests that the difference between "intrinsic" breakdown and "medium field" breakdown or "wearout" is more in the time scale than really bound to localised asperities, metal ions, cracks, voids, stresses or field distortions. However, all of these have of course their influence on the time scale. From the early wearout work of Osburn and Bassous [23], it follows that the anode material is primarily involved in the wearout mechanism and that HCl additon always improves wearout or medium field breakdown.

Mechanical stress induced breakdown

The effect of stress induced by thermal expansion on oxide quality has recently been reported by Eernisse [24] and Kolbesen [25]. The oxidation temperature has a crucial effect on the stress induced effects. At $\sim 950°C$ the glass transition temperature of silica is located. It is a second order phase transition. Far above $950°C$ silica is more like a liquid and induced strains can relax within fractions of seconds. Far below $950°C$ the strains do not anneal within experimental time. The effect of stress and glass transition point is also found in oxidation kinetics [24,26].

189

To demonstrate the effect of stress on the breakdown distributions we have tested large area capacitors. The oxide was grown at 950°C. One lot was withdrawn with normal rate while the other was cooled down slowly to lower temperatures. The active area of a capacitor was 5.25 mm², the gate electrode was poly Si, the oxidation ambient was O_2 with 3% HCl.

Our breakdown curves are given in fig. 13. It is clear that the non-annealed samples have a much higher breakdown probability.

Kolbesen [25] and Isomae et al. [27] report generation of stacking faults below 950°C that would not occur above 1050°C.

THEORIES ON HIGH FIELD BREAKDOWN
The trend in recently published papers is to add arguments which invalidate established breakdown theories; specifically for high field breakdown in thin SiO_2 films [1,3,6].

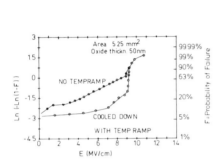

Fig. 13. The breakdown probability for slowly cooled capacitors is significantly smaller than for rapidly cooled capacitors.

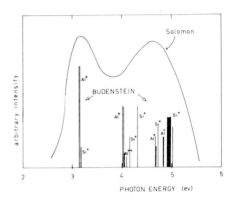

Fig. 14. The correspondence of the light emission at pre-breakdown stress in Al-SiO₂-Si contributed to luminescence by Solomon [10] (curve) and light emission at (partial) breakdown in Al-SiO-Al capacitors contributed to gaseous discharge by Budenstein [1], suggests a common origin.

The relative temperature independence of the conduction mechanism and the independence of duty cycling of high voltage pulses precludes the *thermal runaway*. The relative temperature independence down to 77°K rules out *ionic effects* [6]. *Double injection* and *hole injection* are argued against by the observation [6] "that the breakdown mechanism is intimately related to the generation of electron traps and that these traps do not exist in the oxide before stressing". Holes would become practically immobile at liquid N_2 temperature and breakdown would be expected to occur in shorter time at 77°K than at room temperature. The opposite effect is found (see fig. 8).

Impact Ionisation models are rejected by Harari [6] since in the wide range of breakdown fields (i.e. $\Delta V_{in}/d$) the higher fields would exponentially decrease the time for breakdown. The ionisation rate, α'

190

$$\alpha' = \alpha'_0 \exp{(-H/E)}$$

increases almost exponentially with E (α'_0 and H are constants). This is contradicted by Harari's observation that the time to breakdown was not dependent on the initial voltage V_{in} where injection started. Nor does the constancy of injected charge ($\sim$ 1-2 Coul/cm^2) prior to breakdown fit in the impact ionisation model.

Another argument is that below 12.2 V no holes should be generated in the oxide. High field breakdowns are nevertheless found below 12.2 V [6]. The strong argument in favour of impact-ionisation models has been the predicted thickness dependence. Harari concludes from his observations that below 16 nm impact ionisation did not contribute to breakdown.

Harari's explanation for the high field breakdown mechanism is based on field enhancement in the oxide by space charge caused by generation and filling of electron traps. When local fields of 30 MV/cm are reached, hot electron runaway occurs. Harari decides that without trapping, fields as high as 30 MV/cm (for 4 nm oxides) can be obtained prior to breakdown. Ferry [28] has calculated the electron velocity in SiO$_2$ by iterative techniques and concludes that fields as high as 50 MV/cm are needed for velocity runaway and ionisation rates would become substantial at fields above 30 MV/cm. Ferry [28] admits, however, that these high values are based on equal electron and hole masses and that a hole mass of 3 times the electron mass would reduce the field to 8 - 10 MV/cm.

Harari's model based on field enhancement may be a possible explanation in thinner samples. He finds that electron traps are located near the injecting electrodes. Hence, for thicker samples the field enhancement can never reach the high level of 30 MV/cm. Moreover, if the location of trapped charge is independent of thickness it predicts an increasing breakdown strength for thicker samples.

Gas discharge model
Recently Budenstein [1] has given a survey on breakdown mechanisms in solids covering a large number of investigations on amorphous, crystalline, organic and inorganic substrates. He concludes that the main breakdown mechanism has the nature of a gas discharge. This was based on the following observations.

1. Light emission in the 2-5 eV range was observed at, but also prior to breakdown at high field stressing.

2. Partial breakdowns could be produced by voltage pulses. The breakdowns are called partial since they do not short circuit the two electrodes. The resistance of the dielectric was not altered. The partial breakdowns had tree-like structures. At breakdown they left a hollow main channel and branches with resolidified material. Trees are poor conductors ($\rho \approx 10^6$ Ω cm) and could grow almost perpendicular to the field direction. It could be inferred that they must not be considered as equipotential volumes.

3. Branches had bead–like structures. Beads were found also next to the partial breakdowns.

4. Beads and trees emitted light (prior to breakdown) which consisted of the atomic emission spectra of the constituents of the dielectric. It was probably produced by high pressure gaseous plasmas.

5. When MeV electrons were injected to a dielectric and direct discharge or breakdown was prevented, the tree-like structures could be initiated by mechanical agitation and by choosing the location for grounding the trees could be grown perpendicular to the field produced by the space charge.

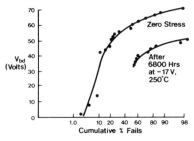

Fig. 15. Long term electrical stressing causes a change in breakdown distribution which is not recoverable (fig. reproduced from [15]).

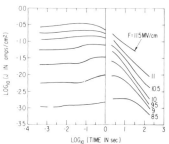

Fig. 16. Trapping at high stress field causes the injection currents to decrease with a $(1/t)^{\alpha}$ law (fig. reproduced from [3]).

Gas discharge in SiO₂?

A number of observations for SiO_2 will be recalled which seem to fit the gaseous discharge model.

1. Light emission in the pre-breakdown stage was reported by Solomon and Klein [10]. The energy of the emitted light was found to be in the range of 2-5 eV, see fig. 14. We have added the relative intensities and energies as found by Budenstein [1] for Al-SiO-Al capacitors at breakdown. Emission was attributed by the latter to the atomic emission spectra of Si and Al. The observations could have the same origin.

2. The recently published breakdown distributions of Anolick [15] with and without low voltage stress show a gradual deterioration of the dielectric under stress, which is permanent. The irreversible change is shown in fig. 15 reproduced from [15]. This is difficult to explain by other effects as ionic migration [15]. This could indicate the existence of "trees".

3. Charge trapping rates are found to decrease by $(1/t)$ laws [6,19,20,29] and therefore trapping itself is probably not the main mechanism for wearout (see above). Even at high field strength injection currents, and probably trapping, decrease with $(1/t)^{\alpha}$ laws with $\alpha \approx 1$, as can be seen in fig. 16 reproduced from Solomon [30]. The total charge Q injected in the dielectric is then given by

$$Q = \int_0^t J dt \propto (1-\alpha)^{-1} t^{1-\alpha} \ . \qquad \text{cf. Eq. (10)}$$

It is unlimited for $t \longrightarrow \infty$ when $\alpha \leqslant 1$. For any $\alpha > 1$ the integral is limited. So if a critical level for Q has to be reached prior to breakdown as suggested by the observations this will certainly happen for $\alpha \leqslant 1$. The situation is just critical for $\alpha = 1$, i.e. $Q \propto \ln t$. It is shown elsewhere that for logarithmic charging, trapping site generation must be assumed [29]. The mechanism for the generation could be the consequence of growing "beads" or "trees" as will be indicated in the following.

4. "The breakdown mechanism is intimately related to the generation of the electron traps" [6]. This generation of electron traps can be explained if it assumed that the contents of a bead are ionised and the relative conductance increases. This will lead to an apparently larger polarisation of the dielectric. Positive and negative charges are accumulated at both sides of a "bead" decreasing the local field. The accumulation is indistinguishable from trapping site generation.

Whether the model of Budenstein for gaseous charges is realistic for SiO_2 films is still a conjecture. However, a number of observations justify further investigations.

CONCLUSIONS
The statistics of extreme values have been used to investigate breakdown and wearout distributions. Area dependence, current density and time dependence have been examined. Most of the existing breakdown theories have to be rejected. A model is proposed based on a charge trapping site generation and conduction in gaseous plasmas.

ACKNOWLEDGMENTS
The author thanks C. Crevecoeur, H.C. de Graaff, P. Hart, F. Vollenbroek, H. Peek, W. Rey, J.F. Verwey, and H.J. de Wit for valuable discussions. T. Hoogestijn for technical assistance, H. Peek, F. Smolders and W. Ruis for sample manufacturing.

REFERENCES
1. Budenstein, P.P., IEEE, E.I. 15, 1980, p. 225
2. Klein, N., Thin Sol. Films **50**, 1078, p. 223
3. Solomon, P., J. Vac. Sci. Technol. **14**, 1977, p. 1122
4. Osburn, C.M., J. of Sol. St. Chem. **12**, 1975, 232
5. Kern, W., RCA Review **31**, 1973, 234
6. Harari, E., J. Appl. Phys., **49**, 1978, p. 2478
7. Johnson, N., Kotz S., Distribution in Statistics (1970) John Wiley and Sons, N.Y.
8. Gumbel, E.J., Statistics of Extremes, 1958, Columbia University Press, N.Y.
9. de Wit, H.J., Wijenberg, C., and Crevecoeur, C., J. El. Chem. Soc. **123**, 1976, p. 231
10. Solomon, P., Klein, N., and Albert, M., Thin Solid Films **35**, 1976, p. 321
11. Kristiansen, K., Vacuum **27**, 1977, p. 227
12. Weber, K.H., and Endicott, H.S., AIEE Trans. **76**, Power App. Syst. p. 393
13. Wolters, D.R., Hoogestijn, T., and Kraay, H., The Phys. of MOS Insulators, Proc. Int. Top Conf. Raleigh N.C. (1980) 349
14. Lenzlinger, M., and Snow, E.H., J. Appl. Phys. **40**, 1969, p. 287
15. Anolick, E.S., and Nelson, G.R., Proceedings 17th annual IEEE, 1979, Reliability Physics p.8
16. Crook, D.L., IEEE, 1979, Reliability Physics, 17th annual Proceedings, p.1
17. Li, S.P., and Maserjian, J., IEEE trans E.D. **23**, 1976, p. 525
18. Metzler, R.A., IEDM (1979) 233
19. Walden, R.H., J. Appl. Phys. **43**, 1972, p. 1178
20. Ushirokawa, A., Suzuki, E., Warashina M., Jap. J. Appl. Phys. **12**, 1973, 398
21. Harari, E., J. Appl. Phys. Lett., **30**, 1977, p. 601
22. Osburn, C.M., and Ormond, D.W., J. El. Chem. Soc. **119**, 1972, p. 591-597
23. Osburn, C.M., Bassous, E., J. El. Chem. Soc. **122**, 1975, p. 89
24. Eernisse, E.P., Appl. Phys. Lett., **30**, 1977, p. 290-293

25. Kolbesen, B., and Strunk, H., Inst. Phys. Cont. Ser. **57,** 1980, p. 21
26. Wolters, D.R., J. El. Chem. Soc. **127,** 1980, p. 2072
27. Isomae, S., Tamaki, Y., Yajima, A., Nanba, M., and Maki, M., J. Electrochem. Soc. **126,** 1014 (1979)
28. Ferry, D.K., Sol. State Comm., **18,** 1976, p. 1051
29. Wolters, D.R., and Verwey, J.F., This proceedings (1981)
30. Solomon, P., J. Appl. Phys. **48,** (1977) p. 3847

Hydrogen-Sodium Interactions in Pd-MOS Devices

Claes Nylander, Marten Armgarth, and Christer Svensson

Department of Physics and Measurement Technology, Linköping University
S-581 83 Linköping, Sweden

The hydrogen induced shift of the flat-band voltage in Pd-MOS structures is shown to increase with sodium contamination of the silicondioxide. Triangular voltage sweep measurements show that the sodium ions become more strongly bound to the Pd-SiO$_2$ interface in the presence of hydrogen.

1. Introduction

In 1975 Lundström et al. demonstrated the effect of hydrogen on a MOS transistor having a gate made out of palladium [1] . Pd is a catalytic metal which dissociates hydrogen molecules on its surface and easily dissolves hydrogen atoms. Upon exposure of such a device to hydrogen a layer of hydrogen atoms will rapidly be formed at the Pd-SiO$_2$ interface. This layer gives rise to an electric potential causing a shift of the flat-band voltage V_{FB} of the MOS structure. The device can be used as a sensitive and selective sensor for hydrogen gas in the ppm range. One problem has, however, made it difficult to use the device as an accurate monitor for hydrogen concentrations. As seen in fig. 1a, the response to hydrogen is not only a rapid shift of V_{FB} but also a slow one possessing several time constants ranging from seconds to hours. There are two reasons to suspect ion migration as an origin for the "slow effect". One reason is that in order to keep the catalytic reactions on the outer Pd surface running satisfactorily, the device has to be run at an elevated temperature where Na$^+$ drift in the oxide is expected to occur. Another reason is that we recently have found that the slow effect can be eliminated through the introduction of a thin layer of alumina between SiO$_2$ and the Pd gate [2]. Such a layer is known from literature to prevent ion migration in the oxide.

In order to find out if the slow effect is related to interactions between hydrogen and sodium ions in the oxide we have performed high frequency CV measurements and TVS measurements (Triangular Voltage Sweep) [3] on both normal and contaminated MOS structures.

2. Experimental

Silicon wafers of p-type (100) were oxidized in oxygen at 1200°C to a thickness of 1000 Å and annealed in forming gas at 500°C for 10 minutes. The oxide surface was then treated in a light HF etch for 15 seconds. Some of the samples were then treated in a 10% solution of NaOH for 5 minutes. On all samples 1 mm dots of palladium and aluminum respectively were deposited through thermal evaporation to a thickness of 1000 Å.

195

High frequency CV measurements were carried out by means of a Boonton 72A bridge. A regulator adjusted the bias voltage in order to retain flat-band condition and this voltage was recorded on a standard X-T recorder. TVS measurements were performed with a Keithley 610C electrometer and a ramp generator applying a triangular voltage with a derivative of 50 mV/second across the sample. All measurements were performed at 150°C.

3. Results

As shown in fig.1 a sample treated with NaOH shows a significantly different response of V_{FB} upon exposure to hydrogen. It should be noted that the "slow" shift is not only increased but the rise time has also become shorter. Some typical results from TVS measurements on both normal and NaOH contaminated Pd-MOS samples are shown in fig. 2. It is clearly seen how the NaOH treatment has increased the number of ions in the oxide.
The upper peak in fig. 2 is due to ion transport from the Si-SiO$_2$ interface and the lower peak is due to ion transport from the Pd-SiO$_2$ interface. The same experiment was performed on ordinary Al-MOS structures. Also in these samples the ion content had increased by approximately the same amount.
One peculiar effect to which we yet have no explanation is the small extra peak occurring to the left of the upper sodium peak. A careful study shows that there are in fact two extra peaks, one hidden in the sodium peak. They occurred on all samples prepared for this investigation. These extra peaks are not affected by sodium contamination fig. 2 . A control experiment with KOH contamination did not affect them either.

We have fitted the lower peak to the theory for ion emission over an image potential barrier presented by TANGENA et al,[4] . In this theory the potential barrier over which the ions are emitted follows the equation

$$E_b = E_{bo} - q\omega\sqrt{V}_{ox} \qquad (1)$$

where E_{bo} is the binding energy for ions at the metal-oxide interface, ω is a factor depending on the actual form of the potential and V_{ox} is the oxide voltage which varies linearly with time. The form of the ionic current

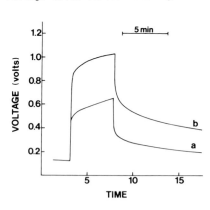

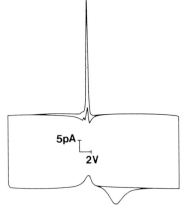

Fig. 1 Typical shift of a) normal and b) NaOH contaminated PdMOS structures upon exposure to H$_2$ in oxygen.

Fig. 2 TVS results for a normal and NaOH contaminated PdMOS structure (see text for further explanation).

196

peak gives us a value for ω which is twice the theoretical value for Schottky barrier lowering. There are two possible explanations for this. First the value for the dielectric constant for SiO_2 used in the theory ($\varepsilon=3.85$) is probably too high since it is the static value. Depending on the velocity of the ions a lower value should be used. Secondly, due to local variations the emission could be something between Schottky and Poole-Frenkel emission, which also gives rise to a higher value for ω. Using the experimentally found value for ω the binding energy E_{bo} can be estimated to 1.4eV under the assumption that the attempt-to-escape frequency is about 10^{12} Hz. The position of the peak varied between different samples (6-10 volts). The corresponding variation in binding energy is, however, only in the range of 0.1 eV.

Fig. 3 shows the result of TVS measurements on a NaOH contaminated sample in pure oxygen and in oxygen containing 500 ppm hydrogen. Both V_{FB} and the upper sodium peak shifts negatively about 1 volt. This should be compared with the result in fig. 1b. The lower peak does, however, show a peculiar behaviour. Instead of following the shift in V_{FB} this peak shifts in the opposite direction about 6 volts relative to V_{FB}. The conclusion must be that, even though the hydrogen atoms at the inner metal surface gives rise to a positive potential in the oxide, Na^+ becomes more strongly bound to the metal surface. This shift of the peak is very drastic on the voltage scale but according to the theory mentioned above the change in binding energy is only 0.1 eV.

The kinetics of the upper peak is about the same as that for V_{FB} shown in fig. 1b. The lower peak seems to follow the slow effect only. Fig. 4 shows the shift of the flat-band voltage and the binding energy, calculated from the position of the lower TVS peak, as a function of time after hydrogen removal from the ambient. The binding energy shows no fast shift but follows the kinetics of the slow shift in V_{FB}.

We also investigated the effect of annealing of the samples in forming gas for 10 minutes at $500^{\circ}C$. In a normal Pd-MOS structure this treatment in-

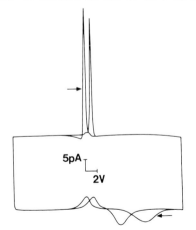

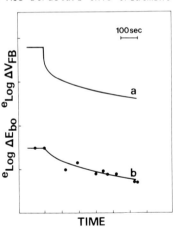

Fig.3 TVS results on a NaOH contaminated PdMOS structure in pure oxygen and in oxygen with 500 ppm H_2 (indicated by arrows).

Fig.4 The response to hydrogen removal from the ambient a) flat-band voltage b) binding energy for Na^+ at the Pd-SiO_2 interface.

creased the number of drifting ions markedly. This was not observed on Al-MOS structures. These did, however, change in another way. The binding strength of Na$^+$ to the metal decreased markedly upon annealing, i.e. the lower TVS peak occurred at a much lower voltage. This phenomena may be the reason for the varying results on ion drift in Al-MOS found in literature [4,5].

4. Discussion and Conclusions

In summary we have made following observations:

a)   premetallization treatment with NaOH makes the hydrogen induced shift of V_{FB} in Pd-MOS larger and faster.

b)   The upper TVS peak follows approximately the shift in V_{FB} and does not change in magnitude significantly when hydrogen is introduced.

c)   The lower TVS peak moves about 6 volts in the opposite direction to the flat-band voltage upon hydrogen introduction.

Observation (a) indicates that more hydrogen atoms are adsorbed at a Na-treated metal-oxide interface than at a nontreated interface *or* that Na$^+$ move towards the SiO_2-Si interface when hydrogen in introduced. The second model is, however, not compatible with the other observations. If hydrogen activates Na$^+$ drift we would expect the upper peak to change in magnitude (area) as a result of hydrogen introduction. Note that the number of ions needed to give 1V shift is about the same as the total number of drifting ions. Observation (c) shows that Na$^+$ becomes more strongly bound to the metal upon hydrogen exposure.

We are thus left with the first model of the slow shift. This model is in fact also supported by observation (c). It clearly shows that the slow effect is located near the metal-oxide interface. Furthermore a simple cluster calculation shows that a layer of positive charges in the oxide, close to the metal, increases the binding energy of alkali ions to the metal. This is due to the form of the image potential. Therefore, if positively charged hydrogen ions are trapped in the oxide, close to the metal, they can cause both the slow shift of V_{FB} and the increase of E_{bo}.

Although we have not been able to understand the origin of the slow hydrogen induced shift completely, we believe we are well on the way. We have also demonstrated that hydrogen-sodium interactions take place in Pd-MOS structures. Such phenomena may be of importance in any type of MOS devices.

We wish to thank Ms Anita Spetz for preparing the samples and Professor Ingemar Lundström for his support. This work has been sponsored by the Swedish Board for Technical Development.

1   I Lundström, S Shivaraman, C Svensson and L Lundkvist, Appl. Phys. Letters 26, 55 (1975).

2   M Armgarth and C Nylander, submitted for publication in Appl.Phys.Letters.

3   N J Chou, J. Electrochem. Soc. 118, 601 (1971).

4   A G Tangena, N F de Rooij and J Middelhoek, J. Appl.Phys. 49, 5576 (1978).

5   P K Nauta and M W Hillen, J. Appl. Phys. 49, 2862 (1978).

Electrical Behaviour of Hydrogen Ions in SiO$_2$ Films on Silicon

Zheng Youdou, We Fengmei, Jiang Ruolian, and Zhou Guangneng

Department of Physics, Nanjing University
Nanjing, China

ABSTRACT

The electrical behavior of the mobile hydrogen ions in the thermally grown SiO$_2$ films were studied by using the C-V measurements,
the thermally stimulated ionic current, and the avalanche injection
techniques. It has been found that the hydrogen ions introduced
into the thermally grown SiO$_2$ films by water vapour or ethanol contamination are high mobility species under low electric field at
room temperature. These mobile hydrogen ions can cause the instabilities of MOS structures at room temperature-bias. It has also
been found that the mobile hydrogen ions located in the SiO$_2$ films
near the SiO$_2$-Si interface may act as electron trap centers.
In addition, the instabilities and electron trap centers due
to the mobile hydrogen ions may be reduced by a low-pressure RF
plasma annealing.

INTRODUCTION

It is well known that hydrogen is the most important extrinsic
species which greatly affects the electrical properties of SiO$_2$-Si
interface in thermally grown SiO$_2$ films. It has been reported
that the hydrogen introduced into thermally grown SiO$_2$ films by
water vapour or ethanol contamination may be mobile positive ions
involved in ion migration through the oxide [1], [2], [3]. We have
studied the electrical behavior of hydrogen ions in thermally
grown SiO$_2$ films by C-V measurements, the thermally stimulated
ionic current, and avalanche injection techniques. This paper
reports the experimental results.

SAMPLES AND EXPERIMENTAL TECHNIQUES

The samples used in the present study were Al-SiO$_2$-Si MOS capacitors with an Aluminium dot of 3.4×10^{-3} cm^2. P-type Silicon
wafers of $\langle 100 \rangle$ orientation and $0.6\,\Omega$-cm resistivity were used.
The oxide layers were thermally grown in wet oxygen with thick-
ness between 1500 to 3000 Å. Aluminium electrodes were evaporated
from a tungsten filament. In order to investigate the effect of
hydrogen ions in thermally grown SiO$_2$ films, the MOS capacitors
were contaminated with water vapour or ethanol.
Three techniques were used to measure the electrical behavior of
the hydrogen ions in thermally grown SiO$_2$ films: (1) High frequency
C-V measurement, (2) Thermally stimulated ionic current (TSIC)
technique [4], (3) Avalanche injection technique [5].

RESULTS

1. Hydrogen ions cause the room temperature-bias instability of
MOS structures. The measurements of C-V characteristics before
and after bias treatment at room temperature were made on a number
of P type MOS capacitors. It has been found that the C-V curve
shifted in the direction of negative voltage after positive bias
treatment and shifted in the direction of positive voltage after
negative bias treatment subsequently, as shown in Fig.1. It is

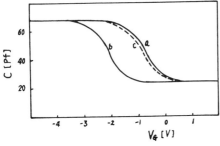

Fig.1 C-V characteris-
tics measured with dif-
ferent bias-voltage treatment at room temperature.
(a) initial C-V (b) +10 V
5 min. (c) -10 V 5 min.

clear that the shifts of C-V curve were caused by mobile positive
ions drifted across the SiO_2 layer when a gate voltage was applied.
The mobile ions drifted in a low electric field (~ 0.3 MV/cm) at
room temperature may be of species having higher mobility than
sodium ions observed in SiO_2 films. It may well be   hydrogen
ions (H^+ or H_3O^+) where released by the hydrolysis of the absorbed
water at the $Al-SiO_2$ interface under applied positive voltage.

2. The dynamic behavior of mobile hydrogen ions in SiO_2 films.
The dynamic behavior of mobile hydrogen ions in thermally grown
SiO_2 films was studied by the measurements of the thermally stimulated ionic current. We have observed three current peaks in
TSIC spectra of the MOS samples, as shown in Fig.2. The peak B

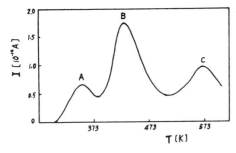

Fig.2 Typical TSIC
curve of $Al-SiO_2-Si$
structure at positive
voltage.

and peak C in Fig.2 are due to sodium ions and potassium ions,
respectively. These are in agreement with several authors. The
peak A is a new peak which lies in the lower temperature region;
it may be due to hydrogen ions introduced by contamination[6].
Fig.3 shows the TSIC curves of hydrogen ions in another MOS samples
for negative (a) and positive (b) gate voltage. The TSIC of hydrogen ions in Fig.3 can be described by the emission theory for a
single level[4]. The current is expressed as

$$I = I_0 \exp(-E_a/KT) \tag{1}$$

where E_a is the activation energy of the trap.

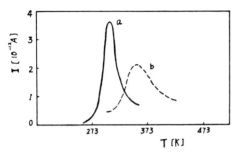

Fig.3 TSIC curves of hydrogen ions in another MOS sample at negative (a) and positive (b) gate voltage.

According to equation (1) and the curves in Fig.3, the activation energy of hydrogen ions can be determined as 0.49 ev for SiO_2-Si interface and 0.57 ev for Al-SiO_2 interface. These results clearly indicate that the motion of mobile hydrogen ions is controlled by emission .from either Al-SiO_2 or SiO_2-Si interface.

3. The hydrogen ion acts as a like-electron trap centre. We have found that the mobile hydrogen ions introduced into thermally grown SiO_2 films by contamination can act as a like-electron trap centre at room temperature. The avalanche injection technique and C-V measurements have been employed to study the electron trapping effects of the hydrogen ions in these SiO_2 films. We have found that the electron trapping effect occured only when hydrogen ions were drifted to the SiO_2-Si interface by applied positive gate voltage at room temperature and not negative one. These indicate that the electron traps are closely related to the appearing of hydrogen ions in the vicinity of the SiO_2-Si interface. Fig.4 shows the typical experimental results of measurements at room temperature. It can be seen that the flat-band voltage shift (ΔV_{FB}) due to this effect is governed by the following equation [7]

$$\Delta V_{FB} = \Delta V_0 \left[1 - \exp(-\sigma j t/q)\right] . \tag{2}$$

Where j is the current density, σ is the trap cross section, q is an elementary electron charge. Using equation (2) the trap cross section σ can be determined as $10^{-15} \sim 10^{-16}$ cm^2. It has also

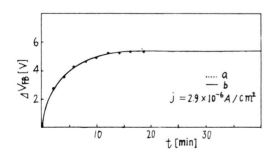

Fig.4 Shift in the flat-band voltage as a function of time during avalanche injection at room temperature. The measured (a) and calculated (b) values.

been found that the amount of negative charge built up at the SiO_2-Si interface is equal to the amount of the mobile hydrogen ions obtained by H. F. C-V measurements before the injection, as shown in Fig.5.

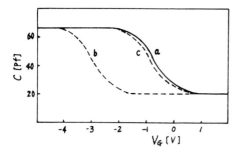

Fig.5 The shifts of C-V curve at room temperature. (a) initial (b) after positive bias treatment (c) after avalanche injection subsequently.

These results suggest that the capture of an electron on a mobile hydrogen ion results in the formation of an hydrogen atom which it is released and diffused out of the samples. There is a posibility that the mobile hydrogen ions were neutralized by the hot electrons injected from silicon substrate.

4. The influence of RF plasma treatments on the instability and like-electron trap centre in MOS structures. We have found that the instability and like-electron trap centre may be reduced by a low pressure RF plasma annealing [8]. The annealing mechanism has not been fully understood at present. However, in the annealing action it seems that the water absorbed on the oxide surface was removed by low pressure RF plasma treatment or the mobile hydrogen ions were neutralized by the hot electrons excited by RF plasma.

ACKNOWLEDGMENTS

The authors would like to thank professor Wu Rulin for valuable discussions.

REFERENCES

1 S. R. Hofstein, IEEE Trans. Electron Devices, 13, 222 (1966) and 14, 749 (1967).
2 E. H. Nicollian, C. N. Berglund, P. F. Schmidt and J. M. Andrews, J. Appl. Phys. 42, 5654 (1971).
3 B. R. Singh, B. D. Tyagl, and B. R. Marathe, Int. J. Electronics, 41, 273 (1976).
4 T. W. Hickmott, J. Appl. Phys. 46, 2583 (1975).
5 E. H. Nicollian, A. Goetzberger, and C. N. Berglan, Appl. Phys. Letters, 15, 174 (1969).
6 Zheng Youdou, Jiang Ruolian, Journal of Nanjing University (Natural Science Edition) 1980.
7 D. R. Young, E. A. Irene, D. J. DiMaria, and R. F. Dekeersmaec, J. Appl. Phys. 50, 6366 (1979).
8 Zheng Youdou, Wu Fengmei, and Su Zonghe, Chinese Journal of Semiconductor, 2, 163 (1980).

Chlorine Implantation in Thermal SiO$_2$

G. Greeuw, H. Hasper

Department of Applied Physics, Groningen State University
9747 AG Groningen, The Netherlands

Abstract

Thermally grown SiO$_2$ layers were implanted with chlorine (10^{11} - 10^{16} cm^{-2}) and in a few cases with argon. The implantation effects were measured on MOS structures by means of the Triangular Voltage Sweep technique, which displays the mobile ion (Na$^+$) motion through the oxide. A strong increase of the number of mobile ions was found with increasing implanted dose. No significant change was found after a 20 min., 900°C anneal in dry N$_2$, from which we conclude that we are dealing with an irreversible process. The implanted chlorine is mobile in the SiO$_2$ film above 700°C. A tentative value for the diffusivity was found:

$$D = 2 \times 10^{-6} \exp(-2.0/kT) \qquad cm^2 s^{-1}.$$

Introduction

A main source of electrical instability of a MOS structure is the motion of mobile ions (Na$^+$, K$^+$) through the SiO$_2$ layer. One way to attack this problem is to perform the (thermal) oxidation step in a chlorine containing atmosphere, as was done by KRIEGLER [1] and others. The resulting oxide has a chlorine rich layer near the Si-SiO$_2$ interface [2], in which mobile Na$^+$ ions can be trapped and neutralized (passivation) [3]. However, the chlorine rich layer has also negative effects, like increase of the interface trap density and of the SiO$_2$ defect density, for chlorine concentrations which cause good Na$^+$ passivation [4]. As the pile-up of Cl near the Si/SiO$_2$ interface during oxidation cannot be avoided, we used ion implantation as a tool to distribute the chlorine more homogeneously over the bulk of the oxide.

A similar experiment was performed by WRONSKI [5]. He reports trapping of Na$^+$ ions by the Cl atoms in the bulk of the oxide after Cl$^+$ implantation with doses ranging from 5×10^{12} to 5×10^{15} cm^{-2}. He also states that the chlorine diffuses out of the oxide during thermal anneal at 900°C. Our experiments support only the latter statement.

Experimental data

All measurements were performed on MOS capacitors, fabricated on <100>, 1 - 30 Ωcm, n-type silicon wafers. The oxide layers were grown at 1200°C in dry O$_2$ to a thickness of about 1000Å. The furnace tube was cleaned by a 2% C$_2$H$_3$Cl$_3$/O$_2$ flow for at least one hour in order to prevent Na contamination during the oxidation. Except for the control wafers, all the samples were implanted with Cl^{35} (dose: 10^{11} - 10^{16} cm^{-2}) and Ar^{40} (10^{11} - 10^{15} cm^{-2}).

The implantation energy was mostly 25 keV and in a few cases 80 keV. Directly after the implantation, most of the wafers were annealed in dry N_2 for 20 minutes at temperatures ranging from 700 - 1000°C. The other samples were heated by a laser pulse.

Aluminum electrodes of 1 mm diameter were evaporated through a metal mask by an electron beam heated evaporation system and sintered for 30 minutes at 450°C in wet N_2 gas.

The triangular Voltage Sweep (TVS) technique [6] was used to study the mobile ion behaviour. In normally prepared oxides, with sodium as the only contaminating species, we expect at 300°C one more or less symmetrical current peak at $V_G \simeq 0.5V$ (see lower curve fig. 1).

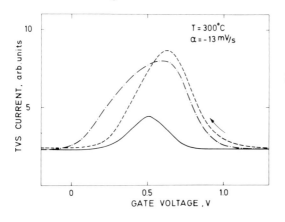

Fig. 1 Three TVS curves
——— reference wafer
----- 25 keV Cl$^+$ implanta-
    tion, 10^{15} cm^{-2}
    700°C anneal
-·-·- *ibid* 900°C anneal

The content of the current peak is a direct measure for the mobile ion concentration N_m (cm^{-2}). The constant current level at $|V_G| > 1.5V$ represents the charging current of the MOS capacitor.

For some of the wafers, the chlorine concentration was measured with long wavelength X-ray analysis ("LWXA") [7].

For some other samples the chlorine profile was determined with the Rutherford Backscattering (RBS) technique.

Results and discussion

A. Increase of mobile sodium
After implantation the TVS curves were different in *shape* as well as in *peak content* (fig. 1). The first effect is not well understood, we suggest shallow bulk trapping to occur as a consequence of the implantation damage.

The second effect means a strong increase of the number of mobile sodium ions (N_m), dependent on implantation dose:
- no increase of N_m for low doses ($< 10^{13}$ cm^{-2}).
- a strong increase of N_m for higher doses (see fig. 2).
- a great *similarity* between the TVS curves of the *chlorine* and the *argon implantation*, except for the implantation energy dependence (fig. 2).
- no significant change in N_m after a 20 min., 900°C anneal in dry N_2 (fig. 1).
To explain the increase of N_m we propose the following model:
We assume that there are a lot of immobile Na$^+$ ions present in the oxide,

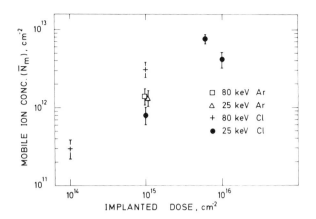

Fig. 2
Mobile Na$^+$ concentration versus implanted dose

at sites were the Si-O bonds are broken [8]. The Na$^+$ ions are trapped here by ionic Na$^+$-O$^-$ bonds. The following reaction may occur during the Cl$^+$ (or Ar$^+$) implantation:

$$\equiv Si - O^-Na^+ + Cl^+ \rightarrow \equiv Si - O + Na^+ + Cl^0$$

This means that the implanted Cl$^+$ ion is neutralized at the position where the Na$^+$ ion was trapped, the latter is thereafter free to move through the oxide. A similar effect is reported by MC CAUGHAN [9] in the case of a 2 keV, 10^{14} cm^{-2} Ar$^+$ implantation. The probability for this process to occur is proportional to the implanted dose; this is in good agreement with our results (fig. 2). The Cl neutralization is supposed to be irreversible, so a high temperature anneal will not change the number of mobile ions.

B. Chlorine diffusion
We have plotted the results of the LWXA [7] measurements in fig. 3. These results are a proof for diffusion of chlorine in the oxide above

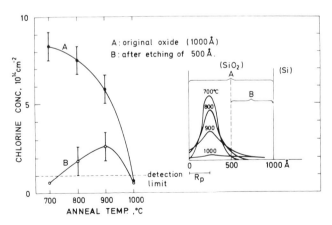

Fig. 3.   Chlorine concentration as function of annealing temperature measured with an X-ray analysis technique (LWXA).

$700^\circ C$. The inset of fig. 3 shows a simple model for this process, starting with a Gaussian implantation profile. A best fit for the diffusivity of Cl in SiO_2 is:

$$D = 2 \times 10^{-6} \exp(-2.0/kT) \qquad cm^2 s^{-1} .$$

C. Laser anneal

The advantage of laser anneal over thermal anneal is the very short high temperature period, in which we expect only minor diffusion of the chlorine. The SiO_2 film is transparant for ruby laser light, but the heat transfer can be accomplished by a thin Al layer (~ 200Å) on top of the oxide. Although about 80% of the laser pulse will be reflected, the remaining energy can be used to heat the oxide. It appeared that the TVS results of laser annealed samples were badly reproducible. The chlorine profile after a 25 keV, 10^{15} cm^{-2} implantation was measured by RBS. Due to the low count rate only the peak position (R) with respect to the Al/SiO$_2$ interface and the peak width (σ) could be evaluated.

Before laser anneal: R = 235 Å ± 40 Å
    (.18 J/cm²) σ = 138 Å ± 20 Å
After laser anneal: R = 392 Å ± 40 Å
                  σ = 210 Å ± 20 Å

The apparent shift of the chlorine profile indicates, in view of the results given in part B, that the oxide temperature must have been high ($\geq 900^\circ C$).

Conclusions

a) A strong increase of mobile Na was found after high dose ($\geq 10^{14}$ cm^{-2}) implantation of Cl$^+$ and Ar$^+$ in thermal SiO_2.
b) The implanted chlorine diffused out of the oxide at temperatures above $700^\circ C$.
c) It is shown by RBS that a ruby laser pulse can be used to heat the oxide for a short period, but the TVS results were ambigious.

Acknowledgements

We are indebted to the following persons:
H. Hermans for doing the LWXA measurements, H. Bezuyen for a lot of TVS measurements, M.W. Hillen and J.F. Verwey for critically reviewing this paper.

References

1. R.J. Kriegler, Appl.Phys.Lett. 20, 449 (1972).
2. B.R. Singh and P. Balk, J.Electrochem.Soc. 125, 454 (1978).
3. A. Rohatgi, Appl.Phys.Lett., 30, 105 (1977).
4. G.J. Declerck, Solid State Devices (ed. H. Weiss), Inst.Phys.Conf.Series 53, 1979.
5. W. Wronski, Phys.Stat.Sol.(a) 53, 659 (1979).
6. M.W. Hillen, G. Greeuw and J.F. Verweij, J.Appl.Phys. 50, 4834 (1979).
7. A.V. Witmer, E. van Meyl, Spectrochimica Acta 34B, 415 (1979).
8. G. Sigel, J.Phys.Chem.Solids, 32, 2373 (1971).
9. D.V. McCaughan, R.A. Kushner and V.T. Murphy, Phys.Rev.Lett. 30, 614 (1973).

Part VII

Technology

Deposition Technology of Insulating Films

E. Doering

'Siemens Research Laboratories, Otto-Hahn-Ring 6
D-8000 München 83, Fed. Rep. of Germany

1. Introduction

This article treats deposition techniques used for secondary
passivation. Secondary passivation refers to oxide layers, nitride layers and glass films which cover either the first
passivation layer or the conductive interconnection levels.

The evaluation of the secondary passivation layers is based
on criteria which include electrical, mechanical and protective
properties. The electrical requirements comprise the dielectric
constants and the dielectric strength.

The mechanical properties include resistance to cracking,
density, pinhole density and step coverage. The intrinsic
stress can be used as a parameter for determining the cracking
resistance,and the density of the films is evaluated by means
of the refractive index and the etching rate. An insufficient
step coverage leads to problems in the electrical isolation.

If the secondary passivation is the final one, it must in
particular provide protection against humidity and contamination caused by the diffusion of impurities (Na^+, K^+ etc.).

2. Deposition techniques

Many deposition techniques have been described for the production
of the secondary isolation layers. Deposition can be affected
by means of mechanisms which are primarily physical, chemical
or physico-chemical in character (Table 1). Excellent summaries
of these methods, based on the literature available up to 1978,
have been published by VOSSEN and KERN [1].

Although some more recent work dealing with the use of
magnetron sputtering for depositing the isolation layer has been
described [2], Chemical Vapor Deposition (CVD) techniques alone
have been examined in the present report. It is in this field
that greatest progress has been made in recent years and these
methods have become widely used in the commercial production of
semiconductor devices, because the CVD-techniques give excellent
results regarding performance, quality and reliability of the
semiconductor devices.

208

Tab. 1: Deposition Techniques

Phys. Methods	Phys.-Chem. Methods	Chem. Methods
Evaporation	Reactive Evaporation	Spin-on
Sputtering	Reactive Sputtering	Spray-on
(rf; magnetron)	Plasma Oxidation	Oxidation
Ionimplantation	Plasma Nitridation	Nitridation
	Anodization	Chemical Transport
	UV-/Laser-CVD	APCVD
	PECVD	LPCVD

CVD: Chemical vapor deposition
AP: Atmospheric pressure
LP: Low Pressure
PE: Plasma enhanced

2.1. Chemical Vapor Deposition

Chemical Vapor Deposition is a deposition technique in which
gaseous substances react together to form a permanent film on
a heated substrate. Various reactions have been utilized in
order to produce the required layers:

1) Pyrolysis: In the early years film deposition was frequently affected by means of a pyrolytic decomposition of
gases. The most common starting materials were organometallic
compounds. A typical example is the decomposition of tetra-
ethylorthosilicate (TEOS) to produce SiO_2 [3].

2) Oxidation: SiO_2 films can also be produced by oxidation
at low temperatures,

$$SiH_4 \text{ (g)} + 2O_2 \text{ (g)} \xrightarrow{400\ °C} SiO_2 \text{ (s)} + 2\ H_2O \text{ (g)}\ .$$

If the doping gas such as phosphine (PH_3) is added to these
starting gases, a silicate glass results [4]. Phosphosilicate
glass (PSG) is the most important type of glass for the
passivation of semiconductor devices:

$$SiH_4 \text{ (g)} + 2\ PH_3 \text{ (g)} + 6O_2 \text{ (g)} \longrightarrow$$
$$SiO_2 \text{ x } P_2O_5 \text{ (s)} + 5\ H_2O \text{ (g)}\ .$$

3) Hydrolysis: A typical example of a hydrolysis reaction
is the deposition of alumina [5],

$$Al_2Cl_6 \text{ (g)} + 3\ CO_2 \text{ (g)} + H_2 \text{ (g)} \longrightarrow$$
$$Al_2O_3 \text{ (s)} + 3\ HCl \text{ (g)} + 3\ CO \text{ (g)}\ .$$

4) Ammonolysis: The ammonolysis reaction is utilized for
the important deposition of silicon nitride [6],

$$3\ SiH_4 \text{ (g)} + 4\ NH_3 \text{ (g)} \longrightarrow Si_3N_4 \text{ (s)} + 12\ H_2 \text{ (g)}.$$

2.1.1. Atmospheric Pressure CVD

Deposition at atmospheric pressure continues to be used principally for the production of phosphosilicate glasses at low
temperatures.

Figure 1 shows an overview of the various reactors for atmospheric pressure CVD (APCVD) at temperatures below 500 °C,
and this has been described in detail by KERN [7]. The various
differences are based on the way in which the wafers are heated
and on the flow relationships present in the reactor.

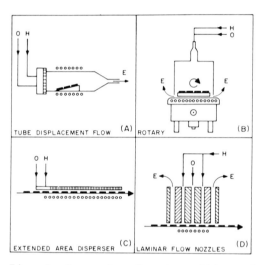

Fig. 1: Apparatus for deposition of films by APCVD
$O : O_2 + N_2$; $H = SiH_4 + N_2$ (+ PH_3) E = Exhaust (from KERN [7])

The deposition and properties of the film are affected principally by the temperature of the substrate and by the O_2/
hydride ratio [8].

Several authors [4, 9, 10] have shown that the composition
of the PSG is an almost linear function of the SiH_4/PH_3 ratio.
In addition, the effects of the carrier gas and of the humidity
on the film deposition have also to be taken into consideration.
KERN and ROSLER [8] have issued a schematic representation of
these relationsships (Fig.2). A notable observation is that the
addition of water vapour and/or phosphorus leads to a reduction
in intrinsic stress, allowing even thick isolation layers to
be deposited without crack formation.

2.1.2. Low Pressure CVD

As early in 1962, SANDOR [11] gave a description of a low
pressure hot wall system for the deposition of SiO_2. But it

210

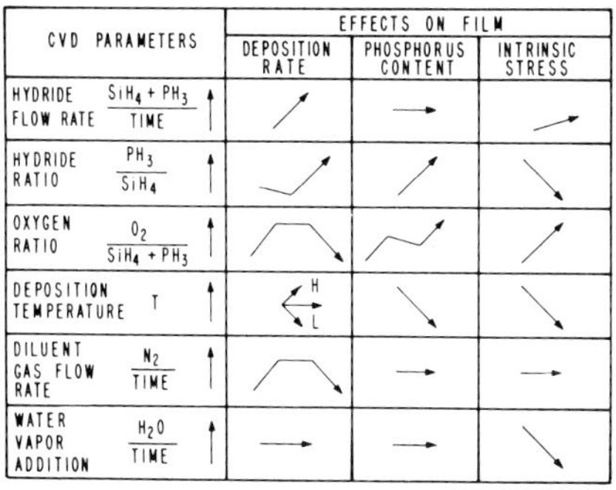

CVD PARAMETERS		EFFECTS ON FILM		
		DEPOSITION RATE	PHOSPHORUS CONTENT	INTRINSIC STRESS
HYDRIDE FLOW RATE	$\frac{SiH_4 + PH_3}{TIME}$ ↑	↗	→	↗
HYDRIDE RATIO	$\frac{PH_3}{SiH_4}$ ↑	⤴	↗	↘
OXYGEN RATIO	$\frac{O_2}{SiH_4 + PH_3}$ ↑	⌒↘	⌁	↗
DEPOSITION TEMPERATURE	T ↑	←H / ←L	↘	↘
DILUENT GAS FLOW RATE	$\frac{N_2}{TIME}$ ↑	⌢	→	→
WATER VAPOR ADDITION	$\frac{H_2O}{TIME}$ ↑	→	→	↘

Fig.2 : Effects of CVD parameters on PSG films (from
KERN and ROSLER [8])

took another 15 years before this technique found its way to
the production lines [12].

The chemical reactions involved are basically the same as
those in the APCVD and the same starting gases can be employed.
Due to its superior uniformity of deposition, however, the use
of SiH_2Cl_2 has proved more popular than SiH_4, so that the
following reactions have to be taken into consideration for
LPCVD [13]:

$$SiH_2Cl_2 + N_2O \xrightarrow{900\ °C} SiO_2 + 2\ HCl + 2\ N_2$$
$$SiH_2Cl_2 + NH_3 \xrightarrow{750\ °C} Si_3H_4 + 6\ HCl + 6\ H_2 \ .$$

The essential difference between APCVD and LPCVD lies in the
fact that the reduction of pressure changes the reaction-
determining step. At low pressure the ratio of the mass transfer
rate of the gaseous reactants to the surface reaction rate
shifts in favour of the surface reaction. According to ROSLER
[14] an increase in the mass transfer of the reactants by an
order of magnitude can be expected. For as the diffusivity of
a gas is increased by a factor of 1000, the stagnant layer
through which the reactants have to diffuse is increased
only by a factor of 3 to 10. Consequently the reactor geometry, which significantly affects mass transfer under normal pressure, plays no essential role under low pressure conditions. The advantage of a low reactor pressure on thickness
uniformity of the deposited layer has been shown by HUPPERTZ
and ENGL [3], who modelled the diffusion current of the reactants.

The LPCVD reactors (Fig.3) are thus optimized with respect
to maximum wafer throughput and temperature stability. Temperature and pressure are the important parameters affecting
the deposition rate [13].

The uniformity of the growth rate along the length of the
boat is modified in an analogous way by temperature and
pressure. With increasing temperature and pressure the deposition at the gas inlet becomes greater, whereas at the gas
outlet a diminution of reactive gases occurs as a result of the

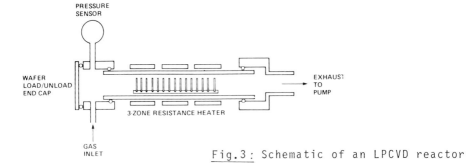

PRESSURE
SENSOR

WAFER
LOAD/UNLOAD
END CAP

EXHAUST
TO
PUMP

3-ZONE RESISTANCE HEATER

GAS
INLET

Fig.3 : Schematic of an LPCVD reactor

increased decomposition at the gas inlet, thus leading to a
low deposition rate. The gas flow affects the deposition at
the gas outlet side in a reverse manner when the reactant
partial pressure is kept constant. With increasing gas flow
more and more reactive gas is transported to the end of the tube
and the growth rate increases.

In an optimised LPCVD system it is currently possible to
produce up to 150 wafers with a deviation of $\pm$ 3 % in layer
thickness, in a single run. This accuracy is comparable with
that of thermal oxidation. This is the significant advantage
of LPCVD technique, for it allows the cost per wafer for CVD
deposition to be drastically reduced by this means combined
with a better film quality.

A disadvantage of the LPCVD technique is that some reactions
require a high deposition temperature. In cases in which the
substrates do not permit any high temperature stresses, the
same chemical reactions can be used if plasma enhancement is
employed in place of thermal energy.

2.1.3. Plasma-Enhanced CVD

The first work which showed the feasibility of a plasma-enhanced chemical reaction in CVD deposition appeared as early
as 1963 [15]. It was not until 10 years later, however, when
REINBERG [16] presented his radial flow reactor, that interest
began to grow in plasma—enhanced CVD (PECVD) [17].

In principle, the same gas sources are used as those for the
other CVD processes in order to synthesize a variety of dielectrics. But the reaction in this case does not occur through
the action of thermal energy but by means of the glow-discharge
effect.

The deposition is determined principally by the RF power
density and the plasma distribution. Sufficient control of the
uniformity of these parameters over a large surface was first
achieved with REINBERG's planar radial flow reactor.
The gases can flow either from the centre to the edge of the
vacuum chamber, as shown in Fig.4 or in the opposite direction.

212

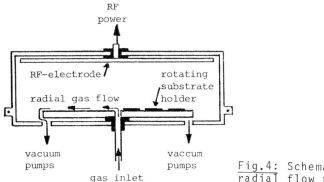

Fig.4: Schematic of planar radial flow reactor for PECVD

The substrates are placed on a heatable plate at ground potential, the RF power electrode being located at a distance of about 5 cm [18]. A disadvantage of the two planar plate reactors is that particles can fall easily upon the horizontally placed wafers.

ENGLE [19] has recently suggested a new reactor configuration, in which the wafers can be layered in a diffusion oven, positioned vertically and closely packed, Fig.5 [20]. This hot wall resistance-heated concept provides the system with a good temperature control. The number of particles on the deposited layers decreases, as in the LPCVD reactor, and in addition the throughput increases.

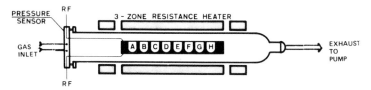

Fig.5: Schematic of vertical diffusion-type reactor for PECVD (from ROSLER and ENGLE [20])

The PECVD reactors are employed principally for the final passivation of semiconductor devices with SiN_x layers. Silicon nitride is known to have better passivation properties than PSG layers in respect of the diffusion of impurities (Na^+, K^+ etc.) and moisture. It has been further suggested that plasma-deposited (PD)-SiO_2 layers be used as the interlayer dielectric [21].

EGITTO [23] has calculated the radial uniformity of deposition in a radial flow reactor. He plotted the non-uniformity as function of power and pressure (Fig.6). From this uniformity surface it becomes clear that there are several conditions under which an optimal uniform deposition is possible. The factors which affect the radial uniformity are residual gas concentration (C), electron density (n_e) and residence time (τ).

213

The gases flow from the outside to the inside and become depleted
in the process. The non-consumed part corresponds to C. Whereas
the electron density is affected by the RF power input, the residence time is a function of the reactor pressure. The residual gas concentration is dependent on both the RF power
and the reactor pressure [24].

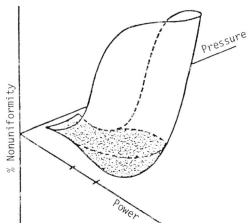

Fig.6: Uniformity surface
(from EGITTO [23])

3. Models

By observing the effect of single parameters on the deposition
rate, on the uniformity of deposition and the step coverage,
the layer formation can be experimentally optimised. However,
in order to obtain statements about the degree of accuracy
with which the respective parameters have to be controlled,
and in order to have a repeatable deposition technique,
detailed knowledge of the respective system is required. This
comprises knowledge of the flow relationships, of the
thermodynamics and the kinetics of the chemical reactions. For
the majority of parameters, it seems that the use of a model
to obtain an analysis and an evaluation of the deposition
can support the efforts a great deal to get an optimum, well-
controlled deposition process.

     Initial attempts of various models have been described in the
literature recently. So by WAHL [25] who used a hydrodynamic
description to obtain a good prediction for the deposition
profile of SiO_2 and Si_3N_4 layers. FISCHER [26] has examined
the heterogeneous reaction at the wafer surface in LPCVD ·
processes.

     SPEAR and WANG [27] started modeling the LPCVD processes
of Si_3N_4 and SiO_2 concerning thermochemical calculations.
Their model uses the temperature, total pressure and the
ratios of the chemical elements in the input gases as thermodynamic variables amenable to experimental control. With the
aid of the SOLGAS and SOLGASMIX programs by ERIKSON [28], the

214

condensed phases which will deposit at equilibrium could be
obtained. The thermodynamic calculations were used to construct
a CVD phase diagram. Figure 7 shows such a diagram for nitride
deposition starting from the gases SiH_2Cl_2 and NH_3. The ratios
of the input gases are depicted as a function of the temperature. It follows from the phase diagram that in a very
wide range both the ratios of the input gases and the temperature
can be changed, the product always remains Si_3N_4. Only when the
silicon content of the input gas is very high is silicon deposited in addition to the nitride phase.

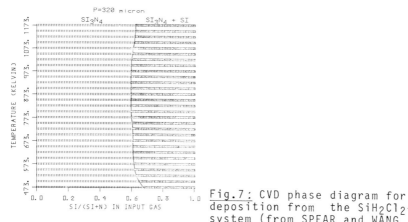

Fig.7: CVD phase diagram for
deposition from the SiH_2Cl_2-NH_3
system (from SPEAR and WANG [27]).

The same model also provides data about the partial pressures
of the gaseous species in heterogeneous equilibria. Figure 8
shows partial pressures vs temperatures for the silicon nitride
system.

The ratio for Si/(Si+N) = 0.43 corresponds to the stoichiometric deposition of Si_3N_4. The gas phase is quite simple and
is almost independent of the temperature. If however, the
composition of the input gases is changed (Fig.9), then the gas
phase composition is strongly affected. It can be clearly seen
that parallel to the decrease of N_2 gas, there occurs a rise in
the silicon-bearing gases, which result in the deposition of
silicon in addition to silicon nitride.

Since this model does not take any mass transport or surface kinetic effects into account at this stage, it can obviously describe only the equilibrium relationships present
before and after the rate-limiting step of the deposition
reaction, but it provides the framework for fundamental understanding of future CVD processes.

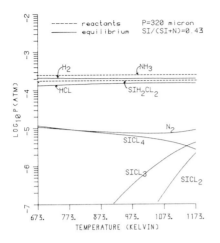

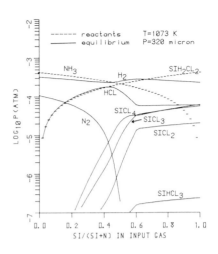

Fig.8: Partial pressures vs temperature for the SiH_2Cl_2-NH_3 system (from SPEAR and WANG [27])

Fig.9: Partial pressures vs Si/Si+N ratio in input gas for the SiH_2Cl_2-NH_3 system (from SPEAR and WANG [27])

4. Film properties

Regarding the deposition technology so far, stress was mainly placed on the uniformity of deposition. In using insulating films for semiconductor devices, however, the film properties mentioned at the outset, such as refractive index, etch rate, stress, density, film composition, dielectric constant and dielectric strength, must be taken into consideration for process optimization.

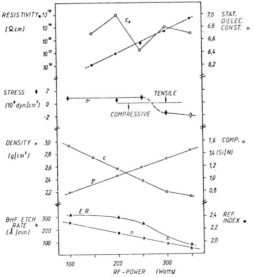

Fig.10: Effect of nominal rf power on properties of reactively plasma deposited Si-N films (after SINHA et al [29,30])

EGITTO [23] has reported that both the RF power and the reactor pressure affect the uniformity of the refractive index in
deposition in a radial flow reactor. Modification of the RF
power, however, changes not only the refractive index, but as
has been reported by SINHA et al.[29,30], a large number of film
properties as well (Fig.10). The production parameters, i.e.
substrate temperature, reactor pressure and total gas flow were
kept constant during these investigations. In general, the refractive index can often serve as one criterion of film quality
and correlate with electrical, mechanical and chemical properties.

The change of intrinsic stress from tensile to compressive
is of importance for the cracking resistance of the films,
since layers under compressive stress have a higher cracking
resistance [31].For the LP Si_3N_4 and LP SiO_2 layers the intrinsic stress, on the other hand, is tensile. Additional
differences between LP layers and PD layers are depicted in
Table 2.

Tab. 2: Film properties (typical values)

	APCVD	LPCVD		PECVD	
	SiO_2	SiO_2	Si_3N_4	SiO_2	Si_3N_4
Gases	$SiH_4 + O_2$	$SiH_2Cl_2+N_2O$	$SiH_2Cl_2+NH_3$	SiH_4+N_2O	SiH_4+NH_3
Dep. temp. [°C]	430	900	750	350	350
Dep. rate [Å/min]	1000	120	40	600	300
Ref. index	1.45	1.45	2.0	1.5	2.2
Density [g/cm^3]	2.15		3.1	2.0	2.8
Diel. const.	4.2	4.2	6.0	4.6	6.5
Step coverage	poor	conformal	fair	conformal	conformal
Resistivity [Ω cm]	10^{16}	10^{16}	10^{14}	10^{16}	5×10^{15}
Stress [dyn/cm^2]	$2 \times 10^9 T$		$1 \times 10^{10} T$	$2 \times 10^9 C$	$5 \times 10^9 C$

5.   Summary

The application of insulating films for secondary passivation
has been discussed. Among the different deposition techniques
the chemical vapor deposition techniques had become the greatest
acceptance for semiconductor device fabrication, because these
techniques, especially with the introduction of low pressure
CVD and plasma enhanced CVD, met the electrical, mechanical,
and protective requirements of the secondary passivation best.
The CVD techniques have been reviewed regarding the basic reactions, the reactor design, the effects of various parameters
on the uniformity of the deposition and models of the deposition
mechanism.

From an optimum deposition technique not only a uniform deposition is required, but also properties such as deposition
temperature, deposition rate, density, etch rate, refractive
index, dielectric constant, step coverage and stress have to
be regarded. These properties have been described for silicon
oxides and silicon nitride, the most important dielectrics.

References

1 Vossen, J.L.; Kern, W. (Eds): Thin Film Processes, New York, Academic Press 1978
2 Serikawa, T.; Yachi, T.: Jap. J. Appl. Phys. 20 (1981) L111
3 Huppertz, H.; Engl, W.L.: IEEE Trans. Electron Devices ED-26 (1979) 658
4 Kern, W.; Schnable, G.L.; Fisher, A.W.: RCA Rev. 37 (1976) 3
5 Tung, S.K.; Caffrey, R.E.: Trans. Metall. Soc. AIME 233 (1965) 572
6 Duffy, M.T.; Kern, W.: RCA Rev. 31 (1970) 742
7 Kern, W.: Solid State Technol. 18 (12) (1975) 25
8 Kern, W.; Rosler, R.S.: J. Vac. Sci. Technol. 14 (1977) 1082
9 Wong, J.; Ghezzo, M.: J. Electrochem. Soc. 122 (1975) 1268
10 Shibata, M.; Yoshimi, T.; Sugawara, K.: J. Electrochem. Soc. 122 (1975) 157
11 Sandor, J.: Electrochem. Soc., Ext. Abstr. No. 96 (1962) 228
12 Kern, W.; Schnable, G.L.: IEEE Trans. Electron Dev. ED-26 (1979) 647
13 Rosler, R.S.: Solid State Techn. 20 (4) (1977) 63
14 Rosler, R.S.: Symp. on Low Pressure Chemical Vapor Deposition W. Kern Chairman; Greater New York Chapter, American Vacuum Society, Murray Hill, N.J., March 15, 1978
15 Alt, L.L.; Ing, Jr., S.W.; Laendle, W.W.: J. Electrochem. Soc. 110 (1963) 445
16 Reinberg, A.R.: Electrochem. Soc. Ext. Abstr. 74-1 (1974) 19
17 Rand, M.J.: J. Vac.Sci. Technol. 16 (2) (1979) 420
18 Rosler, R.S.; Benzing, W.L.; Baldo, J.: Solid State Technol. 19 (6) (1976) 45
19 Engle, G.M.: Plasma Enhanced Chemical Vapor Processing of Semiconductive Wafers, U.S. patent pending
20 Rosler, R.S.; Engle, G.M.: Solid State Techn. 22 (12) (1979) 88
21 v.d.Ven, E.P.G.T.; Sanders, J.A.M.: Electrochem. Soc. Ext. Abtr. 78-2 (1978) 525
22 Mattson, B.: Solid State Technol. 23 (1) (1980) 60
23 Egitto, F.D.: J. Electrochem. Soc. 127 (1980) 1354
24 Reinberg, A.R.: J. Electron. Mater. 8 (1979) 345
25 Wahl, G.: Thin Solid Films 40 (1977) 13
26 Fischer, H.: Z. Phys. Chemie, Leipzig, 255 (1973) 773
27 Spear, K.E.; Wang, M.S.: Solid State Technol. 23 (7) (1980) 63
28 Erikson, G.: Acta Chem. Scand. 25 (1971) 1651
   Erikson, G.; Rosen, E.: Chemica Scripta 4 (1973) 293
   Erikson, G.: Chemica Scripta 8 (1975) 100
29 Sinha, A.K.; Levinstein, H.J.; Smith, T.E.: J. Appl. Phys. 49 (1978) 2423
30 Sinha, A.K.; Levinstein, H.J.; Smith, T.E.; Quintana, G.; Haszko, S.E.: J. Electrochem. Soc. 125 (1978) 601
31 Sinha, A.K.; Smith, T.E.: J. Appl. Phys. 49 (1978) 2756

Very High Charge Densities in Silicon Nitride Films on Silicon for Inversion Layer Solar Cells

R. Hezel

Institut für Werkstoffwissenschaften VI, Universität Erlangen-Nürnberg
D-8520 Erlangen, Fed. Rep. of Germany

Abstract

Different ways of achieving high positive charge densities in Si nitride films on Si for the creation of low resistivity inversion layers for solar cells are demonstrated: (i) optimization of the deposition parameters, (ii) utilizing the Si nitride charge storage effect and (iii) sodium incorporation into the nitride films. Both APCVD and PECVD Si nitride films were investigated. It was shown for the first time, that sodium can be obtained in a positively charged state in Si nitride, and, in contrast to CVD SiO_2, the C-V curves were not distorted and only very small hysteresis was found after the sodium treatment. Very high charge densities up to $1.4 \times 10^{13} cm^{-2}$ with stability at elevated temperatures could be achieved.

1. Introduction

Silicon nitride was recently introduced as a promising dielectric for high efficiency silicon inversion layer solar cells [1,2,3]. In contrast to the MOS field effect transistor, the density of positive charges Q_N in the insulator near the semiconductor/insulator interface should be as high as possible in these solar cells in order to obtain an inversion layer with a high electron density at the semiconductor surface. Furthermore, a low interface state density N_{it} and stability of the charges are required. It is demonstrated in this paper that these conditions can be fulfilled with Si nitride films on silicon by the following means (i) proper adjustment of the deposition conditions for both atmospheric pressure (AP) and plasma enhanced (PE) CVD films, (ii) utilization of the MNOS charge storage effect [4] and (iii) incorporation of sodium into the nitride films.

Since amorphous Si nitride is an excellent barrier against the migration of mobile ions, such as sodium, as well as a getter for impurities, it is widely used as active and passive dielectric in IC technology [5,6]. But despite this importance of Si nitride for sodium passivation no information is available up to now whether sodium is present in Si nitride in a charged or uncharged state. This question is answered and the different behavior of sodium in SiO_2 and Si_3N_4 films is pointed out in the present work.

2. Experimental

The APCVD Si nitride films were deposited on the native oxide of
chem.-mech. polished single crystalline silicon substrates(2-5 Ωcm)
in a hot wall reactor by the ammonia-silane reaction (NH_3/SiH_4 =
1000) with N_2 as a carrier gas [1,7].

PECVD Si nitride were deposited in a parallel plate plasma
reactor as described in [3]. All the films were silicon oxynitri-
des rather than pure nitrides with a refractive index at λ =632.8 nm
ranging from 1.86 to 1.92. The sodium was deposited by the electrolytic decomposition of an aquaeous NaOH solution (20 mg NaOH,
1000 ml H_2O) with the silicon wafer as cathode and platinum as
anode. Standard C-V measurements were performed at 1 MHz, using
a liquid Ga-In alloy as the gate electrode [7].

3. Results and Discussion

3.1 Charge Density of the As-Grown Structures

In Fig.1a the effective interface charge density is plotted as a
function of the nitride deposition temperature for three deposition series on (111) and (100) silicon. As can be seen, Q_N/q
is increasing with decreasing deposition temperature, up to values
of about $4-5 \times 10^{12} cm^{-2}$ at 600 °C (peak values of $7 \times 10^{12} cm^{-2}$ were
achieved in series 1). No definite substrate orientation depen-
dence could be observed. In contrast to thermal SiO_2 on silicon
[8], the high charge densities are associated with a low density
of interface states ($N_{it} \approx 4 \times 10^{10} cm^{-2} eV^{-1}$ was measured for the
nitride films deposited at 640 °C [9]). This is mainly attributed
to the passivation of surface states by the active hydrogen produced during the Si nitride deposition [2,9]. In Fig.1b the interface charge density obtained for PECVD Si nitride is shown as a
function of the SiH_4/NH_3 ratio, with the rf power as a parameter.

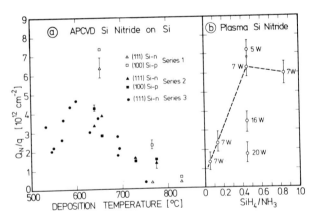

Fig.1   Interface charge density in Si nitride films on silicon
         as a function of nitride deposition parameters

As can be seen, Q_N/q can be increased up to values of 7×10^{12}cm^{-2} either by increasing the SiH$_4$/NH$_3$ ratio or by lowering the rf power. To demonstrate the stability of the charges, in Fig.2 the values of Q_N/q for PECVD and APCVD Si nitride films on silicon are shown as a function of the time of annealing at different temperatures in a N$_2$ ambient. At temperatures below and up to the deposition temperature no remarkable decrease of Q_N/q occurred. The decrease observed for annealing temperatures above the deposition temperature is attributed to hydrogen outdiffusion |9|.

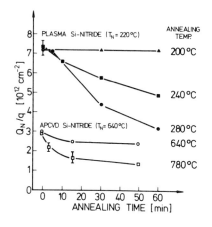

Fig.2 Stability of the interface charges in Si nitride films on silicon

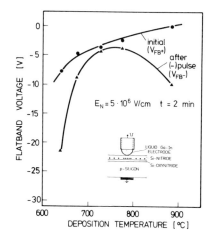

Fig.3 Storage of positive charges in APCVD Si nitride films

3.2 Utilization of the Charge Storage Effect in Si Nitride

In Fig.3 the negative shift of the flatband voltage V_{FB} due to the application of a negative voltage pulse to the APCVD Si nitride film (see insert) is shown as a function of the nitride deposition temperature. As can be seen, at low deposition temperatures (640 °C), where already a high density of positive charges is present in the as grown structures (see Fig.1a) also the shift of the flatband voltage is very large. For the films prepared at 640 °C (d_N = 55 nm) V_{FB} was shifted from -7 volts (Q_N/q = 4.5x10^{12}cm^{-2}) to -22 volts (Q_N/q = 1.4x10^{13}cm^{-2}) by an electric field of 6x10^{6}V/cm applied for 2 min. The low interface state density of the as grown structure is not affected by the storage of additional charges and the stability of the charges is sufficient for solar cell application |2,9|.

3.3 Sodium Incorporation into Si Nitride

Two kinds of MIS structures have been built in order to investigate the effect of the location of the sodium ions in Si$_3$N$_4$ films on the interface properties: (i) deposition of sodium directly on the silicon surface, which was covered only by its native oxide and (ii) deposition of sodium on top of a 6 nm thick Si nitride layer on silicon, followed by the deposition of the final nitride film. The structures were subjected to a bias stress at room temperature in order to study the stability of the ions and the charge tunneling behavior.

In Fig.4 the resulting C-V curves are shown. For the uncontaminated structures (Fig.4a) the typical MNOS memory hysteresis is observed |4,7|. If sodium is incorporated at the oxide/nitride interface, the C-V curves are shifted into the negative voltage direction (Fig.4b). Thus the fact that sodium can be present in Si nitride in a positively charged state was demonstrated for the first time, along with the following new results: (i) no perceptible change of the shape of the C-V curves occurred, indicating qualitatively that the additional charges in the nitride films don't give rise to surface states and (ii) nearly no hysteresis can be observed, showing that the charges are not mobile under the conditions applied and charge tunneling and trapping is inhibited by the sodium incorporation. Similar results were obtained for the structures depicted in Fig.4c. The higher positive charge density (negative shift of the C-V curves) is, as discussed below, due to a higher field strength present during the electrolytic sodium deposition (0.5 V/cm in Fig.4b, 2.2 V/cm in Fig.4c). A small ion hysteresis can be observed, but more severe bias conditions were applied to the structures shown in Fig.4c compared to those of Fig.4b. For comparison, corresponding experiments were performed with CVD SiO_2 films on silicon deposited by the silane-oxygen reaction at 680 °C. After sodium incorporation, in addition to the displacement towards negative voltages, the C-V curves were extremely distorted (an example is given in Fig.4b) and a pronounced ion hysteresis could be observed (without sodium, the C-V curve was nearly identical with curve 1 in Fig.4a). Thus, in contrast to Si nitride, in CVD SiO_2 films on silicon the positive charges and the interface states appear to be correlated, a result similar to that reported for thermally grown SiO_2 films |10|.

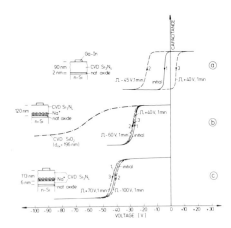
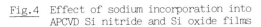

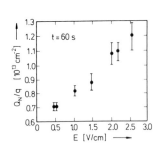

Fig.4 Effect of sodium incorporation into APCVD Si nitride and Si oxide films

Fig.5 Charge density in Si nitride films by electrolytic sodium deposition at various fields

222

In Fig.5 the effective charge density in Si nitride for the structures shown in Fig.4c is plotted as a function of the electric field strength present during the electrolytic sodium deposition process. The current density was kept at a constant value of 0.05 mA/cm^2. As can be seen, the charge density is increasing considerably with the applied voltage and values up to 1.3×10^{13}cm^{-2} were obtained. The Q_N/q values for the structures depicted in Fig.4b were somewhat lower but well within the error bars shown in Fig.5, which represent the scattering of the values both within the 32 mm diameter wafers and from wafer to wafer.

Conclusions

It was shown that for Si nitride films on silicon the highest known and stable positive interface charge densities up to 1.4×10^{13}cm^{-2} together with a sufficiently low interface state density can be achieved by optimization of the deposition parameters, by the Si nitride charge storage effect and by sodium incorporation into the films. Thus Si nitride provides an ideal dielectric for inversion layer solar cells. Up to now such cells with total area AM1 efficiencies of 16% and extremely high UV sensitivity could be fabricated in our laboratory using PECVD Si nitride films.

References

1. R. Hezel, R. Schörner and T. Meisel: Proc. 3rd E.C. Photovoltaic Solar Energy Conf., Cannes, 1980 (Reidel, Dordrecht 1981), p.866
2. R. Hezel: Solid State Electron., 24 (1981),in press
3. R. Hezel and R. Schörner: J. Appl. Phys. 52(3) (1981),in press
4. D. Frohmann-Bentchkowsky and M. Lenzlinger: J. Appl. Phys. 40, 3307 (1969)
5. T.E. Burgess, J.C.Baum, F.M. Fowkes, R. Holmstrom,and G.A. Shirn: J. Electrochem. Soc. 116, 1005 (1969)
6. J. Fraenz and W. Langheinrich: Solid State Electron. 12, 145 (1969)
7. R. Hezel and E.W. Hearn: J. Electrochem. Soc. 125, 1848 (1978)
8. E. Arnold, J. Ladell and G. Abowitz: Appl.Phys.Lett. 13, 413 (1968)
9. R. Hezel:J.Electron. Mater. 8, 459 (1979)
10. A. Goetzberger, V. Heine and E.H. Nicollian: Appl. Phys. Lett. 12, 95 (1968)

Silicon Nitride Layers Grown by Plasma Enhanced Thermal Nitridation

E.J. Korma, J. Snijder, and J.F. Verwey

Department of Applied Physics, Groningen State University
9747 AG Groningen, The Netherlands

Abstract

Silicon nitride layers have been produced by plasma enhanced thermal nitridation of silicon. The growing process has been studied by measuring the layer thickness, ranging from 45 to 70 Å, as a function of time and RF power. A depth profile, obtained by Rutherford Back Scattering, showed that the nitrogen to oxygen ratio was about 2. Current flow through the layers appeared to be controlled by direct tunneling.

Introduction

Through the years silicon nitride layers have become of great interest. These layers, mostly produced by Chemical Vapour Deposition techniques, exhibit outstanding masking and passivating properties due to their chemical inertness, impermeability to metals and dense structure [1]. Unfortunately, attempts to produce CVD nitride films with good silicon-siliconnitride interface properties failed.

Recent publications showed that layers with better interface properties could be obtained by means of thermal nitridation [2]. There are however some disadvantages of this process like its high processing temperature, typically $1200^{\circ}C$, and a selflimiting growth, which keeps the maximum thickness below approximately 100 Å. The thermal nitridation process, however, is expected to be improved by enhancing it by forming an RF plasma in the reactant gas. The concentration of nitrogen radicals will be considerably increased, and as a consequence the processing temperature can be lowered.

Technology

The small experimental reactor used for the plasma enhanced thermal nitridation is schematically drawn in Fig. 1. The heart of the system, which is mounted in a quartz tube, is formed by two planparallel electrodes made of graphite in order to withstand the high temperatures needed. The lower, grounded electrode is connected rigidly to an aluminum lid by means of two stainless steel rods. The upper, "hot" electrode (7 cm long, 3 cm wide) is connected to an RF feedthrough with a tungsten wire encapsuled by a ceramic tube. The two electrodes are separated from each other by ceramic standoffs. The silicon wafer lies between the two electrodes on the grounded one. The quartz tube is vacuum sealed with rubber gaskets mounted in an appropriate copper block. Ammonia of very high purity (99.998 %), in order to avoid oxygen contamination, is used as the reactant gas. The maximum RF power density on the wafer is 5 W/cm^2. Nitridations were carried out on <100> oriented, n-type, 2 Ωcm silicon wafers.

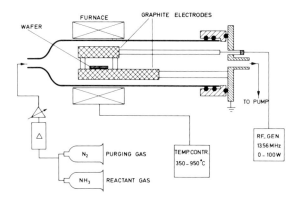

Fig. 1   Schematic drawing of the system for plasma enhanced thermal nitridation.

Results and Discussion

A. Growing process

To characterize the growing process, the layer thickness as a function of
RF power and nitridation time was measured by ellipsometry. The obtained
values, ranging from 45 to 70 Å, should be treated with some reserve because ellipsometric values for such very thin layers can have quite large
errors [3].

The first series of wafers were nitridized for 300 minutes at 920°C and
2 mbar. The RF power was varied from 0 to 32 W. Normal thermal nitridation
(0 W) yielded a layer thickness of approximately 45 Å, whereas plasma enhancement resulted in a thickness ranging from 58 to 65 Å (8 to 32 W respectively), which means an increase of at least 30%. The power dependence
however is small because at these large nitridation times saturation of the
growing process is to be expected [1,4]. This selflimiting growth becomes
evident when we vary the nitridation time and fix the RF power to 32 W.
After 10 min. the thickness has already a relatively large value of about
58 Å, whereas at larger times the growth ratio seems to become nearly zero
($d \approx 65$ Å after 160 min). A recent report from ITO [5] showed that layers
thicker than 100 Å can be obtained with plasma enhancement at 1050°C.

B. Analysis

Analysis of the silicon nitride layers was carried out by Rutherford Back
Scattering. A typical depth profile is shown in Fig. 2. The nitrogen to
oxygen ratio in the bulk of the layer is about 2, which means that the
refractive index is expected to be in the range of 1.8 to 1.9 [6],
which is in good agreement with ellipsometric data. The oxygen contamination might be caused by a small leakage in the reactor, the presence of a
native oxide layer before nitridation starts (typically 10 Å) and a (very)
low concentration of oxygen and moisture in the ammonia gas. There is, however, no evidence that a small oxygen contamination will degrade the desired
properties of the silicon nitride [3,7].

C. Electric properties

Small aluminum dots were evaporated onto the samples to form MIS structures.
Whereas for silicon nitride layers thicker than 200 Å it is commonly found

225

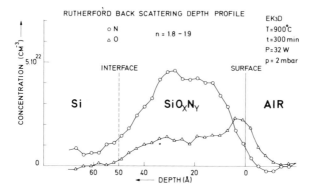

Fig. 2
RBS depth profile of
sample EK3D

that the current through the layer is predominantly controlled by the Poole-
Frenkel conduction mechanism [8], in our case a plot of the logarithm of the
current against the square root of the gate voltage does not yield a
straight line but a slightly curved one. The temperature dependence is, in
contrast to Poole-Frenkel conduction, rather low as can be seen from Fig. 3.
Considering the above mentioned effects it is assumed that the current behaviour is attributed to direct tunneling through the nitride layer. This
is further confirmed by the fact that the current exhibits a strong thickness dependence, a typical tunneling effect, as is shown in Fig. 3. Similar
results have been found before [9].

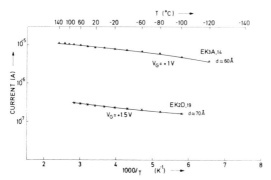

Fig. 3 Arrhenius plot of samples EK2D and EK3A. Both nitridized for 300
min. at 920°C and 2 mbar. RF power 30 W and 8 W respectively.

As a last series of experiments C-V measurements were carried out.
Determination of the interface state density, D_{it}, with the Quasi Static
C-V method appeared to be impossible due to the high tunneling currents.
High Frequency C-V curves, however, could be measured quite reasonably at
1 MHz. An example is shown in Fig. 4. The solid line is the measured curve,
the dashed line is the ideal one, calculated from the measured insulator
capacitance. Some HFCV curves showed a small hysteresis, less than 0.1 V,
which is probably due to shallow traps [10]. The flatband voltage shift is
about -0.46 V which is likely to be a result of fixed charge in the nitride
layer.

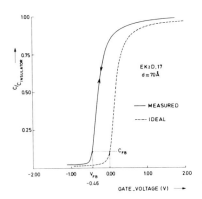

Fig. 4 HFSC curve of sample DK2D

Conclusions

Plasma enhancement proved to be a way to increase the ultimate thickness of a thermally grown silicon nitride layer. Nevertheless, at 920°C the thickness is still limited to approximately 70 Å. Higher temperatures are expected to increase this value.

Oxygen contamination could not be avoided yet. But there was no indication that the performance of the MIS structures was degraded by a small amount of oxygen.

The current flow through the produced layers was clearly controlled by direct tunneling. These high currents made QSCV measurements impossible. Thicker layers are necessary to diminish the current.

References

1. T. Milek, Handbook of Electronic Materials, Vol. 2, IFI/PLENUM (1971).
2. T. Ito *et al*, Appl.Phys.Lett. 32, 330 (1978).
3. D.G. Schueller, Surface Science 16, 104 (1969).
4. S.P. Murarka, C.C. Chang and A.C. Adams, J.Electrochem.Soc. 126, 996 (1979).
5. T. Ito *et al*, to be published.
6. M.J. Rand and J.F. Roberts, J.Electrochem.Soc. 120, 446 (1973).
7. M.L. Naiman *et al*, Proceedings Int.Electr.Dev.Meeting, Washington, Dec. 1980, 562.
8. S.M. Sze, J.Appl.Phys. 38, 2951 (1967).
9. E.H. Sanchez-Lassise and J.R. Yeargan, J.Electrochem.Soc. 120, 423 (1973).
10. E.J.M. Kendall, Brit.J.Appl.Phys. 1, 1409 (1968).

Buried Oxide Layers Formed by Oxygen Implantation for Potential Use in Dielectrically Isolated ICs

K.V. Anand and P. Pang

Department of Electronics, University of Kent
Canterbury, United Kingdom

J.B. Butcher, K. Das, E. Franks, and G.P. Shorthouse

Microelectronics Centre, Middlesex Polytechnic
Enfield, United Kingdom

1. Introduction

The aim is to produce a buried oxide layer by heavy dose implantation ($>10^{17}$vm^{-2}) into 111 Si substrate for use in dielectrically isolated LSI circuits as an alternative to SOS ICs. IZUMI et al.{1} made CMOS devices using this technique and HAYASHI et al.{2} carried out some structural analysis using AES, TEM and XPS techniques. We have extended their work of composition and structural analysis in relating it to practical methodology and assessment procedures.

Figure 1 shows the formation of buried oxide layers with subsequent epitaxial Si deposition for devices.

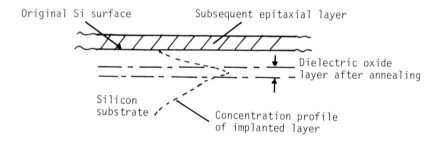

Fig. 1 Fabrication of a Buried Dielectric Isolation Layer

2. Practical Considerations

Oxygen ions were implanted under different conditions of energy, dose and substrate temperature followed by annealing to restore the crystallinity of the top Si layer and to allow chemical reaction between the implant O and substrate. Finally Si is epitaxied by CVD technique.

Table 1 lists the most important practical parameters that were examined.

Table 1 Range of Experimental Variables

For Ion Implantation	
1. Ion source canal material	Ni, Stainless steel, Tungsten, Tantalum
2. Effective energy (for O^+){keV}	50, 100*, 150, 180, 200*
3. Dose{cm^{-2}}	0.5-8.2x10^{17}(at 100 keV), 3-14*x10^{17} (at 200keV)
4. Substrate heating	good thermal contact (less than 200°C), beam induced heating ($\simeq$ 500° to 700°C), external heating

Post Implantation Anneal
1. Gas N_2; Temperature 550°, 1000°, 1150°C; Duration 2hrs to 40hrs
2. Gas H_2;   "   1100°C;   "   5mins* to15mins

Epitaxy
Deposition Temperature 1100°C; Growth rate 0.6 $\mu m.min^{-1}$; Si bearing gas silane; doping gas phosphene

*Most experiments done under this condition

Assessment was carried out both after implantation and after epitaxy. However, only samples that looked promising after post implantation anneal were epitaxied. Table 2 lists the assessment techniques and their inference.

Table 2 Assessment or analytical technique

Technique	Inference
1. Surface light scattering	Surface damage and suitability for epitaxy
2. I.R. spectroscopy	Stochiometry of buried oxide; estimate of oxide thickness; estimate of substrate temperature
3. R.B.S.	Surface crystallinity (χmin); O depth profile and stochiometry (composition - semi quantitatively); Si/O interface location and damage.
4. T.E.M.(cross-sectional method)	Depth distribution of damage type and crystal structure determined by T.E.D. mode
5. Diode characteristics	Breakdown voltage and minority carrier life time

3. Experimental Results and Conclusions

Only the most important results are given below.

3.1 Substrate temperature during implantation critically effects the crystallinity of the top Si surface. At any particular energy, as the substrate temperature increases, the maximum dose required to render top surface amorphous also increases provided the dose was not so high as to oxidise the top surface. The latter can happen when surface O concentration gets to about $3 \times 10^{19} cm^{-3}$.

3.2 A convenient and non-destructive test for surface amorphicity is to anneal the implanted sample in H_2 for a minimum of 5 minutes at 1100° C and take a transmission I.R. spectrum. If the sample is amorphous, H_2 will "getter" the implanted oxygen, and no absorption in the 9.2 to 10.0 μm will be observed.

3.3 I.R. results show that the absorption band minima shifts from around 10.0 μm (associated with SiO) for as–implanted samples to 9.2 μm for SiO_2 after annealing.

3.4 Fig. 2 below shows the effect on R.B.S. of post implantation anneal and dose.

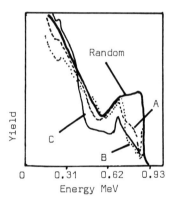

Fig. 2 Implant energy 200 keV, Curve A and B have dose 1.4×10^{18} cm^{-2}, Curve C has dose of 2.5×10^{18} cm^{-2}, Curve A and C before anneal and Curve B after H_2 anneal at 1100^oC for 5 minutes.

χmn reduces from 16% to 5.4% after anneal indicating recrystallisation at the surface. The reduction in yield deeper into the surface and the steeper rise also indicates a reduction in disorder and a better defined interface, respectively. TEM has confirmed that this region is polycrystalline whereas the top surface is single crystal. Curve C shows that with increase in dose the oxide is more homogeneous in its stochiometry and the reduction in yield in the top Si indicates less disorder.

3.5 Both R.B.S. and T.E.M. show that epitaxial layers grown on samples implanted at high temperature are single crystal at the surface. TEM also shows that the polycrystalline interface, however, remains.

230

3.6 Circular P⁺N diodes were formed in the epitaxial layer.
Their median breakdown voltage was 55V as compared to 60V for
controls. Minority carrier lifetime was measured using
reverse recovery method and was found to be 1.2 μ.sec as
compared to 1.8 μ.sec for controls.

Acknowledgements

We would like to acknowledge the measurements done on RBS by Drs.
P.L.F. Hemment and D.W. Wellby of University of Surrey, U.K. and
on TEM by Dr. G.R. Booker and Dr.M.C. Wilson of University of
Oxford, U.K.

References

1. K. Izumi, M. Doken, and H. Ariyoshi, Elect.Lett.14,593(1978)
2. T.Hayashi, S. Maeyama, and S. Yoshi, J.Appl.Phys.(Japan)
   19, 1111 (1980)

Part VIII

Laser Processing

Properties of Patterned and CW Laser-Crystallized Silicon Films on Amorphous Substrates

N.M. Johnson, D.K. Biegelsen, and M.D. Moyer

Xerox Palo Alto Research Centers
Palo Alto, CA 94304, USA

1. Introduction

Semiconducting films on insulating substrates are of technological interest for electronic devices. In particular, a thin-film transistor technology which is compatible with bulk glass substrates would have immediate application in large-area arrays. It has recently been demonstrated that laser-crystallized silicon films can provide the semiconducting material for such a technology [1,2]. As deposited, a polycrystalline silicon film is composed of fine grains (nominally 50 nm in diameter) and is unsuitable as a semiconducting material for active devices due to the dominating effects of grain boundaries on the electrical properties. Laser crystallization is used to increase the grain size to dimensions that are comparable to or greater than those of the device structure, thereby minimizing the detrimental effects of grain boundaries on device operation. It has been shown that heat flow and crystallization during laser irradiation can be controlled to obtain large grains on amorphous substrates [3]. A key advantage of using silicon thin films is the application of conventional silicon microelectronic processing technologies for the fabrication of thin-film devices and circuits.

It has previously been demonstrated that multilayered island structures on amorphous substrates can be used to control nucleation and growth of silicon thin films during laser-induced crystallization [3] and that these structures and laser-irradiation techniques are compatible with device processing [2]. Studies have been initiated on the effects of materials processing on device operation. We have recently demonstrated, with correlated scanning electron microscopy (electron-beam induced conductivity and voltage-contrast imaging) and transmission electron microscopy, that residual grain boundaries intersecting p-n junctions in laser-crystallized silicon films provide efficient channels for dopant diffusion during device processing [4]. In the present paper we highlight recent work on laser crystallization of patterned thin films and the fabrication of thin-film transistors on bulk glass substrates.

2. Silicon Films on Glass

The use of laser crystallization to fabricate thin-film transistors on bulk glass has required the identification and implimentation of new processing principles. Two key materials/laser – processing issues must be addressed in attempting to form large-grain or single-crystal films on bulk glass substrates. First, due to the large disparity of the thermal coefficients of expansion of silicon and glass (e.g., fused silica), laser-crystallized silicon films are highly strained and usually exhibit microcracking, either during laser crystallization or during subsequent device processing. Secondly, with individual islands of silicon, the optical transparency of glass results in edge-enhanced cooling of the islands during laser crystallization, with consequential loss of control of grain growth. It has previously been observed that islands configured for controlled crystallization and tailored for fabrication of individual transistors also provide inherent stress relief in the vicinity of an island [2,3]. It can be more generally stated that a deposited polycrystalline silicon film can be judiciously patterned to control nucleation and growth of grains in a predetermined area in which a device will subsequently be fabricated and that the delineation provides stress relief during scanned laser crystallization. A novel example is shown in Fig. 1. A continuous film of polycrystalline silicon (.65-μm thick) was chemical-vapor deposited on fused silica, photolithographically patterned, and plasma etched to produce the patterned stripe (50-μm wide). The central vertical region is the area designated for device fabrication. The pattern of the stripe was designed to control crystal growth in the central region, provide stress relief, and shunt material transported with the scanned molten zone away from this central region in order to maintain planarity for subsequent device processing. The patterned silicon film was encapsulated with ~20 nm of silicon nitride prior to laser crystallization. The softening as well as the melting temperatures of silicon nitride are above the melting temperature of silicon, and the encapsulant has been shown to minimize mass transport and suppress the tendency for thermal etching or ablation of silicon during laser melting [2,3]. The optical micrograph in Fig. 1 shows the patterned stripe after laser crystallization; an elliptically shaped beam [3] was scanned along the stripe from the bottom to the top of the figure. The regions adjacent to the central area show evidence of mass transport, while the central region is relatively smooth. A TEM bright-field micrograph of the central area is shown in Fig. 2.

Fig. 1 Optical micrograph of patterned stripe of silicon on glass after laser crystallization

Fig. 2 TEM bright-field micrograph of the central region of the patterned silicon film shown in Fig. 1

(The dark borders along the edges of the silicon film are artifacts of specimen preparation.) The region is bisected by a single grain boundary which runs the length of the stripe. The curved dark band bridging the central region is a transmission electron-beam bend contour, which is discontinuous at the grain boundary. The extended structural defects within each grain are stacking faults. The bicrystalline structure is currently being used to evaluate grain boundaries in silicon thin films on glass.

3. Thin-Film Transistors

CW laser crystallization has been used to process silicon thin films on glass as an integral step in the fabrication of thin-film transistors. In Fig. 3 is shown an optical micrograph of a thin-film transistor fabricated in a laser-crystallized silicon island. The transistor is a metal-oxide-semiconductor (MOS) device operating in the n-channel enhancement mode with an alumimum gate electrode. The channel dopant was introduced by boron-ion implantation (e.g., 100 keV, 1×10^{11} cm^{-2}) prior to laser crystallization. Hence, laser processing served to redistribute and electrically activate the dopant over the depth of the silicon film as well as to increase grain size. After laser crystallization, standard photolithographic and plasma-etching techniques were used to remove unwanted material from the vicinity of the island (or patterned region as shown in Fig. 1); it is also possible to redefine the peripheral material for device interconnections. Next, the source and drain regions were implanted with arsenic (e.g., 100 keV, 1×10^{15} cm^{-2}), and the island was thermally oxidized to

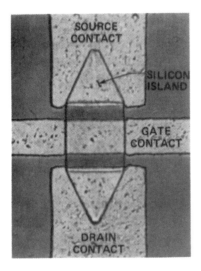

Fig. 3 MOS thin-film transistor fabricated in CW laser-crystallized silicon

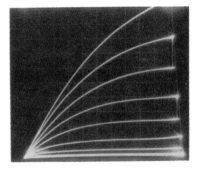

Fig. 4 Current-voltage characteristic for thin-film transistor on glass: vert. axis (I_D): 20μ A/div; horz. axis (V_{DS}): 1 V/div; gate-voltage step: 2 V

form the gate dielectric (e.g., 150-nm thick). After selectively removing the oxide layer over source and drain regions, an aluminum-silicon alloy (1% Si) was vacuum evaporated over the surface and photolithographically patterned into the source, drain, and gate electrodes. In Fig. 3 the channel length is 12 μm (the gate width is 16 μm) and the island width is 24 μm. Test devices received a final sinter at 450 C (30 min in forming gas). The current-voltage characteristic for an MOS thin-film transistor fabricated on bulk fused silica are shown in Fig. 4. For devices as shown in Fig. 3 the silicon islands were found to consist of grains 1 to 5 μm wide, and the channel mobility ranged from 100 to 300 cm^2/V-sec. The results presented here demonstrate the potential of laser-crystallized silicon films for high performance thin-film transistors on bulk glass substrates. However, much materials and device research remains to be performed before the full potential of this emerging new technology will be realized.

References

1. T. I. Kamins and P. A. Pianetta, IEEE Electron Device Lett. EDL-1, 214 (1980)
2. N. M. Johnson, D. K. Biegelsen, and M. D. Moyer, Laser and Electron-Beam Solid Interactions and Materials Processing, eds. J. F. Gibbons, L. D. Hess, T. W. Sigmon (Elsevier, New York, 1981), pp. 463-470
3. D. K. Biegelsen, N. M. Johnson, D. J. Bartelink, and M. D. Moyer, Appl. Phys. Lett. 38, 150 (1981)
4. N. M. Johnson, D. K. Biegelsen, and M. D. Moyer, Appl. Phys. Lett., in press

SiO$_2$ Interface Degradation and Minority Carrier Lifetime Effects of Laser Beam Processing

V.G.I. Deshmukh, A.G. Cullis, H.C. Webber, N.G. Chew

Royal Signals and Radar Establishment
Malvern, England

D.V. McCaughan

G.E.C. Hirst Research Centre
Wembley, England

1. Introduction

In silicon device fabrication, laser beam processing can offer distinct advantages over conventional thermal treatments [1]. However, it is important to determine whether any deleterious consequences accompany the desired results of laser exposure. In view of the crucial role played by the Si/SiO$_2$ interface in MOS-type devices, we have studied the effects of pulsed laser irradiation on the Si/SiO$_2$ interface and present here results of oxide charge (Q$_{FC}$) and interface state density at mid-gap (N$_{SS}$) determined from C-V measurements as functions of laser energy density and pulse number. Minority-carrier lifetimes (MCL) obtained from C-t data [2,3] are also given and the dependence of these parameters on subsequent thermal treatments is discussed.

Sample preparation consisted of the growth of 1000 Å of SiO$_2$ on 10 Ωcm, <100> p-Si substrates at 1150°C in dry O$_2$ followed by an in situ N$_2$ anneal for 30 minutes. The samples were then irradiated through the oxide layer using a Q-switched ruby laser (pulse width ~ 30 nS) fitted with a beam homogeniser. Aluminium evaporation and photolithography to define capacitor/guard ring structures completed the fabrication cycle. Further details are given in [4]. Control values of the MOS parameters are N$_{SS}$ ~ 2.10^{10} cm^{-2}eV^{-1}, Q$_{FC}$ ~ 5.10^{10} cm^{-2} and MCL ~ 40 µS.

2. Single-Pulse Laser Exposure

On irradiation with single laser pulses, N$_{SS}$ and Q$_{FC}$ increase sharply for applied energy densities E $\gtrsim$ 0.75 cm^{-2} and MCL decreases rapidly for E > 0.45 cm^{-2} [4]. The oxide furrowing observed at the higher energy densities has been explained by Hill [5] in terms of the relief of stress in the radiation-transparent dielectric when the underlying silicon melts. In this melting regime, C-V measurements show that the 'passivating' property of SiO$_2$ on Si is progressively worsened [4] though the threshold energy for interface degradation may be less than that necessary for liquefying the substrate. Reductions in MCL are to be expected when trap levels associated with lattice defects are introduced via laser exposure. Representative TEM micrographs of the Si/SiO$_2$ interface in cross-sections are shown in Fig. 1 where a sample irradiated at 1.1 J.cm^{-2} clearly shows a dislocation loop and at 0.75 J.cm^{-2}, possible piping at the melt front is visible and dislocations propagated during regrowth to the surface. Surface ripple is apparent with a period ~ 5 µm and there is possibly an associated variation in oxide thickness with the thinnest oxide occurring at the peaks of the corrugations. These data afford interesting comparisons to be made with other studies, both with CW Ar$^+$ and pulsed ruby lasers on oxide-coated Si [6,7]. Ref. 6 demonstrates reductions in Q$_{FC}$

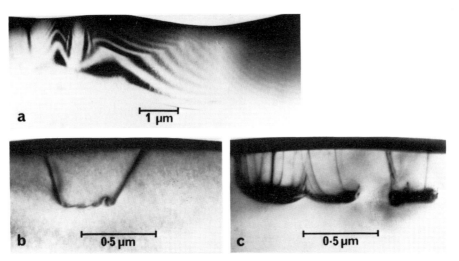

Fig. 1 Cross-sectional transmission electron images of Si laser-irradiated through SiO_2: (a) 1.0 J.cm^{-2} showing good crystal quality and undulating surface, (b) 1.0 J.cm^{-2} as (a) but showing occasional dislocation half-loop, (c) 0.75 J.cm^{-2} showing dense defect network introduced into the Si below the SiO_2.

and only a small variation in N_{SS} with increasing CW laser energy in contrast to our pulsed laser results whilst [7] shows that pulsed irradiation of ion-implanted and thermally annealed Si covered with 250 Å oxide can remove all dislocations for E > 2 J.cm^{-2} and dislocation reorientation at 1.3 J.cm^{-2} is given as evidence of substrate melting. EBIC measurements [8] on pulse-irradiated, bare Si with structural defects give a believed melt threshold ~ 1.2 J.cm^{-2}. Our data, on unimplanted, oxide-coated Si, in conjunction with the comments above suggest that above an energy density threshold that may be identified with melting, defect-free Si can occur. We have been unable however to measure interface properties at high energy densities owing to oxide leakage (see para 3). The effects of thermal annealing on the Si/SiO_2 interface characteristics subsequent to single-shot laser irradiation is given in [4].

3. Multiple-Pulse Laser Exposure

In practical laser-processing, the use of multiple pulses to shape implant profiles as desired is an attractive possibility. We have investigated the effects on the Si/SiO_2 interface of multi-pulse irradiation (5, 10, 50 pulses, repetition rate ~ 0.1 Hz) with energy densities spanning the range 0.4 to 1.5 J.cm^{-2}. Both Q_{FC} and N_{SS} show small monotonic increases with pulse number n though clearly it is the applied energy density that is the more crucial parameter in determining the extent of degradation of the interface (Fig. 2). MCL values shown in Fig. 2 suggest that after the initial reduction caused by a single pulse there is a slightly beneficial annealing effect from subsequent irradiation for small n though again it is energy density that is of prime importance in determining MCL values.

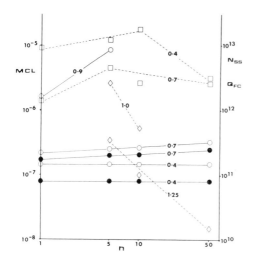

Fig. 2 $Q_{FC}[cm^{-2}]$, $N_{SS}[cm^{-2}eV^{-1}]$ and MCL [sec] versus pulse number n with energy density [J.cm-2] as parameter. $\bigcirc$: Q_{FC}, $\bullet$: N_{SS}, $\square$: MCL, $\diamond$: MCL for reoxidised samples, see paragraph 3.

At $E > 1$ J.cm^{-2}, oxide exhibited total breakdown or leakage that was too severe to allow C-V measurements, but it was observed that the current flow at a set bias was a function of time, increasing over several minutes until a steady-state value was reached. Such current instabilities have been observed previously [9], though at higher applied fields than our case of ~ 2 MV/cm and have been ascribed to positive charge accumulation in the oxide. Surface asperities can lead to locally enhanced electric fields and in view of the severe corruption of the oxide evident at high laser energies, it is unsurprising that breakdown of the SiO_2 occurs at apparently low applied fields. TEM however shows no evidence of asperities.

This weak oxide was chemically stripped and the samples reoxidised to 1000 Å in dry O_2 at 1150°C. Again, however, oxide leakage prevented quasi-static C-V measurement. Experiments on oxide leakage of SiO_2 grown on scratched Si substrates have shown that when surface damage exists, a simple reoxidation to normal thicknesses does not yield hard oxide properties - the surface asperities must be chemically polished prior to oxidation [10]. We therefore restripped the oxide, chemically etched the Si substrates ~ 1500 Å and regrew SiO_2 to 1000 Å in dry O_2 at 1150°C. This oxide was hard except in the region exposed to 50 x 1.5 J.cm^{-2} radiation. Determinations of Q_{FC}, N_{SS} following this thermal treatment show both to be indistinguishable for $E = 1, 1.25, 1.5$ cm^{-2} and n < 50 from unirradiated SiO_2/Si values. MCL results however are dependent on n and E as shown in Fig. 2 implying recombination/generation centres introduced by laser irradiation are not completely annealed out by thermal processing.

4. C-t Transients

In principle, MCL is straightforwardly obtainable from C-t experiment and the values plotted here are taken from such transients characterised by a single value of MCL. At high energy densities (~ 1 J.cm^{-2}) however, the C-t's may show either total surface-dominated behaviour where no unique value of MCL may be extracted or, more bizarre behaviour where dC/dt *increases* with time from depletion to inversion somewhat similarly to the theoretical plots given by [11]. Analysis is rudimentary at present but the shape of the C-t curve and its associated Zerbst plot may be qualitatively reproduced by assuming a trap density that decays exponentially in value from surface to bulk. Further details are to be published elsewhere.

5. Conclusion

Pulsed laser irradiation of Si through SiO_2 leads to significant interface degradation and MCL reductions in the substrate suggesting that careful masking by laser-reflective layers of any gate oxides and their peripheries in real MOS device processing is mandatory. Interface degradation is a strong function of applied energy density with pulse number playing a less important part. At $E \sim 1.1$ J.cm^{-2}, oxide leakage is severe in 1000 Å oxides and chemical polishing of the substrate is necessary (or presumably thick oxide growth) before reoxidising if hard oxides are to be obtained.

References

1. e.g. Laser-Solid Interactions and Laser Processing, ed by S. D. Ferris, H. J. Leamy and J. M. Poate (AIP, New York 1979).

2. F. P. Heiman: IEEE Trans. ED-14, 781-84 (1967).

3. R. F. Pierret: IEEE Trans. ED-25, 1157-59 (1978).

4. V. G. I. Deshmukh, D. V. McCaughan, H. C. Webber: Appl. Phys. Lett. (to be published).

5. C. Hill: Proc. Symp. on Laser and Electron Beam Processing of Electronic Materials, ed by C. L. Anderson, G. K. Celler, G. A. Rozgonyi (The Electrochemical Society, Princeton, NJ 1980) p 26.

6. T. I. Kamins, K. F. Lee, J. F. Gibbons: Solid State Electron 23, 1037-39 (1980).

7. J. Narayan: Appl. Phys. Lett. 37, 66-68 (1980).

8. L. Jastrzebski, A. E. Bell, C. P. Wu: Appl. Phys. Lett. 35, 608-11 (1979).

9. P. Solomon: J. Vac. Sci. Tech. 14, 1122-30 (1977) and refs. therein.

10. J. M. Keen: private communication.

11. L. Manchanda, J. Vasi, A. B. Bhattacharyya: Solid State Electron 22, 29-32 (1979).

Laser-Induced Crystallization in Ge Films and Multilayered Al-Sb Films

L. Baufay, M. Failly-Lovato, R. Andrew, M.C. Joliet, L.D. Laude
A. Pigeolet, M. Wautelet

University of Mons
Mons, Belgium

1. INTRODUCTION

In recent years laser annealing has emerged as a powerful method
for thin film crystallization (1). In the present work, this
technique is applied to the Ge and Al-Sb systems in order to
determine the influence of various experimental parameters
on the crystallization processes, in particular laser beam
inhomogeneties. Results presented here demonstrate that it is
possible to control long range crystal growth by means of
the spatial distribution of light intensity in the incident
laser beam. The choice of Al-Sb is suggested by the remarkably
close melting points of the two metals, by the potential
application of the compound as a photovoltaic material and by
the relative difficulty of its preparation by other means.
Aside from the intrinsic interest in the laser processing
application the results may also shed light upon the nucleation,
fundamentals of laser activated atomic diffusion and crystal
growth mechanisms.

2. EXPERIMENTAL

1000Å thick Ge films were evaporated in vacuum (10^{-5}torr)
by Joule-type evaporation and deposited onto a freshly cleaved
NaCl substrate. Similar experiments were performed on 0.05 to
0.5μm thick mixed Al-Sb films constructed by sequential ultra-
high vacuum e-gun evaporation of alternate Al and Sb layers in
stoichiometric amounts. The metallic films are encapsulated
between two protective SiO layers. After removal of the salt
substrate and deposition on a TEM grid, all free-standing films
were irradiated in air with a medium power pulsed dye laser
with 1 μs pulse duration and 2 eV photon energy.

3. GERMANIUM

After one pulse, transmission electron microscopy observations show that crystalline Ge spots ("stars") are formed
when the laser beam energy density, E, exceeds only 5mJ/cm² (2,3).
Large crystallites are created, the lateral dimension of which
are far greater than film thickness, i.e. up to a factor 100.
Polycrystallites (0.1μm dimension) are present at the boundary of the large crystals. Increasing the pulse energy does
not change the star dimension but their number, although their
concentration is not linearly dependent on the pulse energy.
In order to clarify the origin of these stars, Ge films were
irradiated with a laser beam in which spatial inhomogeneities

were carefully controlled. For that purpose, the laser beam was channeled along optical fibers of a few microns diameter and films irradiated at various distances from the fiber output. Such a beam treatment gives rise to the so-called speckle, effect (4), i.e., light inhomogeneities, the dimension of which depends on the fiber diameter and the fiber-sample distance. Results give clear evidence that the dimension of the stars is directly related to the dimension and geometry of the laser beam inhomogeneities: at a given mean laser energy density, the larger the speckle, the larger the star diameter. This was systematically used to produce well defined crystallized patterns on a sample.

The radial distribution of crystallites in the stars is opposite to that observed using scanned CW laser crystallisation of a-Ge (5). In this case, the process is well interpreted as being due to a conventional thermal annealing. In our case, calculations indicate that temperature is not sufficient to melt or to crystallize Ge in very short times. By comparison with results obtained at low laser power (6), it is believed that metastable states associated with dangling bonds (7) might help the nucleation process and crystal growth, through a modification of the activation energy for the migration of defects.

4. AlSb

4.1 DIRECT BEAM

In the first step, the irradiation is performed with the direct beam. The minimum energy density necessary for the transformation, which is a function of the size of the grid mesh, is in the range of 20-30 mJ/cm^2. A single pulse produces several patterns characterized by:
- an enlargement and orientation of Al and/or Sb crystallites;
- a high density of polycrystallites, their size being less than 0.5μm;
- big crystallites with the maximum observed size of 7X1.5μm;
- a low density of polycrystallites, their size being less than 0.5μm;
- a perforation

Sometimes we note ripples separated by 1.5μm.

4.2 HOMOGENEOUS BEAM

After homogeneizing the laser beam (8), films are irradiated which results in fine and uniform polycrystallisation with small AlSb crystallites.

4.3 STRUCTURED BEAM

Comparison between the crystallizations obtained by a homogeneous beam and a very inhomogeneous beam suggests that the irradiation be performed with a highly structured beam. This can be achieved by introducing into the beam path a grid or an optical grating, thus obtaining a fringe system. The energy density of each fringe depends on the order of the fringe. In this way, it is possible to correlate the appearance of a product resulting of the irradiation with a given value of the energy deposited in the sample.

For a parallel ringe system in which the interfringe distance is greater than 100µm, a single pulse produces different patterns in the sample. The following regions successively appear when the energy density is increased:

- an enlargement of Al and/or Sb crystallites oriented along the perpendicular to the fringe axis;
- a polycrystalline region with a high density of AlSb crystallites, their size being less than 0.5 µm;
- large AlSb crystallites aligned perpendiculary to the fringe axis; these crystallites are several microns long, 7µm being the maximum value noted, and 1 to 1.5 µm wide;
- low density of small AlSb crystallites;
- hole.

Playing with either the geometrical parameters of the fringe system, the regions which are characterized by a poor crystallization (i.e. a low energy density profile) can be successively eliminated.

In particular situations, it is even possible to obtain a self epitaxy (films are not supported), clearly shown by the preferential orientation of the AlSb crystallites.

In these experiments, the large crystallites seem to result from a rapid growth process, whilst polycrystallization appears to result from homogeneous nucleation promoted by either an homogeneous beam or the excess heat of crystallization of larger crystallites. In every case, areas transformed without perforation are greater when each Al layer thickness is small, the best results being obtained when this thickness is about 50 Å.

The temperature distribution in the sample has been calculated with regard to the energy deposited in it. The transformation which occurs in several areas cannot be explained by a classical thermal model, since the calculated value of the temperature is in all cases less than 600°C, in agreement with the lack of globule formation. However, the larger crystallite size seems to be governed by the presence of a thermal gradient.

4.4 OTHER EXPERIMENTAL CONDITIONS

Instead of NaCl substrate, an ordinary glass plate cleaned by ionic sputtering was also used as film substrate.

A ruby laser (15ns pulse duration, 1.8eV photon energy) and the dye laser described previously have been used to irradiate 0.5µm thick glass-supported samples. Despite differences in pulse duration and intensity, the pulse energy (and not power) remains a remarkably constant parameter for achieving complete transformation.

5. CONCLUSION

In this work, large scale crystallite growth in free standing (elemental or compound) semiconducting film is clearly shown to be governed by laser beam inhomogeneities which induce energy (and thermal) gradients, the temperature being less than the melting temperature of the original materials.

This work is supported by project IRIS of the Belgian Ministry for Science Policy. Financial support for M.Lovato is by IRSIA Brussels.

REFERENCES

(1) Laser and Electron Beam Processing of Materials, ed.by P.S.Peerey, C.W.White (Academic Press, New-York,1980)
(2) R. Andrew, M. Lovato, J.Appl. Phys. 50, 1142 (1979)
(3) R. Andrew, L. Baufay, L.D.Laude, M.Lovato, M.Wautelet, Journal de Physique 41, C4-71(1980)
(4) J.C.Dainty (ed.): Laser Speckle and Related Phenomena, Topics in Applied Physics, Vol.9(Springer, Berlin, Heidelberg, New York 1975)
(5) R.L.Chapman, J.C.C. Fan, R.P.Gale, H.J.Zeiger, Appl.Phys. Lett. 36, 158 (1979)
(6) L.D. Laude, M.Lovato, M.Wautelet, Appl.Phys.Lett.34,160(1979)
(7) R. Andrew, L.D.Laude, M:Wautelet, Phys.Lett.77A, 275 (1980)
(8) D.J. Godfrey, C.Hill, Journal de Physique 41, C4-79 (1980)

Part IX

**Transport Properties in
Inversion Layers**

Subband Physics with Real Interfaces

F. Koch

Physics Department, Technische Universität München
D-8046 Garching, Fed. Rep. of Germany

The layer of mobile charges in the semiconductor at the interface to an insulator is a sensitive indicator of the nature of
the atomic bonding and the electric potentials existing at the
junction of the two materials. Special conditions are required
for external control of this charge layer with the metallic
gate of an MOS device. The study of the energies and transport
properties of the quantum levels in the layer is known as subband physics. We survey this field of activities to examine in
particular the kind of information to be derived about the real
interfaces encountered at an insulator-semiconductor junction.

1. Subbands

The semiconductor charge is confined to the layer by the one-
dimensional potential $V(z)$ and occupies a series of quantized
subband levels. The idealized discription, as in Fig.1, assumes

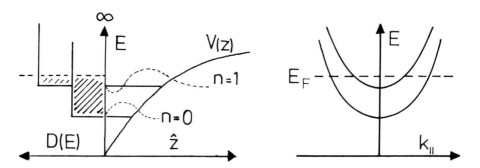

Fig. 1: Subbands at an infinite barrier interface

an abrupt and infinite barrier at $z = 0$, with perfect translational symmetry in $\hat{x}$ and $\hat{y}$. In this picture the wave function
vanishes at the boundary. The motion parallel to the surface
is that appropriate for the bulk semiconductor with effective
mass $m_{\parallel}^*$. One writes the parabolic two-dimensional energy

subbands as

$$E_n(k_{\|}) = E_n + \frac{\hbar^2 k_{\|}^2}{2m_{\|}^*} .$$

The E_n are solutions in the potential well $V(z)$. The potential is self-consistent in that it includes the contribution via Poisson's Equation of the surface charge itself.

For typical interfacial electric fields of 10^5 - 10^6 V/cm the confinement length of the charge is 10 - 100 Å. The effective mass m_{zz}^* plays an essential role for the motion in the potential well. The lowest lying set of subband levels is formed for electrons associated with the lowest conduction band edge. When there are several equivalent minima, as in a multivalley semiconductor, the one with largest m_{zz}^* gives the lowest subbands. The level energies vary with the surface potential and the surface density of charge N_s. Typical N_s values range from 10^{11} - 10^{13} e/cm^2. It is not uncommon to have only a single occupied subband, a situation referred to as the electric quantum limit. Fig.1 has the Fermi level of the degenerate, 2-dimensional electron gas placed such as to partially occupy two subbands.

2. Subband Physics

The deliberate, experimental and theoretical study of subbands is a field of activity only slightly more than 15 years old to date. It has matured and developed into a far-ranging field of fundamental and applied investigations. A thorough sampling of the activities is contained in the series of conference publications [1-4]. I take the liberty in the following to quote and discuss material without specific reference. The interested reader will find documented and properly credited most of the physical ideas and principles that are referred to here in these publications. The portrait sketch of subband physics here follows the natural division of the subject into energy level spectroscopy and subband transport.

a) Spectroscopy of Subbands

Considerable effort in subband physics has been devoted to the determination of binding energies as an essential parameter describing the surface layer system. The energies are a sensitive indicator of the potential, of the occupancy, and in particular also of the interaction with the surface dielectric.

A historical sampling of spectroscopy must begin with tunneling. The thin natural oxide layer formed on such degenerate semiconductors as PbTe and n-type InAs, allowed the observation of tunneling current structures related to the steps in the density of states function in Fig.1. A few years later the resonant excitation of the surface bands was detected in the absorption of far-infrared radiation. In another variant of the experiment, the absorption was registered as a change in the surface layer conductivity. The emission of radiation by a

surface carrier distribution relaxing to equilibrium has also
been observed and has served to determine the level separa-
tions. Most recently inelastic light scattering has been employed as a means of studying the subbands. In Fig.2 are shown
schematically the different excitations and carrier transfers
used in the spectroscopic experiments on subbands. The Raman
excitation involves the simultaneous transfer of an electron
from the filled valence band to an unoccupied subband and the
return of a surface electron to fill the hole. The difference
energy evidently is that of the subband separation.

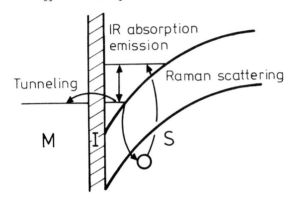

Fig. 2: Spectros-
copic methods used
to determine the
energy levels in the
surface potential
well.

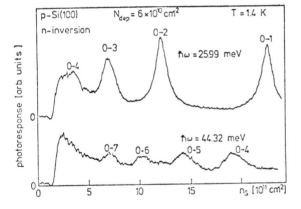

Fig. 3: Subband
transitions ob-
served in Si at
two different
energies of ex-
citation.
[after F. Neppl,
Ph.D. thesis,
TU München
(1979)]

    Fig.3 is a sampling of spectroscopy data on Si(100). Using
the source-drain surface conductivity to indicate resonance
absorption of a fixed-frequency laser source, transition lines
from the occupied $n = 0$ ground-state to 7 higher levels are
resolved.

b) Subband Transport

The surface carriers are free to move in the interfacial layer
and can transport both charge and energy. The electrical conductivity, in particular the fact that it can be externally

250

controlled by the gate voltage, is the feature exploited in
device applications. The conductivity is an important means of
physically characterizing the surface layer. The onset of conduction marked as V_T in Fig.4 signals the first appearance of
mobile carrier and serves as an indication of the band bending.
The differentiated curve $d\sigma/dV_g$ is the field effect mobility
μ_{FE}. The conductivity curve contains specific and detailed information on the scattering processes in the interfacial layer.

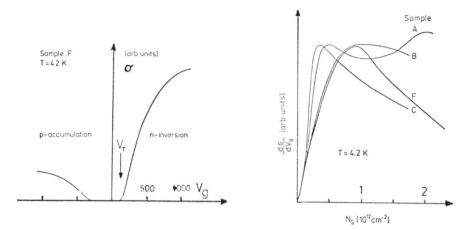

Fig. 4: Conductivity $\sigma(V_g)$ on a Ge(111) surface. Curves
$d\sigma/dV_g$ differ from sample to sample. [after J. Binder,
Ph.D. thesis, TU München (1980)]

A most useful variant of the conductivity experiments is the
magnetoresistance measurement. In a strong magnetic field
applied normal to the surface the parallel motion of electrons
is quantized into Landau levels. The resulting equidistant peak
structure in the gate-voltage dependence of the resistance R_x
(Fig.5) allows one to count the electrons in the subbands,
determines valley and spin degeneracy factor, the g-value, the
scattering rate, the effective mass, and the temperature of the
electron distribution. The steps in the curve R_H have been used
to derive the fundamental constant h/e^2 with great precision.

At far-infrared frequencies several resonant modes of the
parallel conduction have been studied. The most significant is
cyclotron resonance which allows a determination of effective
masses and scattering times. Surface plasma wave modes have
been identified. Resonant absorption from minigaps has been detected. All three of the above processes have been observed in
emission experiments as well as absorption.

Finally, current work on the frequency dependent conductivity
$\sigma(\omega)$ seeks to identify binding of electrons in surface potential fluctuations.

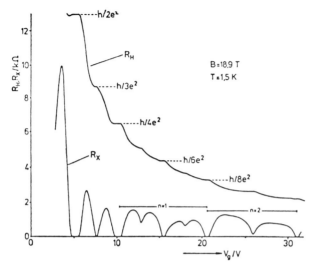

Fig. 5:
The oscillatory
Shubnikov-de Haas
magnetoresistance
observed in a
sweep of the sur-
face electron
density. R_H is
the Hall resis-
tance. [after
K.v.Klitzing et
al. in Phys. Rev.
Letters 45, 494,
1980]

3. REAL Interfaces

The model discussion involved an abrupt, infinite potential
barrier at z = 0. The real insulator-semiconductor interface
that has been thermally grown as a native oxide, produced by
anodic oxidation that has been deposited in a chemical vapor
reaction, grown on epitaxially or physically attached in a
number of different ways, is quite another thing. The specific
interfacial bonding of insulator and semiconductor atoms, the
dielectric polarizability both static and dynamic, the finite
potential barrier heights, random defect structures, periodic
mismatches in the atomic arrangement, interface strain, physical irregularities and roughness, all will have a decisive
and characteristic effect on the subbands. There is such a
great wealth of possibilities that subband physics will require
much time to sort them all out in quantitative detail. I mention here a few of them as they come to mind.

    Dielectric polarizability - The real insulator bounding the
semiconductor at z = 0 is polarizable. Its dielectric constant
ϵ_i is in general smaller than the ϵ_s of the semiconductor. This
figures in the classical electrostatic discontinuity of the
electric field in an obvious way. For the point charge aspects
involved in the exchange and correlation potential corrections,
in the exciton-shift calculation, the polarizable dielectric
must be included. The most detailed subband computations adopt
an image potential contribution of the form

$$V_{image} = \left(\frac{e^2}{4\epsilon_s z}\right)\frac{\epsilon_s - \epsilon_i}{\epsilon_i + \epsilon_s} \ .$$

The image potential is repulsive and acts as a force keeping
the electron away from z = 0 already before it gets there. For
typical Si(100) conditions the effect is about 20% in the subband energy splitting E_{01}. For electrons on the surface of

liquid helium where $\epsilon_i > \epsilon_s$ and the potential is attractive, binding by the image force is a well established fact. For the subbands the point-particle image contribution is still contested.

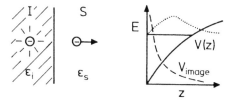

Fig. 6: Image potential and its effect on the subband wave function.

Finite barrier height - The different electron affinities of the insulator and semiconductor provide for a potential barrier at z = 0, but it is not by any means infinite. For the ~ 3.2 eV difference in the conduction bands of SiO_2 and Si, the typical subband energy of 100 meV is small and the idealized description reasonable. For $Ga_{1-x}Al_xAs$ on GaAs, the barrier height can be arbitrarily small. The higher lying states in the quantum-well structures that have been made with molecular beam epitaxy penetrate substantially. There is a marked effect on the energy level scheme of electrons confined to the GaAs in Fig.7. The band-bending at the edges of the rectangular well is the result of a space charge contribution via Poisson's Equation.

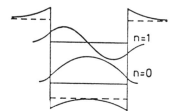

Fig. 7: Subbands in the square well potential of a GaAs layer.

Barrier height also has a sensitive influence on the possibility of tunneling of a subband electron into the insulator where it can occupy a possible defect level. Such missing electrons would be sensitively detected by changes of the magnetoconductivity oscillation period and a shift of V_T. Electrons accelerated parallel to the surface will occupy high $k_{\parallel}$ states and can be injected into the insulator by a scattering event, if the barrier is finite. In a Gunn-type oscillator recently proposed, the electrons from the GaAs layer sandwiched between two $Ga_{1-x}Al_xAs$ layers, enter the latter when heated by a current. The higher $m_{\parallel}^*$ and increased scattering rate in the $Ga_{1-x}Al_xAs$ cools the electron distribution and causes the carriers to return. The result is an oscillator

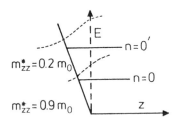

Fig. 8: The graded boundary allows for m_{zz}^*-dependent penetration into the insulator. For Si(100) it decreases the separation $E_{00'}$. [after F. Stern, Sol. State Commun. 21, 163 (1977)]

that depends on the real-space displacement of the carriers between the layers.

Interface grading - A real interface, such as that between the semiconductor and a thermally grown native oxide, cannot be abrupt with mathematical precision. One expects to change over gradually from semiconductor to oxide, perhaps in a distance of a few atomic layers. In Si there is talk of SiO_x, a region where slowly the full stoichiometry of SiO_2 is reached. Such interface grading, as has been discussed in the literature, will change the subband energies because of penetration of the electron wavefunction into the insulator. In particular, it has been shown that grading has a marked effect on the relative separation of corresponding levels on Si(100) formed from different valleys. The mass m_{zz}^* of the two-fold valley-degenerate, lowest-lying subband is $0.9 \, m_0$. The value for the four-fold degenerate state is $0.2 \, m_0$. In a softly graded barrier, the low m_{zz}^* electrons penetrate more deeply into the insulator. The corresponding 0'-level approaches more closely the n = 0 ground-state than for an abrupt barrier. Thus the splitting $E_{00'}$, a quantity that can be measured by thermal population experiments in subband spectroscopy, is an indicator of interface grading.

Interfacial strain - A thermally grown oxide, or any insulating layer firmly bonded to a semiconductor substrate, will exert a force across the interface when there is differential expansion. The resulting strain can be both random, microscopic or large-scale, homogeneous.

For a single-valley semiconductor the result may be small changes in the band parameters, such as m^* and the gap energy. The effect on the subbands is minor. In the multivalley semiconductor Si, however, strain is expected to play a major role in moving different valleys and their associated subbands relative to one another. In the (110) and (111) planes the "wrong" valley degeneracies have been observed and interface strain has figured prominently in searching for an explanation. In related cyclotron resonance experiments, the repopulation of different valleys with uniaxial strain in the (100) Si plane has been convincingly demonstrated. At 1.8 kbar and density $1 \times 10^{12} \text{cm}^{-2}$ the carriers are all transferred to the 0'-subband of the high cyclotron mass valleys (Fig.9). In this way a real, strained interface determines occupancy ratios among subband levels of the different valleys.

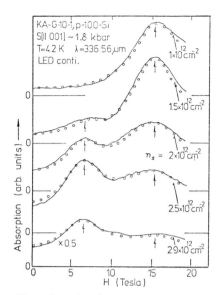

KA-G-10-1,p-100-Si
S[II 001] ~ 1.8 kbar
T=4.2 K λ=336 56 μm
LED conti.

$1\times10^{12}\,cm^{-2}$

$1.5\times10^{12}\,cm^{-2}$

$n_s = 2\times10^{12}\,cm^{-2}$

$2.5\times10^{12}\,cm^{-2}$

$2.9\times10^{12}\,cm^{-2}$

$\times 0.5$

Absorption (arb. units)

H (Tesla)

Fig. 9: Cyclotron resonance
in the presence of interface
strain. High and low field resonances are from different
valleys [G. Abstreiter et al.
in ref. 3]

Interface position - Where
exactly is z = 0? The question
is closely tied to interface
grading, to the image potential and finite barrier discussion. The surface electric
field on Si(100) acts to remove the 2-fold valley degeneracy of the n = 0 ground-
state. The magnitude of the
effect depends sensitively on
the Bloch part of the surface
state wavefunction. The latter
in turn depends on the details
of how the wavefunction is
terminated near z = 0. Calculations have been made for a
model in which the boundary
condition is varied about the
terminating plane of atoms by
± an atomic distance. Such
calculations include the
evanescent wave states encountered in the theory of surface
states. The details of the
atomic bonding at the interfacial plane play a central
role. The valley-splitting
effect is superficially similar to the valley-orbit
splitting encountered in the donor-impurity problem. In Fig.10
the valley splitting is shown, as the Si-SiO2 interface position
is moved about by ± a/8. The sharp resonant structure results
from a surface state that is possible for the particular bounding conditions. At the resonance the subband wavefunction is
considerably changed (Fig.11).

"hkl" interfaces - For Si samples with the interfacial plane
tilted away from [100] by a few degrees, the subband parabolas
have been found to break up into segments. Energy gaps, the so-
called minigaps, appear at quite small values of k (Fig.12).
The basic intervalley gaps at k = ± 0.15(2π/a)sinθ and
± 0.85(2π/a)sinθ are linked with the tilt θ and the position of
the valley in the Brillouin zone. The gaps are another manifes-
tation of the valley splitting discussed above. In this sense
they also are evidence for interaction of the subband electrons
with the interface. Their magnitude, determined in spectroscopic experiments, is a quantitative measure of this interaction.

In addition to the intervalley gap, there is observed also
an intravalley gap that is related to the period of modulation
incurred by cutting an array of atoms at the angle θ. Its
magnitude depends on the interaction of the subband electrons
with the periodic perturbation. One is led to think about the
possibility of tailoring gaps into the dispersion of the surface bands by providing a periodic structuring of the inter-

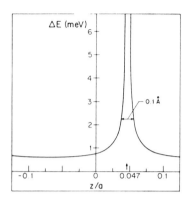

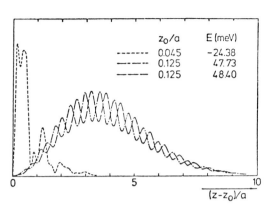

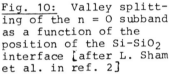

Fig. 10: Valley splitt-
ing of the n = 0 subband
as a function of the
position of the Si-SiO$_2$
interface [after L. Sham
et al. in ref. 2]

Fig. 11: Subband wavefunctions
for different positions of the
interface [after F.K. Schulte,
Sol. State Commun. 32, 483
(1979)]

face or the insulating overlayer. Macromolecules of an insulator or a periodic strain based on a mismatch of lattice constants at the interface could serve for this purpose.

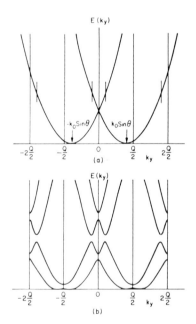

Fig. 12: The subband
structure in the pre-
sence of minigaps (k$_y$
is in the direction of
tilt and Q = $(4\pi/a)\sin\theta$

Interface roughness - Irregularities in the placement of insulator and semiconductor atoms in the
bounding plane will scatter the
subband carriers. This roughness
scattering is responsible for the
decreasing $d\sigma/dV_g$ observed at high
N_S in typical surface layers (compare Fig.4). The effect is a
characteristic of each sample and
depends on the surface preparation.
There have been many models of
roughness scattering. The surface
irregularities are described in
terms of two parameters, one to
characterize the typical spatial
period, the second to describe the
average displacement height. The
subband electrons scatter according
to their probability of being found
in the surface, hence proportional
to the square of the gradient of
the wavefunction. Closer binding to
the surface implies more scattering
and one expects the relaxation rate
to rise with the surface field.
This at once explains the mobility
drop at high N_S and suggests an
experimental method to identify
the roughness contribution to the
scattering. Using the substrate
bias potential the surface field
can be raised at constant N_S. The

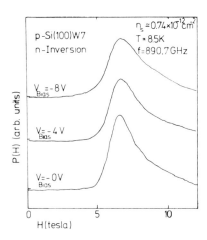

Fig. 13: Subband cyclotron resonance and its dependence on V_{bias}

additional broadening of the cyclotron resonance in Fig.13, for example, is attributed to roughness.

Interface states - Of great interest are states which allow for the exchange of charge with the subbands. Bonding defects, impurity atoms, deviations from stoichiometry, specific structural defects - all belong to this category. It is common to find a V_T far from 0 Volts, because of the interface potential caused by transferred charge. One example is n-doped $Ga_{1-x}Al_xAs$ which will give an electron layer in the neighboring GaAs, because the donor levels lie above the conduction band of the GaAs. Chemical treatment of many compound semiconductor surfaces will lead to the same situation. An interfacial layer of Cs on Si serves the same purpose. One can imagine optically active interface molecules which when illuminated will transfer an electron to the semiconductor. In the sense that subband physics provides accurate ways to measure N_S and to gauge the surface potential, it contributes to a knowledge of the interface states.

Interface charge - A realistic interface can contain charged impurities or defect centers. Na^+ in the SiO_2 at the interface is a well known example. As mentioned above, charge causes band bending and a change of V_T. It also provides for a breakdown of the translational symmetry of the interface and causes potential fluctuations. The result is binding and localization of the carriers at the interface. One generally finds at low values of N_S an activation barrier for surface conduction. Deviations from Drude behavior for $\sigma(\omega)$, negative magnetoresistance, changes in the cyclotron resonance position and linewidth, and a marked effect on the subband resonance have all been observed when Na^+ is moved to the $Si-SiO_2$ interface.

Interface energy transfer - The real interface that is polarizable, that contains impurities and defect states, will have various modes of excitation. A most obvious one are the polar phonons in SiO_2. The excitation of such modes by subband carriers at the interface has been considered in a number of publications. For the present, it suffices here to point to a number of possible ways that the real interface will serve to exchange energy with the semiconductor subbands.

... etc., etc., etc. There are still many more examples of how the subband carriers interact with the interface, but the limitations of time and space weigh heavily. I have tried to give the reader a portrait of what current subband physics is all about. By focussing on those experiments and results, where the real interface plays a role, I have inadvertently succeeded in sketching the current trends of the field. No longer is the primary effort in subband physics the self-indulging study of the fascinating physical properties of an ideal, 2-dimensional electron gas. The field has matured to start looking at the small differences, the deviations from the idealized description, and in the process it has become a useful means to examine the complex physical system known as the REAL insulator-semiconductor interface.

References

The most useful means to acquaint oneself with the field of subband physics are the proceedings of the four conferences, three of them held in '75, '77 and '79, a fourth to be held this year, on the Electronic Properties of Two-Dimensional Systems. In these publications the reader will find documented much of the material that I have cited and discussed here.

1. Surface Science, Vol. $\underline{58}$ (1976)
2. Surface Science, Vol. $\underline{73}$ (1978)
3. Surface Science, Vol. $\underline{98}$ (1980)
4. Surface Science, to be published.

Transport Properties of Carriers at Oxide-Hg$_{1-x}$Cd$_x$Te Interface

J.P. Dufour, R. Machet, J.P. Vitton, J.C. Thuillier, F. Baznet

Laboratoire Diélectriques, Faculté des Sciences Mirande
Université de Dijon, B.P. 138
F-21004 Dijon Cedex, France

Cadmium mercury telluride (CMT) was recently the subject of many publications in the field of surface properties in view of its applications 1-3. One of the fundamental difficulties encountered in device technology is that the surface of the material has to be passivated before the deposition of an insulating film, and the anodic oxidization is the most commonly used process 4-6. Our work is mainly oriented towards the study at low temperature of degenerate electron layers forming a two dimensional gas of carriers. Several publications were devoted to these topics 7-8 but no correlation is possible between the different results. We report here the approach to get a set of data suitable to make a systematic study.

1. Experimental Procedures

In our experiments, CMT single crystals (*) are n or p type with a cadmium concentration of ~20% and ~28%. Samples with different compositions do not give significantly different results. The sample carrier concentration has a mean value of 4.10^{15}cm^{-3} for n type, and $2-3.10^{16}$cm^{-3} for p type.

Crystals are cut and polished in a conventional manner. But, generally, after the 5% bromine-methanol etching the surface of CMT is strongly degenerated with an electron concentration of ~10^{12}cm^{-2} and moreover a pinning of the bands occurs which does not allow the modulation of the surface charge. A passivation is then necessary. Oxidization of CMT was made with two different solutions : the first one is a saturated solution of ADP in diethylenglycol with a constant current of 0.1mA cm^{-2} up to a compliance bias of 90V. The second one is an anhydrous methanol-KOH (5.6g/l) solution. The current is kept constant : 0.5mA cm^{-2}. The oxidization is stopped when the bias on the electrolyte cell reaches 20V.

In each case the films formed are uniform. By the first process films of 2000Å in thickness are obtained with quite good insulating properties. Permittivity is ~20, close to that of tellurium dioxide. Breadkown occurs at about 0.8MV/cm. By oxidization in methanol-KOH solution, the films are always leaky. The thickness is limited to 700-1000 Å to avoid large, macroscopic, pinholes. The deposition of an insulator is then necessary (ZnS in our case).

(*)We want to express here our thanks to the SAT Company, in Paris, for the samples supply.

The presence of the oxide layers plays two parts:
- it allows the deposition of a second layer of insulator by increasing the adherence of the film,
- it allows the modulation of the surface charge. In the study of the quantum effects, for the sake of simplicity, we generally applied the electric field across a mylar film (6µm) onto the oxidised surface.

Capacitance, conductivity measurements are classical. Magneto-resistance was measured up to 20T (at Grenoble, SNCI, CNRS Laboratory).

2. Results.Discussion

2.1 Capacitance

Measurements of capacitance and conductance of our samples were made mainly on samples oxidised by the first technique. In any manner C(V) curves exhibit a very weak hysteresis effect (0.5V maximum).

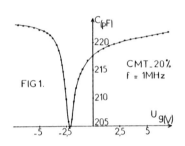
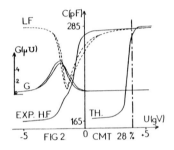

Two typical behaviours are noted:
- 20% cadmium content samples
They always exhibit low frequency (LF) C(V) curves for n and p type independent of doping level. It is easily unterstood by considering that the minority carrier charge in the inversion layer is generated by tunelling across the band gap. Associated with the narrowness of the gap, the influence of the non-para-bolicity of the conduction band appears clearly on the C(V) by a lagging along the V axis (Fig.1). A study of this effect will be published elsewhere.
- 28% cadmium contents samples
Samples have normal C(V) curves with a transition of LF-HF type curves for the best samples at 10kHz, however, mainly between 50-100 kHz (Fig.2 a,b) indicating a high surface recombination velocity.

TERMAN's method [7] was used to estimate the interface state density: $N_s \sim 10^{12} cm^{-2} eV^{-1}$ is found in the band gap, with a strong in-crease up to $10^{13} cm^{-2} eV^{-1}$ comparable to the results of LEONARD [1]. Inclusion of non-parabolicity effects will probably reduce the measured density near the conduction band edge.

2.2 Surface Conductivity

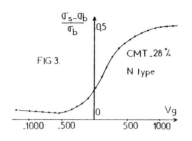

FIG 3.

CMT _28 %

N type

We performed essentially surface conductivity measurements to check the possibility of observation of quantum effects. Fig. 3 shows the variation of the surface conductivity for a sample n type 28% of cadmium. A marked saturation occurs indicating a strong reduction of the surface mobility as the number of induced carriers increases linearly with the bias. Mobility up to $50000 \text{cm}^2/\text{V.s}$ were measured at $4.2°\text{K}$: one half to one order of maqnitude lower than bulk for zero field plate bias or 10^{12}cm^{-2} charges. The drop in mobility may be attributed to two dominant facts : the spatial confinement of electrons of the highest subband which contains 60% of the mobile charge as shown in Fig.4 and the most probable cause is the increase in the effective mass of these carriers (Kane model dependence and surface effects). Up to now no model gives the description effects surely play a role at $77°$ K.

2.3 Quantum Effects

The application of a magnetic field on samples with pinned bands reveals clearly the degeneracy of the surface with $N_s \sim 10^{12} \text{cm}^{-2}$.

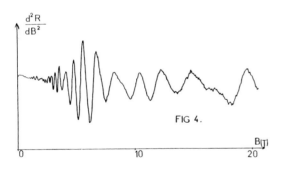

FIG 4.

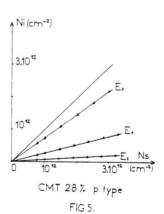

CMT 28 % p type

FIG 5.

Fig. 4 shows the second derivative of the magnetoresistance. We have to note several facts before proceeding further

261

With our anodization technique, the modulation of the surface
charge was possible.
-The surface fixed charge density is constant all along the
experiments (stable in time),
-Thermal cycles from $300°K$ to $4.2°K$ tend to induce a decrease of
the surface mobility,
-A reduction of the amplitude of the Shubnikov de Haas oscillations
(and mobility) occurs for $N_s > 1.5 \cdot 10^{1.2} cm^{-2}$. It corresponds to
the saturation of the field effect conductivity curves shown
before (Fig. 3).
-At high magnetic field : $B > 10T$ a severe decrease of the
amplitude of SdH oscillations is observed. The cause is unknown
up to now. It may correspond to the transition toward the
magnetic quantum limit of carriers of the higher subband,
expelled toward deeper ones with a higher mass.
-Some beat patterns are observed independent of the superposition
of the SdH oscillations. It may correspond to slight deviation
in spatial homogeneity of our samples.

The analysis of the SdH periods reveals the existence of at least
3 subbands for charge densities $N_s < 3 \cdot 10^{12} cm^{-2}$. A fourth one is
suspected for narrower band gap materials. Figs. 5a, b show
the variation of the population in the different subbands versus
the total charge. A comparison with the model of Y. TAKADA and
Y. UEMURA [9] indicates a fairly good agreement between the theoretical values for inversion layers and the measured one, specially for the higher subband. An attempt to apply the model developed by ANTCLIFFE et al. [8] was made but does not give a realistic
explantation of the problem.

1. W.F. Leonard, Techn. Report AFAPL-TR 76-4
2. R.A. Chapman, M.A. Kinch, A. Simmons, S.R. Borrello, H.B.
    J.S. Wrobel, D.D. Buss, Appl.Phys.Lett. 32, 434 (1978)
3. M.A. Kinch, R.A. Chapman, A. Simmons, D.D. Buss, S.R.
    Borrello, Infrared Physics, Vol 20, p.1.20 (1980)
4. Y. Nemirowsky, E. Finkman, J. Electrochem.Soc. 125, 481(1978)
5. P.C. Catagnus, C.T. Baker, U.S. Patent 3977-018 (1976)
6. T.S. Sun, S.P. Buchner, N.E. Byer, J.Vac.Sci.Technol. 17 (5),
    1067 (1980)
7. L.M. Terman, Sol. St. Electronics 5, 285 (1962)
8. G.A. Antcliffe, R.T. Bate, R.A. Reynolds, Proc.Conf.Phys.
    Semimetals and Narrow Gap. Semic., Dallas 1970, Ed. D.L.
    Carter, J.Phys. Chem.Sol 32, p. 499 (1971) Supp. 1
9. Y. Takada, private communication

Role of Interface States in Electron Scattering at Low Temperatures

A. Yagi

Semiconductor Div. Sony Corporation
Atsugi, 243, Japan

1. Introduction

Experiments in the inversion layer on a silicon surface at low temperatures indicate that the effective mobility μ_{eff} increases with the density of surface carriers N_s in the low N_s region. Fang and Fowler [1] assumed that the increase in μ_{eff} with N_s arises from Coulomb scattering of carriers in the inversion layer caused by the charged scattering centers, such as (i) fixed oxide charges in SiO_2 near the Si-SiO_2 interface, (ii) charged interface states and (iii) charged donors and acceptors in the depletion layer. They pointed out that the observed increase in mobility is too steep, however, to be explained by the k_F-dependence of screened Coulomb scattering in two-dimensional (2D) systems.

The present author and Kawaji [2] proposed a variable screening effect on 2D Coulomb scattering due to a tailing at the 2D subband edge, and interpreted the behaviour of μ_{eff} in the low N_s region in terms of the width of the band tailing σ and the effective density of Coulomb centers at the Si-SiO_2 interface, N_{coul}. They also showed that the width of the band tailing reflects the extent of microscopic disorder in the interface and correlates with N_{coul} such that $\sigma \propto N_{coul}{}^{\gamma}$, where $\gamma \cong 0.5$. The present paper reports results of further investigation of the relation between N_{coul} and the interface state density.

2. Experimental Results

2.1 Fabrication of Samples

Samples were fabricated on (100) surfaces of p-type Si substrates doped with boron atoms to a concentration $N_A \cong 2 \times 10^{14} cm^{-3}$. The boron concentration is so low that the effect of charged acceptors in the depletion layer is very small.

Two groups of samples were fabricated: Group A and Group B. Samples in Group A have a layer of 700 Å thick Si_3N_4 on a ~1000 Å thick SiO_2 layer. Si_3N_4 films are deposited by the chemical reaction of SiH_2Cl_2 and NH_3 at 700°C. One of the samples is annealed in N_2 gas at 1000°C after the deposition of Si_3N_4 film on thermally grown SiO_2 film, and the other is not annealed. Samples in Group B have a single layer of thermally grown SiO_2 of ~1000 Å thick. One of the Group A samples is subjected to post-oxidation annealing in N_2 gas at 1000°C and the other is not. Table I summarizes the fabrication process for each group of samples.

2.2 Flat-Band Voltage and Interface States Density

Flat-band voltages (V_{FB}) at room temperature are shown in Table I. As a simple and practical method to evaluate interface state density $N_{ss}(E)$ $(cm^{-2}eV^{-1})$, we employ the shift in the threshold voltage of conductivity between room temperature and 77K. This is similar to the method of Gray and Brown {3}. Thus an integrated interface state density defined by

$$\bar{N}_{ss} = [\int_{-\infty}^{\infty} N_{ss}(E)f(E,T)dE]_{77K} - [\int_{-\infty}^{\infty} N_{ss}(E)f(E,T)dE]_{RT} \cong \int_{E_F(RT)}^{E_F(77K)} N_{ss}(E)dE$$

will be used in the present paper, where $f(E,T)$ is the Fermi distribution function and $E_F(RT)$ and $E_F(77K)$ are the surface Fermi levels at which the band bending equals twice the bulk Fermi potential at room temperature and 77K, respectively. We can calculate that $E_F(RT) = 0.83eV$ and $E_F(77K) = 1.08eV$ relative to the valence band edge for $N_A=2\times10^{14}cm^{-3}$. The values of $\bar{N}_{ss}$ thus obtained are listed in Table I.

Table I

Sample #	Post-Oxidation Annealing (1000°C)	Post-Deposition Annealing (1000°C)	V_{FB} (V)	$\bar{N}_{ss}$ (10^{10} cm^{-2})	N_{coul} (10^{10} cm^{-2})
Group A					
K521	yes	yes	−1.9	73	90
K511	yes	no	−2.0	2.5	6.5
Group B					
6825	no	—	−2.0	6.8	20
6861	yes	—	−1.1	2.8	7.0

2.3 Effective Mobility at Low Temperatures

Experimental log μ_{eff} versus log N_s curves at 4.2K are shown in Fig. 1 for samples in Group A and in Fig. 2 for samples in Group B. Here μ_{eff} is determined by the conductivity $G(=eN_s\mu_{eff})$ where N_s is given by $N_s=C_{ox}(V_g-V_t)/e$. Threshold voltage V_t at 4.2K is determined by the conductivity threshold voltage at 77K. Our model {2}, which takes into account the variable screening effect caused by the band tailing by employing appropriate parameters σ and N_{coul}, can accurately predict the behaviour of μ_{eff} for respective samples. Table I lists the values of N_{coul} employed.

3. Discussion

3.1 Effect of Acceptor-Like Interface States

Group A samples indicate that the as-deposited Si$_3$N$_4$ film on SiO$_2$ has no appreciable effect on $\bar{N}_{SS}$ and N_{coul}, but subsequent high temperature annealing (post-deposition annealing) results in a great increase in these numbers. The increment in $\bar{N}_{SS}$ and the corresponding increment in N_{coul} yield $\Delta\bar{N}_{SS} = 7.1\times10^{11}$cm^{-2} and $\Delta N_{coul}=8.4\times10^{11}$cm^{-2}, respectively. Thus we have $\Delta\bar{N}_{SS}\cong\Delta N_{coul}$, which implies that μ_{eff} of sample K 521 is lowered by the Coulomb scattering caused by the charged interface states. In addition, $\Delta\bar{N}_{SS}$ originates from acceptor-like interface states, since V_{FB} of sample K521 shifts little from that of the reference sample K511; that is, $\Delta\bar{N}_{SS}$ is neutral at flat-band condition and negatively charged as E_F approaches the conduction band edge. The origin of the acceptor-like states observed in the annealed sample with double layers of insulating films (SiO$_2$-Si$_3$N$_4$) might be broken or distorted bonds at the Si-SiO$_2$ interface induced by compressive stress in Si$_3$N$_4$ films {4,5} when the films are annealed at temperatures higher than deposition temperatures. This is not inconsistent with a recent theoretical study of Laughlin, Joannopoulos and Chadi {6} which shows that bond-angle disorder, dangling Si bonds or Si-Si bonds in SiO$_2$ near the interface introduce a band tailing or trap states near the conduction band edge. At this point, we have no knowledge whether there is any correlation between the stress-induced tailing and the width of the band tailing σ(σ=19meV for sample K521 and 5.2 meV for sample K511).

3.2 Effect of Donor-Like Bound States

The difference in V_{FB} between samples in Group B probably comes from the net change in positive fixed charges (ΔN_{ox}^{+}) near the interface caused by the post-oxidation annealing, as proposed by Deal {7}. We obtain $\Delta N_{ox}^{+} = (C_{ox}/e)\Delta V_{FB}=1.46\times10^{11}$cm^{-2}. A positive center at the interface gives rise to a donor-like

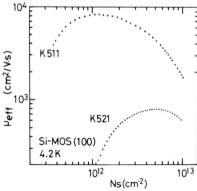

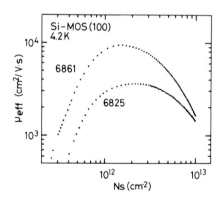

Fig. 1 Effective mobility versus electron density for samples in Group A at 4.2K (see Table I)

Fig. 2 Effective mobility versus electron density for samples in Group B at 4.2 K (see Table I)

bound state {8}. Thus, the difference in $\bar{N}_{ss}$ between sample 6825 and 6861, which equals $\Delta \bar{N}_{ss}$ of $4 \times 10^{10} cm^{-2}$, is considered to be same as the density of bound states produced by the corresponding ΔN_{ox}^{+}. Note that only a part of positive centers existing very close to the interface gives rise to bound states since $\Delta \bar{N}_{ss}$ is smaller than ΔN_{ox}^{+} by a factor of 3.7. Moreover, the difference in N_{coul} between samples of Group B ($\Delta N_{coul}= 1.3 \times 10^{11} cm^{-2}$) is close to the difference in N_{ox}^{+}. Because of this, the effect of the bound states on the Coulomb scattering can not be clearly observed in the present experiment. When a donor-like bound state binds an electron, it should act as a neutral scattering center whose scattering probability is small and independent of k_F {9}.

4. Summary

The effective density of Coulomb centers at Si-SiO$_2$ interface (N_{coul}) is compared with the interface state density integrated over the energy in the upper portion of the Si band gap. N_{coul} has been evaluated from d.c. conductivity at 4.2K by taking into account variable screening in two-dimensional systems.

It is found that, for the sample with a large density of acceptor like interface states, the Coulomb scattering which is proportional to N_{coul} is predominantly determined by the density of negatively charged interface states. For the sample with a large density of positive fixed centers near the interface, the Coulomb scattering is found to be predominantly determined by the density of positive fixed charges. The effect of donor-like bound states on the electron scattering can not be clearly observed.

Acknowledgements

I would like to thank my colleagues in particular M. Nakai for his help in computer calculations and M. Kamada for providing some samples. I am indebted to Professor S. Kawaji for his helpful discussion. I also thank T. Shimada and H. Yagi for their support of this work.

Reference

{1} F.F. Fang and A.B. Fowler: Phys.Rev. 169(1968) 619
{2} A. Yagi and S. Kawaji: Japan, J.Appl.Phys. 20 (1981)
{3} P.V. Gray and D.M. Brown: Appl.Phys.Lett. 8 (1966) 31
{4} S. Isomae, M. Nanba, Y. Takami and M. Maki: Appl.Phys.Lett 30(1977)564
{5} A.Bohg and A.K. Gaind: Appl.Phys.Lett.33(1978) 895
{6} R.B. Laughlin, J.D. Joannopoulos and D.J. Chadi: Phys. Rev. B21(1980) 5733
{7} B. Deal: J.Electrochem.Soc. 121(1974)198C
{8} F. Stern and W.E. Howard: Phys.Rev. 163(1967) 816
{9} A. Yagi and S. Kawaji: Appl.Phys.Lett. 33(1978) 349

Neutral Scattering Centers Near the Si/SiO$_2$-Interface of MOSFET Devices Prepared by TCE Oxidation

D. Kohl

2. Physikalisches Institut der RWTH Aachen, Sommerfeldstraße
D-5100 Aachen, Fed. Rep. of Germany

Abstract

Hall mobility at 13 K was studied on MOSFET devices. The SiO$_2$
layers were thermally grown in oxygen. Samples fabricated
show a constant difference in reciprocal mobility, Fig. 1 a.
For devices with a trichloroethylene admixture during oxidation
(TCE) a contribution to reciprocal mobility (increasing
linearly with electron density) was found, Fig. 1 b and c.

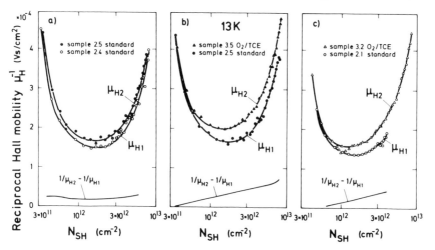

Fig. 1 Reciprocal Hall mobility and differences of reciprocal
Hall mobility as a function of the electron density [2].

The first case can be ascribed to homogeneously distributed
neutral scattering centers [1], whereas the TCE devices should
contain additional neutral scattering centers located within
2 nm from the interface [2]. The scattering behaviour of
neutral centers is calculated assuming different sizes. A
linear response theory orginally developed for bulk centers

[3] is applied to a box potential. It turns out that centers with a diameter above 1 nm are strongly screened by electrons of the inversion layer, Fig. 2. This predicts a decrease of reciprocal mobility with increasing electron density. The observations are explained with the assumption of neutral centers less than 1 nm in size. The scattering cross section of such small centers grows fast with the diameter. Therefore an estimate of the center density is not reasonable without knowledge about their nature.

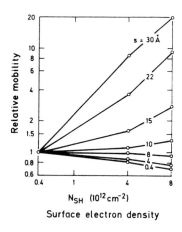

Surface electron density

Fig. 2 Calculated relative mobility for scattering at neutral centers (box potential), normalized to 1 for $N_{SH}=4 \times 10^{11} cm^{-2}$. The diameter s of the potential is given as parameter. The values are calculated only at the marked points.

References:
[1] A. Yagi und S. Kawaji, Appl. Phys. Lett. 33 (1978) 349; Solid-State Electronics 22 (1979) 261.
[2] D. Kohl, D. Becker, G. Heiland, U. Niggebrügge and P. Balk, J. Physics C, 14 (1981), 553.
[3] E. Gerlach and P. Grosse, Festkörperprobleme 17 (1977),157.

Part X

Films on Compound Semiconductors

Native Oxide Reactions on III-V Compound Semiconductors

G.P. Schwartz

Bell Laboratories
Murray Hill, NJ 07974, USA

1. Introduction

Thermally grown native oxide films on materials such as GaAs
[1,2], GaSb [2], InAs [2], InSb [3,4], and InP [5] have been
found to contain elemental inclusions of the respective Group V
metalloids As, Sb, or P which tends to limit the utilization of
these films in a variety of device-related applications. Recent
studies of native oxides grown via electrochemical or plasma
oxidation techniques suggest that these metalloid deposits can
be substantially reduced or eliminated under specific growth
conditions [6-8]. However, fundamental thermodynamic considerations implicit in the condensed phase form of the respective
III-V-O phase diagrams indicate that the films which result from
anodic or plasma oxidation are not necessarily stable to subsequent chemical reactions under conditions associated with thermal
processing or aging.

Two general classes of reaction are anticipated. The first
class involves interfacial reactions which occur between specific
Group V oxides and their respective substrates. Reactions of
this type evolve deposits of the Group V metalloids in native
oxide films which were initially free of these species. The
second general class involves reactions between the oxide constituents within these films to generate new oxide products.

Evidence for both reaction classes are reviewed from available
data pertaining to the identification of thermally induced solid
state reactions in native oxides on III-V substrates. Thermally
induced oxide-substrate reactions are illustrated for the cases
of anodic films on GaAs, (Al,Ga)As, InAs, and GaSb. Data for
rf and dc plasma oxidized films on GaAs are also presented. Recent estimates of the In-P-O and Ga-P-O phase diagrams taken in
conjunction with thermal aging studies of anodic films on InP
and GaP suggest that oxide-oxide reactions may predominate on
these substrates. The general reaction patterns observed on the
arsenides, antimonides, and phosphides will be shown to be consistent with recent estimates of their respective condensed phase
ternary (III-V-O) diagrams.

2. Experimental

Surface reflection Raman scattering has been the primary technique used to identify the presence and growth of elemental inclusions of As, Sb, and P. Spectra were obtained using 5145 Å excitation from an Ar^+ ion laser. An I.S.A. monochromator (Model HG-2S) equipped with holographic gratings and f/1.8 collection optics was used to analyze the scattered light. For the work reported here, the electric vector of the incident light was polarized in the scattering plane (H), and the polarization of the Raman scattered light was unanalyzed (U). This combination of incident and scattered electric field vectors is designated as (HU).

Samples were prepared by polishing using bromine-methanol and were then electrochemically anodized at room temperature in an ethylene glycol based electrolyte. This electrolyte consisted of 3% H_3PO_4 adjusted to a pH of 6.2 with NH_4OH; this solution was then diluted in a 1:2 volume ratio with ethylene glycol. Anodizations were carried out under constant current conditions (typically 1 mA/cm^2) up to voltages of 50-125V. Thermal aging was performed in evacuated (5×10^{-7} torr) quartz ampules.

Estimates of the condensed phase portions of the respective III-V-O ternary phase diagrams were constructed through a combination of thermodynamic calculations and binary mixture reaction experiments. Details of these experiments appear in [9] and [10].

3. Results and Discussion

3.1 General

In the last several years Raman scattering has been used rather extensively for detecting elemental deposits of As, Sb, or P in native oxide films on III-V semiconductors. Its principal advantages lie in its non-destructive application and its high detection sensitivity [10,11] for the Group V metalloids (20-50 Å deposits can be detected). The detection sensitivity for these materials is strongly enhanced by resonance Raman processes [12] which occur for materials which are optically absorbing at the excitation wavelength. On the other hand, the oxide constituents in these films are transparent at 5145 Å, and their normal Raman cross sections are usually insufficient to allow the detection of these species unless very thick films are grown or special waveguiding techniques [13] are employed. It should also be noted that the spatial location of the metalloid deposits within the transparent oxide matrix cannot be specified. In order to obtain this information, controlled film·removal is necessary.

3.2 Oxide-Substrate Reactions

For electrochemically anodized films on the gallium and indium based arsenides and antimonides, currently available data suggest oxide compositions composed primarily of (Ga_2O_3, As_2O_3), (Ga_2O_3, Sb_2O_3), (In_2O_3,As_2O_3), and (In_2O_3,Sb_2O_3) mixtures respectively [14]. These mixtures are not necessarily found in a 1:1 ratio due to selective dissolution into the electrolyte. Traces of either elemental As or Sb can also be present depending on the choice of certain growth parameters such as current density. For plasma oxidized GaAs, the major film components are also Ga_2O_3

271

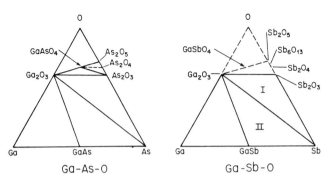

Fig.1 Ternary phase diagrams for the Ga-As-O and Ga-Sb-O systems. The upper region of the latter diagram has not been fully determined; the presence of As_2O_4 on the Ga-As-O diagram is open to question.

and As_2O_3 [7] with elemental arsenic typically present as a minor constituent [7,8].

The specification of the oxide components in a particular film allows one to utilize the appropriate phase diagram in order to predict which oxide-substrate reactions would be expected to occur. For example, consider the Ga-As-O and Ga-Sb-O diagrams in Fig.1 [9,10]. Oxide phases which are connected to either GaAs or GaSb by a tie line will not react with those materials. One notes that neither As_2O_3 nor Sb_2O_3 are connected to their respective substrates with a tie line; both are therefore expected to be subject to interfacial oxide-substrate reactions according to (1) and (2) below.

$$As_2O_3 + 2GaAs \rightarrow Ga_2O_3 + 4As \qquad .(1)$$

$$Sb_2O_3 + 2GaSb \rightarrow Ga_2O_3 + 4Sb \qquad (2)$$

These equations represent a mass balance expression for the crossing point between the stable tie lines (Ga_2O_3,As), (Ga_2O_3,Sb) and the initial reactants $(As_2O_3,GaAs)$, $(Sb_2O_3,GaSb)$. For the typical situation in which one has a thin anodic film on the base substrate (GaAs for example), there is always an excess of the substrate relative to the quantity of the Group V oxide present, and the final equilibrium phase field will be that bordered by GaAs, Ga_2O_3, and As assuming that the reaction (1) has gone to completion. Considerably more complex reaction patterns are also possible if the as-grown films contain some of the other possible oxide phases. For the In-based systems, data for the In-As-O [15] and In-Sb-O [16] diagrams suggest that the form of the phase fields in the lower portion of the diagrams are similar to their gallium counterparts so that similar interfacial reactions with InAs and InSb will be expected analogous to (1) and (2).

The process of interfacial oxide-substrate reaction can be monitored using Raman scattering as the native oxide-substrate

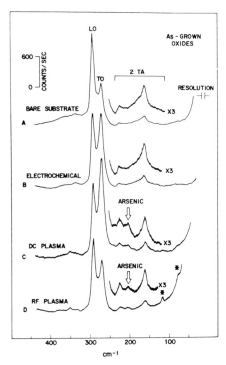

Fig.2 Raman spectra from as-grown anodic and plasma films on GaAs. Asterisks denote laser plasma lines

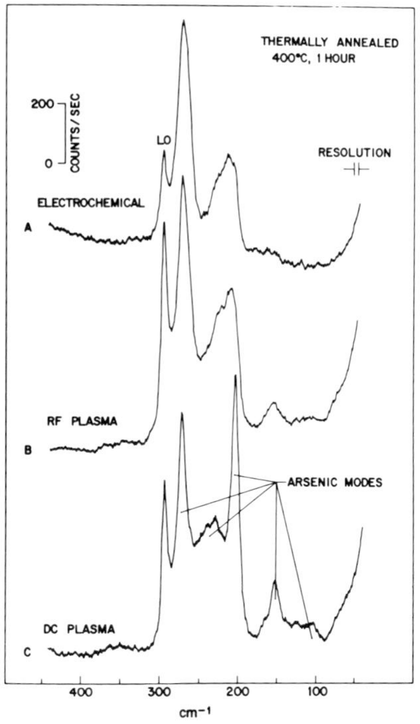

Fig.3 Raman spectra from thermally aged anodic and plasma grown films on GaAs. Modes associated with elemental arsenic are labeled

system is driven towards thermodynamic equilibrium by thermal aging. Figures 2 and 3 illustrate the process in 1000 Å anodic and plasma films on GaAs. Raman scattering from the substrate (Fig.2A) displays contributions from longitudinal and transverse optic (LO,TO) modes at 292 and 269 cm^{-1}, respectively, as well as two-phonon scattering from transverse acoustic (2TA) modes at lower frequencies. Crystalline arsenic has sharp A_{1g} and E_g modes at 257 and 195 cm^{-1} [17], whereas amorphous As (a-As) shows a broad band centered between 220-240 cm^{-1} and weaker structure below 150 cm^{-1} [18,19]. As can be seen in Fig.2, in the as-grown state the electrochemically anodized film shows little if any evidence for elemental arsenic inclusions, whereas the plasma grown films appear to contain a few (∼4) volume percent of this metalloid. After thermal aging at 400°C for 1 hour however, both amorphous and crystalline allotropes of arsenic are detected in both sets of films. Although the reaction and crystallization rates for these films vary somewhat, all three films are incipiently unstable to thermal aging due to the presence of As_2O_3 and its reaction with the underlying substrate according to (1).

Similar observations have recently been reported for the ternary alloy $Al_xGa_{1-x}As$ (x=0.25) [20]. The phonon mode structure

273

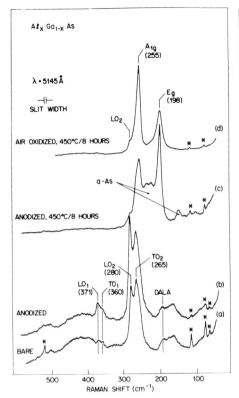

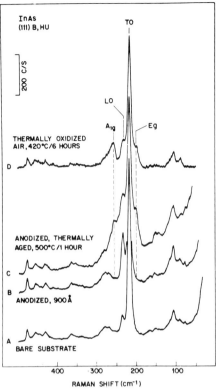

Fig.4 Raman spectra from samples
of Al$_x$Ga$_{1-x}$As with anodic and
thermally grown native oxide
films

Fig.5 Raman spectra from InAs
samples with anodic (B,C) and
thermally oxidized (D) films

of the substrate is characterized by two-mode behavior (Fig. 4)
in which separate AlAs-like (LO$_1$,TO$_1$) and GaAs-like (LO$_2$,TO$_2$)
branches develop. A defect activated longitudinal acoustic (DALA)
mode has also been identified in these alloys [21]. The Raman
spectra of the bare substrate and substrate plus 1000 Å anodic
film (Fig.4) provide no indication for the presence of elemental
arsenic in the as-grown film. Once the film is aged at 450°C,
however, the A$_{1g}$ and E$_g$ modes of crystalline arsenic become readily apparent. As mentioned previously, the direct thermal oxidation leads directly to the thermodynamically stable interfacial
phases consisting of Ga$_2$O$_3$, Al$_2$O$_3$, and elemental As. The exclu-
sion of the latter product in the as-grown condition in anodic
and plasma generated films reflects the fact that oxide growth
has occurred under conditions far from thermodynamic equilibrium.
This result is not surprising in view of the rapid film growth
rates which can be achieved with the latter growth techniques.

Preliminary data for InAs [15] are presented in Fig.5. As
observed previously for the gallium compounds, there is little

274

evidence for the incorporation of elemental arsenic inclusions in the film in the as-anodized condition. Thermal aging at 420°C is seen to generate arsenic, although it should be noted that the detected quantities are typically much less than what was seen with GaAs. We suspect that the composition of the anodic film contains much less As_2O_3 and we are currently investigating the role of current density and electrolyte pH and composition in order to see how the film composition is influenced. Consistent with the data reported by FARROW et al. [2], thermal oxidation leads directly to elemental arsenic inclusions as anticipated from the form of the In-As-O phase diagram. For oxidation at 520°C, these inclusions are sufficiently thick (250 A) to totally absorb the excitation beam and mask the underlying substrate signal.

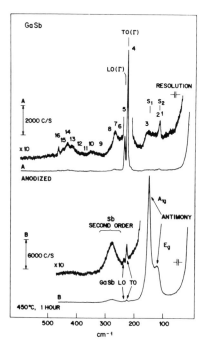 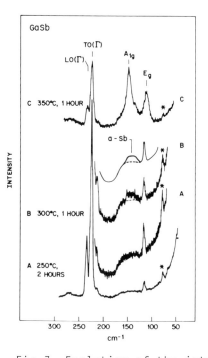

Fig.6 Raman scattering from as-anodized and thermally aged films on GaSb. S_1 and S_2 are crystalline Sb modes; the numbered (1-16) features correspond to substrate scattering

Fig.7 Evolution of the interfacial reaction. Initially amorphous Sb is detected; for more elevated reaction temperatures the crystalline A_{1g} and E_g modes develop

Both gallium and indium antimonide are also subject to interfacial reactions involving Sb_2O_3 (in anodic films). Figures 6 and 7 demonstrate the reaction (2) on GaSb and the facility with which it proceeds at temperatures on the order of 300°C. Raman spectra of the bare substrate (not shown) and the anodized (1000 Å film) substrate are essentially indistinguishable. Suffi-

cient crystalline Sb is generated by thermally aging the film
at 450°C that optical absorption of the excitation beam is nearly
complete and the substrate LO and TO can barely be distinguished.
One notes in Fig.7 that the initial low temperature thermal aging
has produced amorphous Sb; distinct crystalline modes are observed
at somewhat higher (350°C) temperatures. This is analogous to
what has been observed during the low temperature (300-366°C)
thermal aging of anodic films on GaAs [22], i.e. the initial reaction evolves the amorphous product which can then crystallize.

Within the last year, NAKAGAWA et al. have reported on thermally induced reactions in anodic films on InSb [4]. In that
system the evolution of elemental Sb is readily detected after
thermal aging of the film at 260°C for 10 minutes. The facility
with which these oxide-substrate reactions occur on the antimonides
should be carefully considered in any device-related application
which proposes to utilize either anodic or plasma generated native oxide films on these materials.

3.3   Oxide-Oxide Reactions

The identification of the oxide components in anodic films on
InP and GaP is still an open question. On InP, In_2O_3 and P_2O_5
have been suggested based on photoemission studies [23]. The
In/P ratio within these films changes as a function of electrolyte pH [24], thereby ruling out certain single phase products
such as $InPO_4$ and $In(PO_3)_3$. In general, the In/P ratio is
greater than one, so that mixtures of the above products would
also be excluded. Anodic films on InP are also known to rapidly
getter water from the atmosphere [25] which would be consistent
with the presence of a hygroscopic material such as P_2O_5. For
GaP, Rutherford backscattering has been used to analyze the
Ga/P ratio within the film. As with InP, this ratio is variable
and dependent on the electrolyte, pH, and current density. At
the present time, Ga_2O_3 and P_2O_5 are implicitly considered to
be the most likely film products.

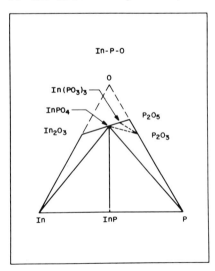

Fig.8 Estimate of the In-P-O
phase diagram; P_2O_3 and $In(PO_3)_3$
are questionable but cannot be
rigorously excluded based on the
available data.

276

Figure 8 shows a recent estimate of the In-P-O phase diagram. For an anodic film composed of In_2O_3 and P_2O_5, one notes that in addition to possible oxide substrate reactions, two intra-film oxide-oxide reactions given by (3) and (4) are also possible.

$$In_2O_3 + P_2O_5 \,--\, 2InPO_4 \qquad\qquad (3)$$

$$In_2O_3 + 3P_2O_5 \,\rightarrow\, 2In(PO_3)_3 \qquad\qquad (4)$$

Only one of these products ($InPO_4$) is connected to InP by a tie line; the other product in principal should also react with the substrate. Reaction of P_2O_5 with an excess of InP should yield products which lie within the $InP-InPO_4-P$ phase field, i.e. elemental phosphorus would be generated if the oxide-substrate reaction dominates over the depletion of P_2O_5 via direct volatilization or consumption according to reactions (3) and (4). However, kinetic rather than thermodynamic considerations may play the dominant role in these systems at least in the temperature range likely to be accessed for device-related applications. For example, binary mixture experiments have indicated that reactions (3) and (4) proceed between 650-700°C, whereas reactions between either In_2O_3 or P_2O_5 and InP require higher temperatures. Even this information is difficult to acquire due to the thermal decomposition of InP at elevated temperatures.

The basic difficulty in studying or observing intrafilm oxide-oxide reactions using Raman scattering is one of detection sensitivity. Only part of the problem is related to the cross section for Raman scattering of the oxide species. Other factors should also be considered. For instance, oxide films of arbitrary thickness cannot be grown due to breakdown at high applied anodization voltages. Also, the composition of as-grown films is generally depleted in P_2O_5 due to selective dissolution into the electrolyte. In addition, the high temperatures which are necessary to drive either the oxide-substrate or oxide-oxide reactions are sufficient to produce direct volatilization of P_2O_5 from these native oxides.

Raman data for GaP are shown in Fig.9. Panels A and B show the bare substrate and substrate with an anodic film of roughly 1500 Å. Within our signal-to-noise, we can find no evidence for elemental phosphorus or either of the oxide products Ga_2O_3 or P_2O_5. Studies of the low temperature thermal oxidation of InP [26] have demonstrated the low that elemental phosphorus can be readily detected if present; there is essentially no evidence for this product in the as-grown anodic films. Thermal aging at 850°C for 6 hours (Fig.9C) similarly shows no evidence for $GaPO_4$ which we know forms at comparable temperatures in the gallium analogue of reaction (3). $GaPO_4$ can be detected, however, after thermal (air) oxidation of GaP at 850°C for 2 days or after 950°C for 15 hours (panels D and E). The Raman spectrum of pure $GaPO_4$ is shown in panel F to indicate the position of the strongest Raman active modes. The identification of $GaPO_4$ in thermally oxidized material is consistent with the presence of a $GaPO_4-GaP$ tie line in the phase diagram and has been identified previously at high oxidation temperatures [27] using x-ray dif-

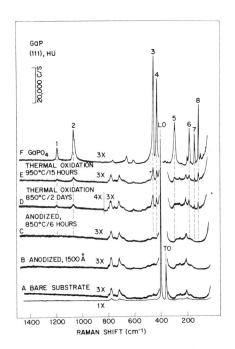

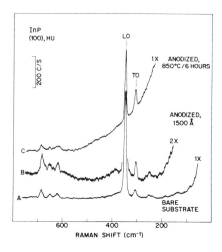

Fig.9 Raman spectra for (A) bare GaP, (B) anodized, (C) anodic film aged at 850°C, and (D,E) air oxidized films at 850 and 950°C. The top panel (F) shows the spectrum for pure $GaPO_4$

Fig.10 Raman spectra of (A) bare InP and with (B) as-grown and (C) thermally aged anodic films

fraction and at lower temperatures by x-ray photoemission [28]. No evidence for either oxide-substrate (P generation) or oxide-oxide ($GaPO_4$ formation) reactions is apparent, however, in the thermally aged anodic film spectra.

A similar situation has been observed for anodic films on InP as seen in Fig.10. No substantial differences are observed between the bare and anodized spectra (panels A and B) except for a weak fluorescence tail present in the anodized spectrum which is believed to result from impurities in the electrolyte. Thermal aging of the film to 620°C (Fig.10C) does not produce any identifiable signal associated with the Raman bands of either $InPO_4$ or elemental phosphorus. Although it appears that at the present time reactions within anodic films cannot be easily identified using Raman scattering with excitation sources in the visible, it may prove possible in the future to examine these reactions using uv or deep uv sources.

4. Summary

Oxide-substrate reactions in thermally aged anodic films on Ga and In based arsenides and antimonides have been observed using surface reflection Raman scattering to detect the evolution of elemental As and Sb. The reactions are consistent with the form of the respective III-V-O phase diagrams and data concerning the initial composition of the native oxide films. Considerations of the In-P-O and Ga-P-O diagrams suggest that both oxide-substrate and intrafilm oxide-oxide reactions should occur in thermally aged anodic films on InP and GaP. Kinetic considerations and detection sensitivity limitations have so far hindered the direct verification of oxide-oxide reactions in thin film structures.

Acknowledgments

It is a pleasure to acknowledge J. E. Griffiths, G. J. Gualtieri, C. D. Thurmond, B. Schwartz, and W. A. Sunder of Bell Laboratories and Takuo Sugano of the University of Japan for their active contributions and collaboration on the studies reported in this text.

References

1. J.A.Cape, W.E.Tennant, L.G.Hale: J. Vac. Sci. Technol. 14, 921 (1977)
2. R.L.Farrow, R.K.Chang, S.Mroczkowski, F.H.Pollak: Appl. Phys. Lett. 31, 768 (1977)
3. A.J.Rosenberg, M.C.Lavine: J. Phys. Chem. 64, 1135 (1960)
4. T.Nakagawa, K.Ohta, N.Koshizuka: Jpn. J. Appl. Phys. 19, L339 (1980)
5. G.P.Schwartz, W.A.Sunder, J.E.Griffiths: Appl. Phys. Lett. 37, 925 (1980)
6. D.E.Aspnes, G.P.Schwartz, G.J. Gualtieri, A.A.Studna, B. Schwartz: J. Electrochem. Soc., in press
7. G.P.Schwartz, B.Schwartz, J.E.Griffiths, T.Sugano: J. Electrochem. Soc. 127, 2269 (1980)
8. J.B.Theeten, R.P.H.Chang, D.E.Aspnes, T.E.Adams: J. Electrochem. Soc. 127, 378 (1980)
9. C.D.Thurmond, G.P.Schwartz, G.W.Kammlott, B.Schwartz: J. Electrochem. Soc. 127, 1366 (1980)
10. G.P.Schwartz, G.J.Gualtieri, J.E.Griffiths, C.D.Thurmond, B.Schwartz: J. Electrochem. Soc. 127, 2488 (1980)
11. G.P.Schwartz, J.E.Griffiths, B.Schwartz: J. Vac. Sci. Technol. 16, 1383 (1979)
12. For a review of resonance Raman scattering see R.M.Martin, L.M. Falicov: In Light Scattering in Solids,ed. by M.Cardona, Topics in Applied Physics,Vol.8 (Springer, Berlin, Heidelberg, New York 1975)
13. F.L.Galeener, W.Stutius, G.T.McKinley: Proc. Intern. Topical Conf. Phys. MOS Insulators, ed. G.Lucovsky, S.T.Pantelides, F.L.Galeener (Pergamon Press, New York, 1980) p.77
14. For a review see C.W.Wilmsen, S.Szpak: Thin Solid Films 46, 17 (1977)

15. G.P.Schwartz, W.A.Sunder, J.E.Griffiths: unpublished
16. T.P.Smirnova, A.N.Golubenko, N.F.Zacharchuk, V.I.Belyi
    G.A.Kokovin, N.A.Valisheva: Thin Solid Films $\underline{76}$, 11 (1981)
17. R.N.Zitter: <u>The Physics of Semi-metals and Narrow Gap
    Semiconductors</u>, ed. D.L.Carter, R.T.Bate (Pergamon Press,
    New York, 1971) p.285
18. J.S.Lannin: Phys. Rev. B $\underline{15}$, 3863 (1977)
19. R.J.Nemanich, G.Lucovsky, W.Pollard, J.D.Joannopoulos:
    Solid State Commun. $\underline{26}$, 137 (1978)
20. G.P.Schwartz, B.V.Dutt, G.J.Gualtieri: unpublished
21. R.Tsu, H.Kawamura, L.Esaki: <u>Proc. Intern. Conf. Phys. Semi-
    cond.</u>, Warsaw, 1972 (Elsevier Press, Amsterdam, 1972) p.1135
22. G.P.Schwartz, G.J.Gualtieri, J.E.Griffiths, B.Schwartz: J.
    Electrochem. Soc. $\underline{128}$, 410 (1981)
23. C.W.Wilmsen, R.W.Kee: J. Vac. Sci. Technol. $\underline{15}$, 1513 (1978)
24. C.W.Wilmsen: Workshop for Dielectrics on III-V Materials
    (San Diego, 1980) unpublished results
25. A.A.Studna, G.J.Gualtieri: unpublished
26. G.P.Schwartz, W.A.Sunder, J.E.Griffiths: Appl. Phys. Lett.
    37, $\underline{925}$ (1980)
27. M.Rubenstein: J. Electrochem. Soc. $\underline{113}$, 540 (1966)
28. Y.Mizokawa, H.Iwasaki, R.Nishitani, S.Nakamura: Jpn. J.
    Appl. Phys. $\underline{17}$, 327 (1978)

MISFET and MIS Diode Behaviour of Some Insulator-InP Systems

D.C. Cameron, L.D. Irving, G.R. Jones, and J. Woodward

RSRE, St. Andrews Road
Malvern, Worcestershire, United Kingdom

1. Introduction

Recently there have been an increasing number of publications describing the operation of n-channel inversion mode InP MISFETS [1-4]. Nevertheless there remains a considerable lack in understanding of the interfacial phenomena which control the device characteristics. This paper reports a study of three dielectrics deposited on to InP, and attempts to correlate the performance of transistors with interface parameters deduced from capacitance and conductance measurements on MIS diodes.

2. Insulator Deposition

Silicon nitride was deposited by a plasma enhanced CVD technique using an inductively coupled r.f. plasma system operating at 400 kHz. An optimised deposition process was found to require a silane and ammonia gas mixture (volume ratio of NH_3 to SiH_4 was 200 to 1), a substrate temperature of between 350 and 400°C, and a plasma power of less than 50 watts. This procedure gave insulators with a resistivity of 10^{16} ohm cm, a dielectric constant of 7 with no dispersion over the frequency range 1 to 10^6 Hz, and a breakdown field of about $5 \times 10^6 Vcm^{-1}$.

The plasma enhanced decomposition of tetraethoxysilane in an O_2 atmosphere was used to deposit silicon dioxide. The plasma was excited by a 400kHz r.f. generator capacitively coupled to the deposition system. The best bulk dielectric properties have been achieved with plasma powers of less than 50 watts and with a substrate heated to 300°C. The insulator resistivity has been found to be 5×10^{14} ohm cm, with a dielectric constant of 5.5 at 10^6Hz increasing by about 5% as the frequency was reduced to 1Hz. Typical breakdown fields of $5 \times 10^6 Vcm^{-1}$ have been observed for these silicon dioxide films.

Aluminium oxide films were produced by the pyrolytic decomposition of aluminium isopropoxide on a heated substrate in a low pressure N_2 atmosphere. This method has produced glassy films for substrate temperatures between 300 and 450°C. At 400°C the films were found to have a resistivity of 5×10^{12}ohm cm and a breakdown field of $2 \times 10^6 Vcm^{-1}$. The dielectric constant was 8.5 at 1MHz, but increased by up to 25% as the frequency was reduced to 1Hz.

3. MIS Diode Studies

The insulators were deposited on bulk p-type substrates of InP with carrier densities ranging from 1×10^{15} to $5 \times 10^{16} cm^{-3}$. Alloyed Ni/Zn/Au was used as a back contact to the samples with a circular gate electrode of Al (0.5mm diameter) defined on the dielectric by shadow mask techniques. Typically dielectric layers 100nm thick were used in this work. Parallel capacitance and conductance measurements were made simultaneously using a small signal a.c. field applied during a slow bias voltage sweep in the usual manner.

The C/V curve shapes from the three dielectrics were readily separated into two types. Silicon nitride diodes reproducibly showed features attributable to accumulation and depletion, but did not show an increase in capacitance for positive biases up to the breakdown strength. A number of variations to the basic deposition process were examined including surface treatments both prior to and after loading into the deposition chamber. However the gross features outlined above were largely unaltered. Modelling studies of the shape of the C/V curves indicated that a shift in surface potential from the valence band to just beyond mid-gap was the maximum change which could be induced for a silicon nitride MIS structure.

In contrast both oxides showed markedly different C/V curves to those displayed by silicon nitride. Silicon dioxide MIS diodes have been found to exhibit an increase in capacitance to the value of the oxide capacitance at around zero bias for frequencies less than about 11 Hz indicating the onset of strong surface inversion. Under negative bias the device shows depletion but not strong accumulation. Structure is also observed in plots of parallel conductance against both voltage and frequency (Fig.1). Surface state densities (N_{SS}) calculated from the latter peaks indicate values in the range 10^{11} to 10^{12} states $cm^{-2}eV^{-1}$ with time constants around 0.2secs. This value of N_{SS} is comparable to that reported by other workers on n-type material using the same measurement technique.

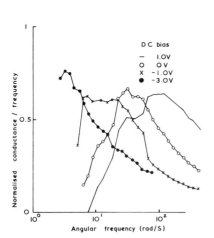

Fig.1 G_p/ω against ω plot for Al-SiO$_2$-pInP diode

However the time constant of these states differs by many orders of magnitude from those reported by other workers [5,6]. In order to make a complete calculation of the N_{SS} variation across the InP bandgap, the relationship between applied voltage and surface potential must be established. The more reliable techniques for obtaining this relationship require a classical low frequency C/V curve. Because of the lack of such a curve for silicon dioxide diodes, a full N_{SS} analysis has not been possible to date.

Aluminium oxide films have produced the classical C/V characteristics which have allowed a complete analysis by the Berglund technique of the N_{SS} variation throughout the bandgap (Fig.2). The range of N_{SS} values obtained by this method is in reasonable agreement with that obtained from peaks in the conductance-frequency curves. This technique indicates N_{SS} values between 10^{11} and 10^{12} states cm^{-2} eV^{-1} in the weak inversion region. The time constants of these traps lie in the same range as those in the silicon dioxide structures, namely 0.2 to 1 sec.

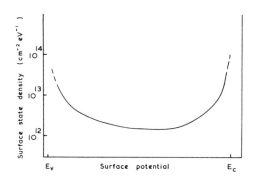

Fig. 2 Variation of surface state density with surface potential for Al_2O_3 - InP

4. MISFET Device Characteristics

In order to assess transistor performance, arrays of fairly large single FET devices have been fabricated on bulk p-type InP. Source and drain contact pads (200 by 300 microns) of alloyed Ni/Ge/Au were defined using standard optical photolithography and float off techniques. The channel length of these devices is 5 microns. An aluminium gate pad is finally defined on the dielectric surface again using photolithography. The low frequency device performance (measured on a curve tracer) has shown good correlation with the MIS diode studies reported in the previous section as shown in Table 1.

TABLE 1 MISFET Device Performance

Gate dielectric	Transconductance (mS/mm)	Channel Mobility $(cm^2V^{-1}sec^{-1})$	Inversion in C/V measurements
Si_3N_4	0.5	7	No
SiO_2	15	220	Yes
Al_2O_3	63	2100	Yes

The channel mobility in this table is calculated from the device characteristics in the saturation region. The best device result we have achieved to date is shown in Fig. 3. This result is comparable to the best InP MISFET performance previously reported in the literature. [1]

5. Conclusions

The formation of a surface inversion layer is an obvious prerequisite for the observation of FET action in a device. Hence the observed correlation between MIS diode inversion and improved device characteristics is not surprising. However such a correlation has not been clearly demonstrated for InP previously. Sufficient evidence is not available at this time to make any further deductions about the quality of the interface which is presumably reflected in the ultimate level of device performance. Samples with Al_2O_3 as the gate oxide have also shown some tendency to display a lack of accumulation for negative biases as shown by the SiO_2 samples. However it is not yet possible to relate this observation to device results. DLTS studies have provided supporting evidence for Fermi level pinning in the bandgap of the SiO_2

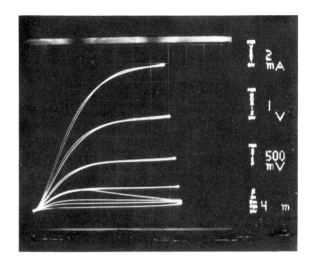

Fig. 3 I-V characteristics
of Al₂O₃-pInP MOSFET (maximum gate voltage = 2V)

Fig. 3 I-V characteristics of Al$_2$O$_3$-pInP MOSFET (maximum gate voltage = 2V)

samples which may prevent hole accumulation. Preliminary measurements on Al-SiO$_2$-nInP structures have identified two states, one at 0.3eV below the conduction band (approximate density 10^{12} states cm^{-2}eV^{-1}) and one at 0.6eV below the conduction band (10^{13} states cm^{-2}eV^{-1}) [7].

It is believed that the ultimate level of device performance and reproducibility with an oxide gate insulator has not yet been achieved. In contrast silicon nitride deposition would seem to involve some intrinsic feature which limits device performance. The most likely explanation of this result is that gross surface phosphorus depletion occurs during the initial stages of nitride deposition by the PECVD technique. This reaction would be caused by the large concentration of reactive hydrogen species present in the plasma causing phosphine formation. This conclusion is in agreement with other recent studies of this deposition process. [8]

References

1. T Kawakami, M Okamura: Electron.Letts. 15, 503 (1979)
2. D Fritsche: Inst.Phys.Conf.Ser. 50, 258 (1979)
3. D L Lile, D A Collins, L G Meiners, L Messick: Electron. Letts. 14, 657 (1978)
4. A J Grant, D C Cameron, L D Irving, C E Greenhalgh, P R Norton:
   Inst.Phys.Conf.Ser. 50, 266 (1979)
5. D Fritsche: Electron.Letts. 14, 266 (1978)
6. G G Roberts, K P Pande, W A Barlow: Solid State & Electron.Dev. 2, 169 (1978)
7. P Tapster (Private Communication)
8. L G Meiners: J.Vac.Sci.Technol. (to be published)

Plasma Anodised Alumina Films in GaAs and InP MIS Structures

W.S. Lee, C.V. Haynes, and J.G. Swanson

Department of Electronics, Chelsea College, University of London
London SW6 5PR, England

1. Introduction

The complete plasma oxidation of a metal layer on a semiconductor substrate offers a versatile way of forming metal
oxide layers at temperatures as low as 60°C {1}. The experimental procedures have been detailed previously {1,2}. The
method avoids the possibility of contamination from the electrolyte and dissolution during growth which can occur with 'wet'
anodisation.

2. The Formation of Alumina Layers on InP and GaAs

The aluminium layers were vacuum deposited and oxidised in situ.
The substrates were (100) chemically polished wafers which had
been free etched immediately before loading into the vacuum
system. The end-point of the metal oxidation was detected
during the anodisation by observing the sudden change in the
anodisation voltage which was necessary to sustain a constant
anodisation current density of 1.3mA.cm^{-2}. In the case of n-
type substrates the end-point was made more obvious by the
sensitivity to light flashes. The incident photons which had
passed through the oxidising structure were able to generate

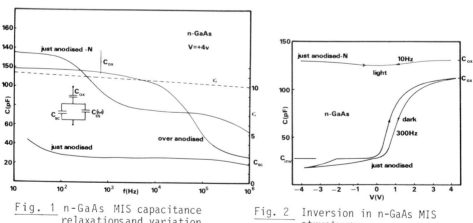

Fig. 1 n-GaAs MIS capacitance
relaxations and variation
of ε_r' for an MIM structure

Fig. 2 Inversion in n-GaAs MIS
structures

holes in the substrate which would otherwise have had to be
provided by impact ionisation, necessitating a slightly
greater anodisation voltage without light. The films had
thicknesses in the range 200Å to 1000Å and exhibited breakdown
electric fields of about $4 \times 10^6 V.cm^{-1}$. The electrical resist-
ivities were about $10^{14} \Omega.cm$ at a field of $1 \times 10^6 V.cm^{-1}$. The
variation of dielectric permittivity with frequency obtained
from MIM structures is shown in Fig. 1. It reduces with increasing frequency at about 2% per decade having a relative value
of 10 at 1MHz.

3. Electrical Properties of Alumina-GaAs MIS Structures

In this case the final chemical treatment of the GaAs surface
was a 4 min. etch in molar NaOH followed by a 1 min etch in
0.5% Br-methanol. The structures were completed by depositing
Al counter-electrodes through a mask although photolithographic
delineation with acidic and alkaline etches has been used
without degrading the alumina. The samples were then annealed
at 340°C in $90\%N_2$-$10\%H_2$ for 30 minutes. The C-V characteristics
of n-type structures show the usual frequency relaxation of
capacitance under positive bias {3}. Fig. 1 shows these
variations at a bias of +4V for samples which were just-anodised
and over-anodised, with and without nitrogen glow discharge
exposure of the GaAs before Al deposition.

Except for the just-anodised nitrogen treated sample for which
deep depletion was not observed it was possible to directly
deduce the substrate doping concentration. This was used to
calculate the potential at which the surface was pinned under
positive bias. In each case this occurred when the Fermi level
was 0.9eV + 0.1eV below the conduction band edge. The capacitance
plateau, Fig. 2, at the onset of deep depletion corresponded
to a Fermi level position within 0.2eV of the valence band edge
indicating that carrier inversion had been achieved. This was
confirmed by the tendency of the capacitance under negative
bias to approach the oxide value at low frequencies or under
illumination. This was particularly marked in the just-anodised
nitrogen treated samples which did not show deep depletion.

The frequency dependence of the device capacitance under positive
bias may be explained in terms of the equivalent circuit elements
in Fig. 1 in which the surface state capacitance, C_{ss}, is
frequency dependent. Since the surface potential is pinned
it follows that C_{sc} is fixed and that the observed frequency
dependence is caused by states at the pinning level. Surface
treatment of the GaAs surface before Al deposition has an important effect on this frequency dependence. In just-anodised
samples which have not been nitrogen treated the pinning states
respond predominantly below 100Hz. Over-anodisation always
results in extra states responding in the range 10Hz to 100kHz.

Although the curve for just-anodised nitrogen treated samples
indicates pinning states which respond up to 300Hz, it actually
conceals a more significant difference in the number of very
slow states. This is apparent on comparing the constant high
frequency capacitance voltage transients shown in Fig. 3. These

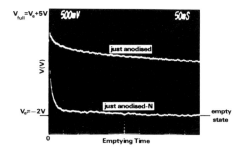

$V_{full}=V_o+5V$ 500mV 50mS

V(V)

just anodised

$V_o=-2V$ just anodised-N empty state

0 Emptying Time

Fig. 3 n-GaAs CCDLTS
transients

were recorded after pulsing the metal electrode to +3V for
20μs and measuring the voltage required to set the capacitance
to the value for the pinned surface measured at 20MHz. The trace
commences 20μs after removing the filling pulse. In the case
of the just-anodised nitrogen treated sample the number of
filled states at 20μs was $1.9 \times 10^{12} cm^{-2}$ compared with 3.3×10^{12}
cm^{-2} without nitrogen, but at 50 ms only $6 \times 10^{10} cm^{-2}$ states
remained filled in the former case compared with $2.8 \times 10^{12} cm^{-2}$
without nitrogen. The total number of electronic charges induced
by the filling pulse was $6 \times 10^{12} cm^{-2}$. It is not possible to
infer how many, if any, were accumulated in the conduction band
although 8×10^{11} electrons cm^{-2} would have been needed to fill
the depletion region if this had happened.

Measurements made on p-type MIS structures prepared in a similar
way did not show inversion. The behaviour was very similar to
that reported by HAYASHI {4} for electrolytically over-anodised
alumina-GaAs structures. In our experiments slight over-
anodisation was necessary to reveal the end-point as there was
no optical sensitivity in the case of p-type material.

4. Electrical Properties of Alumina-InP Structures

In the case of InP the final chemical treatment was a 30sec
etch in 0.5% Br-methanol. Fig. 4 shows typical C-V characteristics of just-anodised n-type structures after annealing as
above. Inversion is clearly apparent. The frequency dependence
of the capacitance under positive bias matches the variation

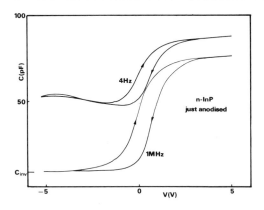

100

C(pF)

4Hz

n-InP
just anodised

50

1MHz

C_{inv}

−5   0   5
V(V)

Fig. 4 Inversion in n-InP
MIS structures

287

of insulator permittivity observed in the MIM structures and may be taken as an insulator effect. If pinning by surface states which respond at frequencies much higher than 1 MHz is discounted, then this oberservation would be consistent with majority carrier accumulation.

The behaviour is modified considerably by over-anodisation and contrasts with that of similarly prepared GaAs structures. Pronounced trapping hysteresis occurs and is exaggerated by illumination, see Fig. 5. The charge traps retain charge when the illumination is removed under negative bias, eventually discharging with time constants of tens of seconds. They are immediately emptied on re-applying a positive bias. Notice that the dark negative bias capacitance matches the inversion value closely.

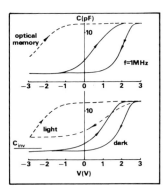 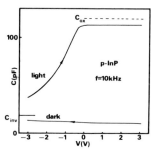

Fig. 5 Hysteresis due to over-anodisation on n-InP

Fig. 6 Inversion in p-InP MIS structures

Fig. 6 shows the C-V characteristic of a just-anodised p-type MIS structure. The insulator capacitance was inferred from measurements on other anodised structures since it could not be observed directly. In the dark the structure shows a constant capacitance close to the expected high frequency inversion value for positive bias. The value rises only slightly under negative bias and majority carrier accumulation could not be observed. On illumination the inversion capacitance rises to approach the expected oxide value confirming that minority carriers have accumulated and can now respond at the signal frequency.

5. Conclusions

All of the structures formed on n-GaAs showed pinning at a level 0.9eV ± 0.1eV below the conduction band edge. This agrees reasonably well with values for other oxide-GaAs systems quoted by SPICER {5}. Exposure to a nitrogen glow discharge before the Al deposition marginally reduced the number of surface states at the pinning level but dramatically reduces their ability to retain charge for longer than 50mS. Auger analysis did not reveal any chemical change at the interface arising from this surface treatment.

Comparison of our p-type GaAs observation with those of HAYASHI
{4} suggests that the extent of anodisation may be critical
in yielding carrier inversion. Our structures were slightly
over-anodised and did not exhibit inversion.

In the case of InP, inversion was clearly evident in n-type
structures which were just-anodised. Over-anodisation produced
a considerable increase in slow charge trapping which was
increased further by illumination. Auger analysis revealed clear
evidence of the ingress of In & P into the alumina centres in
the oxide which can be ionised optically. Inversion was also
observed on just-anodised p-type InP but the surface was depleted
for all practical values of negative bias.

6. Acknowledgements

The authors gratefully acknowledge the financial support
provided by SRC, DCVD, GEC and the University of London Central
Research Fund.

7. References

1. A. Saad and J.G. Swanson, Thin Solid Films, 61, 3,1979,p.355

2. W.S. Lee and J.G. Swanson, Proc. 8th Int.Vac.Congress, Cannes
                              1980, p.590

3. E. Kohn and H.L. Hartnagel, Solid State Electronics, 21,
                              1978, p. 409

4. H. Hayashi et al., Appl.Phys.Letts., 37, 4, 1980, p. 404.

5. W.E. Spicer et al., Insulating Films on Semiconductors,1979,
                       Inst.of Physics Conf.Series No.50, p.216

Composition Changes During Oxidation of $A^{III}B^{V}$ Surfaces

M. Somogyi, M. Farkas-Jahnke

Research Institute for Technical Physics of the Hungarian Academy of Sciences
1325 Budapest, Ujpest 1, P.O. Box 76, Hungary

As the interface region determines the properties of the semi-conductor-insulator layer system, the transitional oxidation products of the substrate are of interest as possible consti-tuents. Investigations concerning the formation of these con-stituents may help to understand the behaviour of the inter-face. In this work the problem is investigated structurally, i.e. oxidation products of well defined crystalline structure were identified. The native oxide systems of GaAs, GaP and InP were studied in two ways: anodic oxide layers by electron dif-fraction and thermal oxidation process with the aid of "in situ" high temperature X-ray technique.

Experimental

1. Surface oxidation   The slices used were Czochralsky grown, (100) oriented, mechanically polished and chemically cleaned as follows: GaAs with $3H_2SO_4/H_2O_2/H_2O$; GaP with $3HCl/HNO_3$; and InP with 2% Br_2-methanol. The etched surfaces were also in-vestigated by electron diffraction after heating them in vac-uum for 5 min at 150°C. The anodization of etched samples was carried out under typical conditions {1} in water-glycol mix-ture using constant current density /0.1 - 0.2 mA/cm²/, sub-mitted subsequently to a heat treatment in N_2 gas at 250°C for 2 hours. The layer thicknesses measured by Talystep were about 100 nm and showed the characteristic interference col-ours. Further samples having uneven oxide layers were prepared by anodizing rough surfaces. A great variety of interference colours were observed and at some points the electron reflec-tion patterns obtained were characteristic to the substrate. The surfaces were investigated by RHEED technique using 100 kV electrons in a JEOL type 100kV electron microscope. The observations made on different types of surfaces are summed up in Table 1.

2. Thermal oxidation "in situ" investigated.   Powders of the same slices fixed on a platinum mesh were heated in air to 1000°C with an average rate of 32°/hour in a Guinier-Lenné type high temperature camera. Cu K_α radiation was used. The changing of the X-ray patterns as registered during slow oxi-dation of GaAs, GaP and InP are presented on Figs. 1a, 1b, 1c, respectively. To facilitate identification Guinier X-ray patterns of powdered material were taken at room temperature, after heating the powders in air at certain fixed temperatures /e.g. 500°C, 600°C, 700°C etc./ for 2 hours. The line systems of these samples correlated well with the lines appearing at the corresponding temperature during continuous heating.

290

Table 1 Phases identified by electron diffraction

Preparation	GaAs	GaP	InP*
anodized, heat treated	GaOOH, β-Ga_2O_3 traces: α-Ga_2O_3 amorphous phase Fig.2.	$GaPO_4$ low cristobalite, α-Ga_2O_3 traces: β-Ga_2O_3 X component	$In_4/P_2O_7/_3$ in large crystallites amorphous In_2O_3 $InPO_4$ "calcite-like" structure
anodized, no heat treatment	β-Ga_2O_3 X component amorphous phase	"calcite-like" structure α-Ga_2O_3, β-Ga_2O_3 amorphous phase	$In_4/P_2O_7/_3$ traces: In_2O_3 Fig.3.
etched, heat treated	β-Ga_2O_3 "calcite-like" structure	"calcite-like" structure traces: α-Ga_2O_3 $GaPO_4$ low cristobalite	In_2O_3

* ASTM diffraction lines were used to identify $In_4(P_2O_7)_3$. Some doubt about the correct identification and synthesis of this product exists however.

Discussion

As shown by Fig. 1. the GaAs oxidation is the most complicated process. In agreement with literature {2} arsenic appears first. About 800°C the GaOOH→β-Ga_2O_3 transition is accompanied by a characteristic line system which could be reproduced on GaAs powder heated at 800°C. As the electron diffraction pattern of this powder agreed with patterns observed on Ga containing sur-faces, it is supposed to be an oxidation product of Ga. On Table 1 it is designated as X component. The main thermal oxidation products were identified as β-Ga_2O_3, $GaPO_4$ and orthorhombic $InPO_4$, respectively. $GaPO_4$ is known in three forms {3} α-quartz, low and high cristobalite. We have observed that the α-quartz to cristobalite ratio depends on the rate of oxidation.

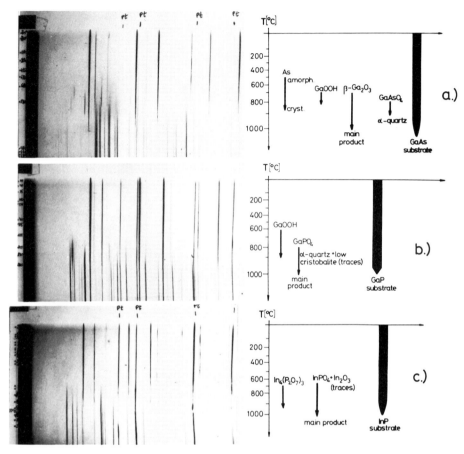

Fig.1. "In situ" high temperature X-ray patterns of
a/ GaAs   b/ GaP   c/ InP

According to Table 1 a great variety of crystalline phases
were detected by electron diffraction. No essential difference
was observed between the patterns of evenly thick and uneven
oxide layers. Heating promotes crystallization not only in the
anodized layers but on etched surfaces too.

Besides the X component mentioned above, another intermediate phase showing calcite-like structure was found on each
substrate. Each attempt to prepare it separately has failed.

Fig.2 Polycrystalline αGa_2O_3 and $\beta-Ga_2O_3$ on anodized and heat treated GaAs substrate / RHEED pattern /

Fig.3 Polycrystalline $In_4/P_2O_7/_3$ and amorphous In_2O_3 on freshly anodized InP substrate / RHEED pattern /

Conclusions

In native oxide layers of GaAs, GaP and InP a strong tendency to crystallization was observed, the presence of numerous crystalline compounds could be stabilized. Of the two unknown structures the so-called "calcite-like" appears in layers or on surfaces always in contact with the substrate.

1. Hasegawa, H.L. Hartnagel
   J.Electrochem.Soc. 123,713 /1976/

2. G.P.Schwartz, G.J. Gualtieri, J.E. Griffiths, C.D. Thurmond,
   B. Schwartz
   J.Electrochem.Soc. 127,2488 /1980/

3. A.Perloff
   J.Am.Ceram.Soc. 39,83 /1956/

RF-Sputtering of Silicon Nitride Layers on GaAs Substrates: Characterization of an Intermediate Layer Between the Substrate and the Deposited Film

L.M.F. Kaufmann, R. Tilders, K. Heime

University of Duisburg, Solid State Electronics Department, P.O.Box 101629
D-4100 Duisburg, Fed. Rep. of Germany

H.W. Dinges

Research Institute of the German Post Office, FTZ, P.O. Box 5000
D-6100 Darmstadt, Fed. Rep. of Germany

(100)-oriented GaAs wafers are coated with a silicon nitride film by rf-sputtering. The technology for depositing uniform, electrically insulating layers without inclusions or defects, suitable for high-temperature processes ($800^{\circ}C$) is presented. Reflection ellipsometry reveals an intermediate layer between the silicon nitride film and the GaAs substrate, which cannot be explained by gallium oxide or nitride. This layer, a damaged layer between the substrate and the deposited film, is found to be the origin for this effect. Similar results were reported from ion implanted GaAs [1], ion implanted GaP [2] and sputter etched silicon surfaces [3] by other authors.

1. Introduction

Silicon nitride films are often used as a protective cap for GaAs integrated circuits. As they are deposited as one of the final processing steps during the IC-fabrication, the film should not influence the characteristics of the active FET-devices. Reflection ellipsometry cannot only be used as a fast, damage-free method to control the thickness and the refractive index of the deposited film itself, but it is a sensitive tool to characterise the interface between the silicon nitride layer and the substrate.

2. Technology

In this work silicon nitride films were deposited by rf-sputtering from a hot-pressed silicon nitride target using standard MRC Sputtershere equipment. By optimizing the sputtering conditions, using: 1. low rf-power (typical 50 W, turret voltage 250 V),

2. a mixture of argon ($4 \cdot 10^{-3}$ Torr) and nitrogen ($1 \cdot 10^{-3}$ Torr)

3. sputter-etching of the substrate prior to the deposition of the silicon nitride film,

4. bias-deposition,

homogeneous films without inclusions or defects can be obtained [4]. These films can be used as a diffusion mask for the selective diffusion of Sn into GaAs at 800°C from spun-on emulsions [5].

3. Ellipsometry

A Rudolph Research Co. ellipsometer, type 43603-200 E [6], in a
dark laminar flow box is used for the measurements. The angle of
incidence was 70°, the wavelengths used, chiefly 546.1 nm, were
selected by interference filters of 2 nm halfwidth from the spectrum of a halogen lamp.
The difficulties, arising by the determination of the optical properties from ellipsometric measurements are well known from literature [6,7]. A sample consisting of a substrate covered with a
homogeneous film has five optical constants, i.e. the index of
refraction of the substrate $\tilde{n}_s = n_s - ik_s$, the index of refraction of the film $\tilde{n}_f = n_f - ik_f$ and the thickness of the film d_f.
The system of the ellipsometric equations is never fully determined
by ellipsometry only, since there are only two measured quantities
Ψ and Δ. Thus three parameters have to be determined independently.

4. Experimental results

Chemically cleaned and etched GaAs slices were therefore examined
prior to the sputter process to determine the substrate properties.
It is known, that immediatly after chemical etching [8] GaAs is covered with a thin natural oxide layer, which is absorption-free at
546.1 nm. The values for $k_s = 0.304$ and for $n_{ox} = 1.9$ (α -Ga$_2$O$_3$)
at 546.1 nm were taken from literature [9]. We found, that the real
part of the refractive index of GaAs is $n_s = 4.05 \pm 0.01$.
The thicknesses of the thin natural oxide layer were in the range
of 0.9 to 1.5 nm depending on the exposure time to air after etching
(5 - 60 min). The same values were reported in literature [8].
It is known that pure silicon nitride films are absorption-free for
546.1 nm. With the optical constant $\tilde{n}_s = 4.05 - 0.304i$ at 546.1 nm
and with values between 1.8 and 2.0 for the refractive index of the
sputtered silicon nitride films a Ψ, Δ - network was drawn in order
to insert the measured Ψ, Δ - values of the silicon nitride covered
GaAs-slices.

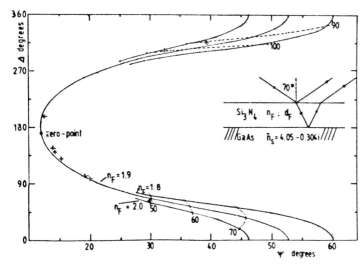

Fig.1
Ψ, Δ -graph for
GaAs-substrates
($\tilde{n}_s$=4.05-
0.304i)
covered with
homogeneous,
plan parallel
and absorption-
free films with
refractive in-
dices n_f be-
tween 1.8 and
2.0, the thick-
ness of the
film d_f is
curve parameter.

The measured ψ, Δ-values of our films cannot be fitted by one unique value of the refractive index of the silicon nitride film for thicknesses between 5 and 190 nm (Fig.1).

Also, the curve through the measured ψ, Δ-values does not fit the zero point (Fig.1) given only by the optical constants of the bare GaAs substrate and for film thicknesses d_f equal zero. The curve passes the zero point shifted by 0.9 degrees towards greater ψ values. This shift cannot be explained by absorption in the silicon nitride film. Absorption would shift the measured ψ, Δ-curve to smaller ψ values than the zero point [11].

Assuming homogeneous and absorption-free films on a substrate with changed optical constants we found a very good fit with the numerical data of the substrate $\tilde{n}_s$(changed) = 4.2 - 0.195i and a refractive index of the deposited film n_f = 1.98 (Fig.2).

Earlier investigations on anodic oxide films on GaAs showed similar effects [7]. As the measured data for films with thicknesses in the order of the repeating thickness fit the new theoretical ψ, Δ-graph very well, the assumption that the silicon nitride film is absorption-free was right.

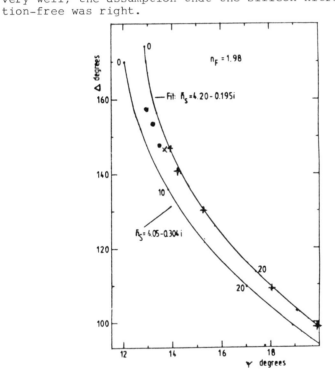

Fig.2 Partial ψ, Δ-graph for silicon nitride covered GaAs substrates with a fit for $\tilde{n}_s$(changed) = 4.2 - 0.195i ; in both cases the refractive index of the film is n_f = 1.98 . The measured data for sputter-etched samples (●) and for films deposited without bias and without sputter-etching the substrate (x) are also included.

5. Discussion

As the optical constants for the chemically cleaned and etched
GaAs substrates are in agreement with data from literature [7,8,9]
the origin for the change of the refractive index of the substrate
after a sputtering process has to be found in the process itself.
The measured optical constants of the sputter-etched GaAs substrates are not in agreement with the numerical data based on the refractive index of the bare GaAs substrate $\tilde{n}_S = 4.05 - 0.304i$.
As reported all sputter processes were performed in a pure argon/
nitrogen atmosphere. The shift of the measured Ψ, Δ -values cannot
be explained by the formation of an gallium oxide or gallium nitride layer with well-known refractive indices on top of a sputtered
GaAs sample.
The shift of the measured Ψ, Δ -values of sputter-etched substrates
is larger using higher sputter-power. Using constant power we found
the shift is independant of the sputter-time (15 sec - 60 min).
This means, that a disturbed substrate surface region is formed
within the first few seconds of the process.
The optical constants of silicon nitride films sputter-deposited in
an argon/nitrogen atmosphere without sputter-etching of the substrate and without using the bias-mode of the equipment also show a
shift compared to the calculated Ψ, Δ -graph for $\tilde{n}_S = 4.05-0.304i$
(Fig.2). This means a disturbed GaAs substrate surface region is
also formed during sputter-deposition even using low power.
The superposition of both effects, the formation of a disturbed
substrate surface region during sputter-etching as well as during
sputter-deposition explains the large shift of the measured Ψ, Δ
values, resulting in a calculated, changed refractive index of the
substrate $\tilde{n}_S$(changed) = 4.2 - o.195i . From dispersion measurements
we hope to get the information necessary to calculate the thickness
and the refractive index of the disturbed GaAs substrate surface
region.

Acknowledgements

The authors are indebted to E.Hesse, FTZ for advice, G.Joel, FTZ for
chemical preparations and the members of the Solid State Electronics
Department, University of Duisburg for their helpful discussions.

1   Q.Kim and Y.S.Park, J.Appl.Phys., 51 (1980) 2024-2029
2   K.Watanabe, M.Miyao, I.Takemoto and N.Hashimoto, Appl.Phys.Lett.,
    34 (1979) 518-519
3   B.C.Dobbs, W.J.Anderson and Y.S.Park, J.Appl.Phys., 48 (1977)
    5052-5056
4   H.Öchsner, Appl.Phys., 8 (1975) 185-198
5   N.Arnold, K.Heime, 18th annual Solid State Physics Conf.,
    January 1981, York, GB
6   F.L.McCrackin, E.Passaglia, R.Stromberg and H.L.Steinberg,
    J. Res. Nat. Bur. Stand. A67 (1963) 363-377
7   H.W.Dinges, Thin Solid Films, 50 (1978) L 17
8   A.C.Adams and B.R.Pruniaux, J.Electrochem.Soc., 120 (1973) 408
9   M.D.Sturge, Phys.Rev., 127 (1962) 768
10 C.J.Mogab and E.Lugujjo, J.Appl.Phys., 47 (1976) 1302-1309
11 H.W.Dinges, to be published in Thin Solid Films

Surface Analytical and Capacitance-Voltage Characterization of Anodic Oxide Films on $Hg_{0.8}Cd_{0.2}Te$

H.J. Richter, U. Solzbach, and M. Seelmann-Eggebert

Fraunhofer-Institut für Angewandte Festkörperphysik
D-7800 Freiburg, Fed. Rep. of Germany

H. Brendecke, H. Maier, J. Ziegler, and R. Krüger

AEG-Telefunken Serienprodukte AG, Geschäftsbereich Halbleiter
D-7100 Heilbronn, Fed. Rep. of Germany

1. Introduction

Surface passivation is an essential technological process for the production of IR quantum detectors. In the case of n-type $Hg_{0.8}Cd_{0.2}Te$ an anodic oxide proved to be an adequate passivant. However, the chemical and physical properties of the anodic oxide are still subject to discussion.

2. Sample Preparation

The $Hg_{0.8}Cd_{0.2}Te$ samples to be oxidized were grown by the solid state recrystallization technique. The single crystalline n-type samples had a carrier concentration $n(77K) \approx 5 \times 10^{14} cm^{-3}$ and a band gap energy $E_g(77K) \approx 0.1$ eV. They were randomly oriented and homogeneous in composition and electrical properties.

The native oxide was grown by anodization in a KOH electrolyte (0.1M KOH, 90% ethylene glycol, 10% H_2O) in accordance with a standard technique [1]. Before oxidation the $Hg_{0.8}Cd_{0.2}Te$ surface was mechanically polished and etched with 0.5% bromine in methanol and cleaned in pure methanol. Anodic oxide films up to 2200 Å were grown and analyzed.

3. Surface Analytical Characterization

X-ray photoelectron spectroscopy (XPS) with MgKα radiation (1253.6 eV) was employed for the surface analysis and, in combination with Ar^+ sputtering, for the depth profile analysis of the anodic oxide films as well. The qualitative analysis was based on XPS signature data of Te, O, and Cd which exhibit characteristic chemical shifts in the process of oxidation [2]. These XPS signature data as measured for the anodic oxide and for potential constituents of the anodic oxide are listed in Table 1.

298

Table 1 Binding energies of the $Cd3d_{5/2}$, O 1s, and $Te3d_{5/2}$ core electrons and peak energy of the $M_5N_{45}N_{45}$ Auger electrons of Cd as appearing in the binding energy spectra of the compounds listed. The underlined data match or nearly match the corresponding signature data of the anodic oxide.

	$Cd3d_{5/2}$ [eV]	O 1s [eV]	$Te3d_{5/2}$ [eV]	$CdM_5N_{45}N_{45}$ [eV]
Anodic oxide	405.0	530.2	576.0	878.4
Cd (metal)	404.9			875.6
CdO	404.4	529.2		876.8
Cd(OH)$_2$	405.0	531.3		878.6
TeO$_2$		530.2	576.2	
CdTeO$_3$	405.0	530.2	576.0	878.3

For the data comparison in Table 1 a single crystal of CdO was examined by XPS. The crystal was grown by oxidation of Cd vapor and kindly supplied by R. Helbig (University of Erlangen). The Cd(OH)$_2$ and the TeO$_2$ samples were prepared by chemical standard procedures as thin layers on Cd and Te substrates, respectively. For this study CdTeO$_3$ was available as a powder sample only (CERAC, Milwaukee, USA). For XPS examination it was formed into a flat pill by application of moderate pressure.

All samples were subjected to sputter cleaning treatments prior to each measurement. It was found that Cd(OH)$_2$ was partially transformed into CdO, and that some CdTeO$_3$ was reduced to CdTe due to sputtering.

In view of previous results [2] CdTeO$_3$ was expected to be a constituent of the anodic oxide on $Hg_{0.8}Cd_{0.2}Te$. Also the phase diagram of the CdO-TeO$_2$ system as presented by ROBERTSON et al. [3] suggests that CdTeO$_3$ and/or CdTe$_2$O$_5$ might be formed by anodization. A CdTe$_2$O$_5$ sample was not available for this study, but CdTeO$_3$ and CdTe$_2$O$_5$ are expected to have qualitatively identical XPS signatures because of similar bonding configurations of their components. Although TeO$_2$ is included in Table 1 it is not considered a candidate constituent of the anodic oxide. It is attacked by alkalis, i.e. it is not stable in the used electrolyte [4].

Since only CdTeO$_3$ completely matches the XPS signature data of the anodic oxide we conclude with respect to Cd that predominantly CdTeO$_3$ and/or CdTe$_2$O$_5$ should form the anodic films on $Hg_{1-x}Cd_xTe$, which is in accordance with a previous suggestion by DAVIS et al. [5]. From a recent study by SAKASHITA et al. [6] concerning anodic oxide films and electrochemical reactions on HgTe one can infer that the anodic Hg is also likely to be bound in tellurites. The quantitative analysis was based on the weighting of the photoelectron response (i.e. the integral of the $Cd3d_{5/2}$, $Hg4f_{7/2}$, $Te3d_{5/2}$, or O 1s XPS signal above background) with the effective photoionization probability (or sensitivity factor) of the respective core level. Using the same sensitivity factors as SUN et al. [7] we found a composition of the anodic oxide as given in Table 2. Also with respect to the quantitative depth profile analysis a close agreement with the previous results by SUN et al. was observed except for the $Hg_{0.8}Cd_{0.2}Te$ bulk value. This is simply related to the fact that our XPS data were taken with a sample temperature of 300 K.

Table 2 Composition of the anodic oxide on $Hg_{0.8}Cd_{0.2}Te$ as obtained by XPS examination of a sputter cleaned sample.

	SUN et al.	THIS STUDY
O	59 at%	60 at%
Te	24 at%	25 at%
Cd	14 at%	14 at%
Hg	3 at%	2.5 at%

The composition of a sputtered $Hg_{1-x}Cd_xTe$ surface deviates from the bulk stoichiometry, and this deviation is a function of the energy of the Ar ions [2] and of the temperature of the sample [8]. An intact, however thin, anodic oxide film acts as a protective layer against preferential sputter effects on the $Hg_{0.8}Cd_{0.2}Te$ interface. Hence, that part of the sputter profile in which the thickness of the anodic film is comparable to the XPS probing depth is of particular interest for an assessment of the stoichiometric quality of the $Hg_{0.8}Cd_{0.2}Te$ interface region. As far as we can tell from XPS measurements this interface region has a rather stoichiometric composition without a substantial Hg depletion.

In conclusion of the surface analytical and chemical characterization we interpret the anodic oxide on $Hg_{0.8}Cd_{0.2}Te$ as a mixture of cadmium-mercury di- and mono-tellurite with an approximate composition of 44 mol% $CdTe_2O_5$, 8 mol% $HgTe_2O_5$, 41 mol% $CdTeO_3$, and 7 mol% $HgTeO_3$. This assignment is consistent with only 60 at% analyzed oxygen since the analyzed Te was composed of 23 at% Te^{4+} (i.e. Te with four oxygen bonds) and 2 at% Te^{2-} (i.e. Te with two bonds to Cd or Hg) as a result of the above mentioned reduction caused by sputtering.

4. Electrical Characterization

The electrical characterization of the oxide layer and the interface was obtained by C-V measurements. As metal contacts for the MOS structure chromium-tin dots with a diameter of 0.6 mm were evaporated through a metal mask directly onto the oxide. All C-V experiments were carried out at 77 K.

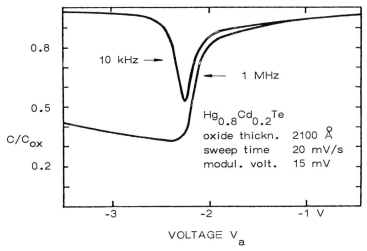

Fig. 1 Typical experimental C-V curves of n-type $Hg_{0.8}Cd_{0.2}Te$

Figure 1 presents typical experimental data for our n-type samples with measuring frequencies of 10 kHz and 1 MHz. At 1 MHz the sample shows the expected high frequency response. However, at 10 kHz the sample still exhibits a low frequency response, i.e. the inversion layer is still able to follow the applied frequencies. This peculiar behavior has been previously observed [9, 10] . The insulator capacitance per unit area is about 8×10^{-8} Fcm^{-2} for an insulator thickness of 2100 Å. Although not indicated in Fig. 1 a slight hysteresis is observed in the C-V curves. From the hysteresis shift the slow surface state density can be calculated to a value of 1.5×10^{10} cm^{-2}. The position of the flat-band voltage indicates that the density of fixed charges in the oxide and/or at the interface is positive, i.e. in absence of an applied voltage the surface of the n-type $Hg_{0.8}Cd_{0.2}Te$ is in an accumulated condition. From measurements on several MOS structures an average fixed charge density of 8×10^{11} cm^{-2} can be calculated.

The fast surface-state density N_{ss} was determined from the low frequency response by the integration method of BERGLUND [11]. In the upper part of Fig. 2 the normalized capacities of the experimental and theoretical curves are plotted over the surface potential γ_s. γ_s was calculated by integrating the measured C-V curve. By comparing the experimental with the ideal curve the distribution of the fast surface-state density N_{ss} per unit energy as a function of γ_s is obtained and shown in the lower part of Fig. 2. Integrating the surface-state density over the 0.1 eV band gap of $Hg_{0.8}Cd_{0.2}Te$ gives N_{ss} 5×10^{11} cm^{-2}.

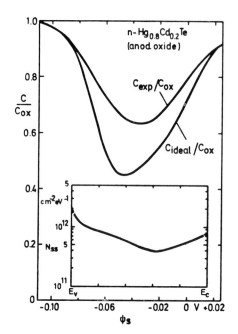

5. Conclusion

The anodic oxide obviously leads to the desired accumulation of the n-type $Hg_{0.8}Cd_{0.2}Te$ interface. The low density and the nearly flat distribution of the fast surface states over the band gap indicate a high quality $Hg_{0.8}Cd_{0.2}Te$-oxide system which is a prerequisite for the production of high performance IR detectors and devices.

An unambiguous correlation of the chemical and electrical properties of the $Hg_{0.8}Cd_{0.2}Te$-oxide system was not found to be possible as yet but it remains the ultimate goal of further investigation.

Fig. 2 Analysis of the C-V curves by the BERGLUND method. Insert: Distribution of the fast surface-state density over the band gap energy.

References

1   P.C. Catagnus, C.T. Baker, U.S. Pat. No. 3 977 018 (1976)
2   U. Solzbach and H.J. Richter, Surface Sci. 97, 191 (1980)
3   D.S. Robertson, N. Shaw, and I.M. Young, J. Mat. Sci. 13, 1986 (1978)
4   M. Pourbaix, Atlas d'Equilibres Electrochemiques à 25°C, Paris (1966), p. 567
5   G.D. Davis, T.S. Sun, S.P. Buchner, and N.E. Byer, J. Vac. Sci. Technol. 19, to be published (1981)
6   M. Sakashita, H.-H. Strehblow, and M. Bettini, J. Electrochem. Soc., to be published (1981)
7   T.S. Sun, S.P. Buchner, and N.E. Byer, J. Vac. Sci. Technol. 17, 1067 (1980)
8   H.M. Nitz, O. Ganschow, U. Kaiser, L. Wiedemann, and A. Benninghoven, Surface Sci. 104, 365 (1981)
9   M.A. Kinch, R.A. Chapman, A. Simmons, D.D. Buss and S.R. Borrello, Infrared Physics 20, 1 (1980)
10 Y. Nemirovsky, I. Kidron, Solid-State Electr. 22, 831 (1979)
11 C.N. Berglund, IEEE Trans. Electron. Dev. ED-13, 701 (1966)

Surface and In-Depth Analysis of Anodic Oxide Layers on Cd$_{0.2}$Hg$_{0.8}$Te

A. Benninghoven, O. Ganschow, U. Kaiser, J. Neelsen, L. Wiedmann

Physikalisches Institut der Universität Münster
D-4400 Münster, Fed. Rep. of Germany

H. Brendecke, H. Maier, U. Ziegler

AEG-Telefunken
D-7100 Heilbronn, Fed. Rep. of Germany

1. Introduction

Cadmium mercury telluride, Cd$_{0.2}$Hg$_{0.8}$Te (CMT), is well suited for detectors in the far infrared because of its band structure. The control of its electrical properties, however, poses some severe problems to both the semiconductor and the surface scientist. Anodic oxide layers on CMT grown in KOH ethylene glycol solution to different thickness are of special interest, because they are used for passivation of the surface.

In the present paper, such oxide layers with thickness 200 Å, 900 Å and 2300 Å have been characterized by quasisimultaneous secondary ion mass spectrometry (SIMS), X-ray photoelectron spectroscopy (XPS), thermal desorption mass spectroscopy (TDMS), and Auger electron spectroscopy (AES). The latter method was for the first time successfully applied to CMT samples.

2. Thermal stability

Bare CMT surfaces show appreciable changes of elemental concentrations with time, which are due to mercury evaporation /1/. From the kinetics of surface concentration and evaporation we have found the diffusion constant of mercury in CMT to be

$$D = D_o \exp(-\Delta E/kT)$$

with $D_o = 3.6 \times 10^{31}$ cm^2/s and $\Delta E = 0.34$ eV.

During sputter depth profiling, the simultaneous occurrence of preferential sputtering and mercury diffusion results in depletion profiles of this element below the actual surface. In a detailed discussion of these effects /1/, the experimental results have been treated quantitatively within the framework of an appropriate diffusion theory /2/. Fig.1 shows calculated depletion profiles at sputter equilibrium for different temperatures. At 180 K, the composition changes are confined to the uppermost three monolayers. Therefore our experiments were performed in the 170 - 180 K range. Apart from the specimen temperature, the experimental arrangement was as described in /1/.

303

The mercury evaporation rate from oxide coated CMT samples has been estimated by TDMS. Even at maximum thickness of the coating (2300 Å), the Hg atom flux density through the surface was reduced by at most one decade as compared to the bare substrate, where it is in the order of 10^{13} to 10^{16} atoms/cm^{2}s depending on temperature and pretreatment.

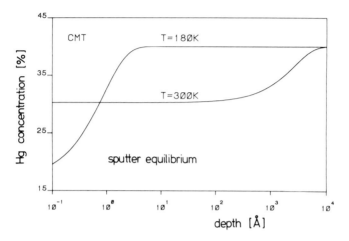

<u>Fig.1</u> Calculated   mercury   depth   profiles   in   sputter equilibrium for different temperatures.

3.   <u>Preferential sputtering and erosion rate</u>

Fig.2 shows the elemental AES and XPS signals from the 900 Å oxide layer as a function of depth. Similar results have been obtained for the 2300 Å and 200 Å layer. Throughout the oxide layer, the signals remain reasonably constant. As compared to the XPS data, the concentrations determined by AES are smaller for Hg and O and larger for Cd. In view of the smaller information depth of the AES transitions, these differences indicate concentration gradients of these elements due to preferential sputtering in such a way, that for the total sputter yields $S_{Hg} > S_{Te} > S_{Cd}$ must hold, as has been found for the bare CMT /1/. From the same argument it can be concluded that oxygen is sputtered preferentially as well. This can additionally be seen from the appearance of an unshifted Te 3d doublet due to non-oxidized Te soon after the onset of sputtering.

The mean erosion rate of the oxide has been determined from the interface position in the AES data as $(5.3 \pm 0.3)\times10^{-22}$ cm^3 per incident primary ion (Ar$^+$, 5 keV), similar to the value of $(5.8 \pm 0.3)\times10^{-22}$ cm^3 per primary ion for the bare substrate at room temperature /1/.

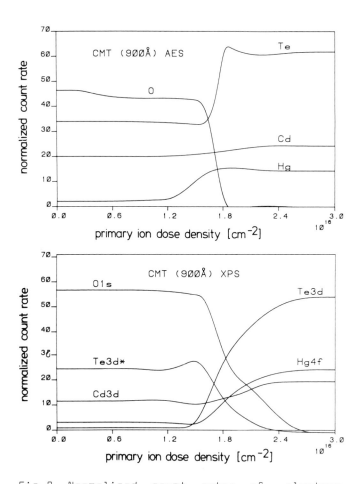

Fig.2 Normalized count rates of electron spectroscopic
       signals as a function of primary ion dose density for
       the 900 Å layer.
       a) AES signals,
       b) XPS signals (Te 3d* denotes the shifted Te 3d line
       from oxidized Te).

4. Chemical information

From various XPS investigations of anodic oxides on CMT /3-5/,
chemical shifts of the Cd MNN Auger transition and the Te 3d
photoelectron lines with respect to their energy in the CMT
matrix are known. In this study, we additionally observed
shifts of the Te MNN, Hg NVV, Te 4d, and Cd 4d lines. The
most prominent feature is the shifted Te 3d doublet, the
intensity of which is indicated in fig.2 as Te 3d*.

The secondary ions emitted from the samples consisted of
atomic and molecular species characteristic for the matrix and
the oxide such as $TeO_m^\pm$, $Te_2O_m^-$ (m=0,1,2...), $Cd^\pm$, O^- etc.

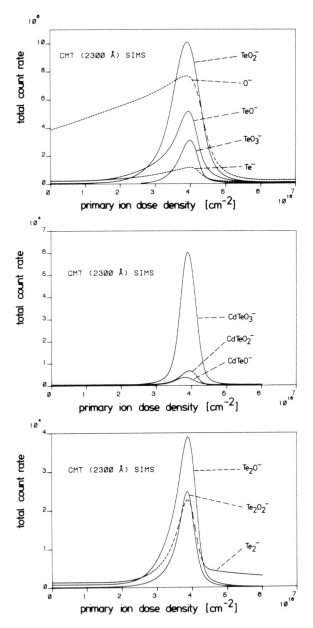

Fig.3   Integrated count rates of the oxide-specific secondary
        ions as a function of primary ion dose density for the
        2300 Å layer.
        a) O^-, Te^-, TeO_n^- (n=1-3),
        b) $CdTeO_n^-$ (n=1-3),
        c) $Te_2O_n^-$ (n=0-2).

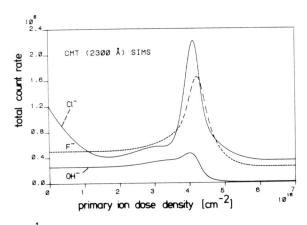

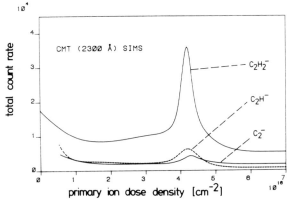

Fig.4 Integrated count rates of impurities as a function of primary ion dose density for the 2300 Å layer.
a) Cl^-, F^-, OH^-,
b) $C_2H_n^-$ (n=0-2).

as well as a number of contaminations like C^-, C_2^-, C_2H^-, $C_2H_2^-$, F^-, Cl^-, OH^-, K^+ and Na^+. The depth distribution of the most prominent of these species is given in figs.3 and 4 for the 2300 Å layer. In contrast to the AES and XPS results, the SIMS data indicate an appreciable inhomogeneity of the layer. The changes of the secondary ion intensities with depth must be attributed to an inhomogenity with respect to the oxidation number of the elements and the ionicity of chemical bonds, which influence the ionization probability very strongly. This evidence is supported by an increase of the Te 3d* signal in XPS near the interface, which corresponds to the maximum signal of the oxide specific secondary ions.

The problem of interference of different secondary ion species (e.g. TeO_m^-, CdO_{m+1}^- and $CdO_{m+1}H^-$, or $Te_2O_m^-$, $CdTeO_{m+1}^-$ and $CdTeO_{m+1}H^-$) has been solved by numerical separation of the intensities of the constituent molecules according to their natural isotopic abundances. This analysis yields an upper

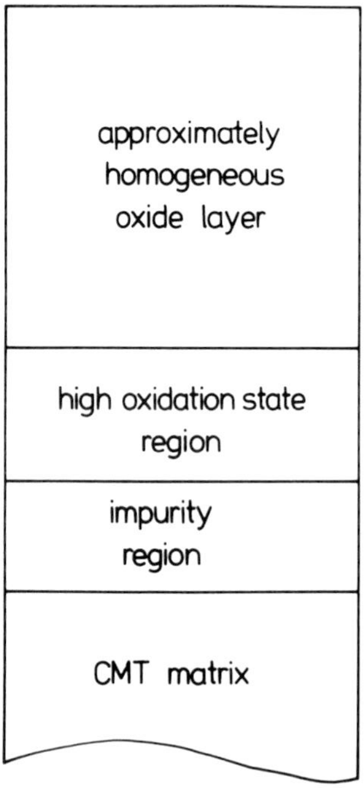

| approximately homogeneous oxide layer |
| high oxidation state region |
| impurity region |
| CMT matrix |

Fig.5 Composition of the 900 Å and 2300 Å layers as a function of depth (schematically).

limit for CdO_m^- and CdO_mH^- of 1 % of the TeO_{m-1}^- signal, so that there is no evidence for Cd oxidation from these data.

The 200 Å layer shows a similar composition as the interface region of the thicker layers. The structure of the thicker layers, as indicated by our experimental data, is shown schematically in fig.5.

The total amount of the contaminations accumulated in the interface region is estimated to be in the order of 3×10^{14} atoms/cm^2, independent of the layer thickness within a factor of 2. We therefore suggest that these contaminations are present at the surface from the very beginning of anodization and are not continuously incorporated into the layer during this process. Because of their large electronegativity, they may well represent the reason for the accumulation layer observed in CMT MOS structures /3/.

5. Conclusion

Anodic oxide layers on $Cd_{0.2}Hg_{0.8}Te$ show a rather complicated structure. Their quantitative composition analysis, as shown in our investigation, requires an elaborate surface analytical equipment, as well as detailed models for the interpretation of the experimental data. We hope that the results of such investigations will lead to improved coating techniques, better CMT devices and better understanding of the anodization process.

References

/1/ H.M.Nitz, O.Ganschow, U.Kaiser, L.Wiedmann and A.Benninghoven, Surface Sci. 104 (1981), 365.
/2/ R.Webb, G.Carter and R.Collins, Radiat.Eff. 39 (1978), 129.
/3/ H.J.Richter, M.Seelmann-Eggebert, U.Solzbach, H.Brendecke, H.Maier, J.Ziegler, R.Krüger, these proceedings.
/4/ U.Solzbach and H.J.Richter, Surface Sci. 97 (1980), 191.
/5/ T.S.Sun, S.P.Buchner and N.E.Byer, J.Vac.Sci.Technol. 17 (1980), 1067.

We thank the Fraunhofer Gesellschaft for financial support.

308

Impact of Insulator Charge Trapping on I.R.C.I.D. Transfer Efficiency

G. Boucharlat[1], J. Farré[2], A. Lussereau[1], J. Simonne[2], M. Sirieix[1],
J.C. Thuillier[3]

[1]S.A.T. 41, rue Cantagrel, F-75624 Paris, France
[2]L.A.A.S-C.N.R.S. 7, avenue du Colonel Roche, F-31400 Toulouse, France
[3]Laboratoire des Diélectriques ERA CNRS no. 19, Faculté des Sciences
 Mirande, Université de Dijon, F-21004 Dijon, France

A direct effect of the increase of sensing sites in a CID array,
has been to enhance the number of charge transfers involved
in each cell, during the read out of the whole matrix. As a
result, the CID properties - namely transfer efficiency - can
be affected by a higher charge loss.

Several factors contribute to a loss of information when a charge
packet is transferred back and forth under the two adjacent
electrodes of a cell; geometrical dimensions and actual carrier
mobilities are considered as mainly responsible; but transfer
inefficiency measurements give often worse results than those
expected even when clock timing and pulse shapes are ajusted
to optimize the procedure.

Among these factors: insulator charge trapping - a phenomenon
often noticed in an Infra Red Charge Injection Device when an
insulator of high enough quality cannot be achieved - may degrade
the sensitivity of the device through a decrease of the signal
and an increase of the fixed pattern noise, and brings about a
major contribution to transfer inefficiency.

The analysis of the impact of insulator defects on IRCID
efficiency is the purpose of this paper.

Trapping and release mechanism in an elementary IRCID cell

The charge: Q_{ph} stored during a time : T_i in a MIS structure
which is deeply depleted and exposed to a photon flux F,
consists of a free carrier layer at the surface of the semiconductor : Q_t, and in a part : Q_d trapped in the first layers
of the dielectric through a tunneling mechanism, which can be
significant when the technological process used is a deposition
of the insulator film upon the narrow band gap semiconductor.

$$Q_{ph} = Q_t + Q_d \qquad (1)$$

Q_d is a function of the free carrier stored in the inversion
layer, and of the trap density per energy unit in the insulator,
N_T, according to the time dependent expression.

$$Q_d = \gamma q\, N_T \, \log \left(1 + \frac{T_i}{\tau} \right) \frac{\Delta E}{kT} \qquad (2)$$

where ΔE is the energy interval concerned with carrier trapping
{1} {2}; τ and γ are constants related to the tunneling mechanism {3}. This leads after some algebra {2} to:

$$Qd \cong \lambda q\ N_T Q_t^{4/5} \ Log\ (\frac{T_i}{\tau}) \qquad (3)$$

where λ is a constant related to the semiconductor material.

When half or full selection is operated for read out procedure,
the free carrier charge packet is transferred from one to the
adjacent electrode of the cell. At the same time, the charge
trapped is released from the insulator according to a similar
law depending on the amount of charge previously trapped, and
on the transfer time T_t.

$$Q_r \cong \lambda q\ N_T\ Q_{t_o}^{4/5}\ Log\ (\frac{T_t}{\tau}) \qquad (4)$$

where Q_r is the charge released during T_t and Q_t the free carrier
charge at $t = T_i$ given by (2) and (3).

Charge injection read-out mode of a C.I.D. array

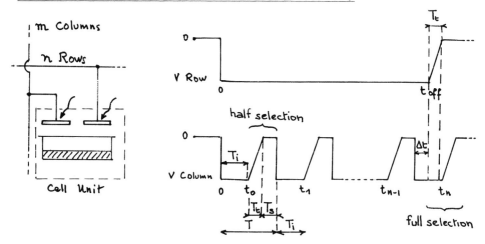

Fig. 1

In Fig. 1, the clock timing controlling the row and column voltage
of a cell during read out of a m x n CID matrix, exhibits n half
selections on the column electrode, and one full selection on the
row/column electrodes when injection takes place.

Following each transfer, the column electrode is "OFF" during T_S.
The amount of charge released during this period will be lost
by recombination if we consider that the interval of time T_S is
much higher than the minority carrier lifetime.

Assuming that the free carrier charge Q_{t_0} stored during T_i and the charge released during T_t are entirely transferred, we get, through Eq. (4), the charge lost during the first half selection

$$Q_{L1} = \lambda q\ N_T\ Q_{t_0}^{4/5}\ Log\ (\frac{T-T_i}{T_t})\ . \qquad (5)$$

After each period T, both electrodes store charge again which becomes equally distributed at $T + T_i$.

After n half selections, the charge lost will be

$$Q_{L_n} = \sum_{k=1}^{k=n} Q_{L_k} = \lambda q\ N_T \sum_{k=o}^{k=n-1} Q_{t_k}\ . \qquad (6)$$

After read-out, the global loss will be

$$Q_L = Q_{L_n} + Q_{Loff} \qquad (7)$$

where $Q_{L\ off}$, for the row electrode, is computed in the same manner as Q_{L1} for the column electrode, using the appropriate time interval : t_{off} and corrected for the n transfers.

Charge loss through trapping mechanism in the read-out procedure

A program allowing the evaluation of Q_L, when the dielectric is SiO_2, and the semiconductor InSb, has been organized according to the block diagram of (Fig.2).

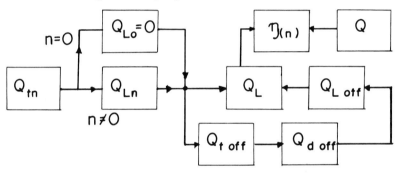

Fig. 2

In Fig.3, transfer efficiency: $\eta = \frac{Q-Q_L}{Q}$, is presented for any number of n transfers (half selection), followed by the read-out (full selection), up to n=32. Q is the total charge due to the photon flux integrated under both electrodes.

The trap density, running from $N_T = 10^{17} cm^{-3} eV^{-1}$ to $10^{20} cm^{-2} eV^{-1}$, is used as a parameter to show the influence of the insulator quality.

These results have been checked by performing a measurement on a cell after one half selection, plus a full selection; then after 32 half selections plus a full selection, for an equivalent storing time in both cases. The trap density is evaluated through a method already published (1).

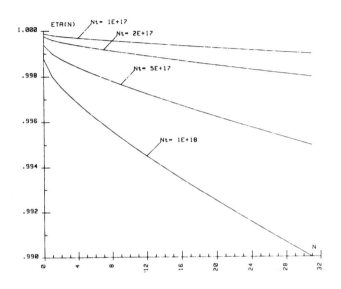

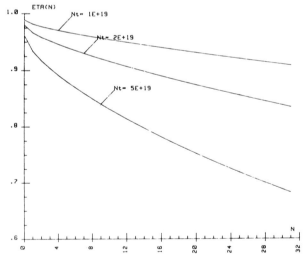

The curves were obtained using the following values:

$F = 10^{15}$ ph. $s^{-1}cm^{-2}$; $T_i = 12\mu s$; $T_t = 100ns$;

$T_s = 2\mu s$; $T \cong 14 \mu s$; $T° = 77°K$;

InSb dielectric constant : 17,9; $\tau : 10^{-13}s$; $\lambda = 7.6 \cdot 10^{-5}cm$ eV.

312

Acknowledgments
We wish to thank J. Baudet for the computing program and N. Causero for her assistance.

References
1　J. Buxo et al.- Appl. Phys.Lett. 33, 969 (1978)
2　J. Farre Sc. D. Thesis n° 939 Université P. Sabatier
　　Toulouse (1980)
3　I. Lundström, C. Svenson - J.Appl.Phys. 43,12 (1972)

Index of Contributors

Inelastic Particle-Surface Collisions

Proceedings of the Third International Workshop on Inelastic Ion-Surface Collisions Feldkirchen-Westerham, Federal Republic of Germany, September 17–19, 1980
Editors: E. Taglauer, W. Heiland
1981. 194 figures. VIII, 329 pages
(Springer Series in Chemical Physics, Volume 17)
ISBN 3-540-10898-X

Contents: Electron Emission. – Electron and Photon Impact. – Electron Transfer. – Polarized Light Emission. – Excited Particle Emission. – Index of Contributors.

Secondary Ion Mass Spectrometry SIMS-II

Proceedings of the Second International Conference on Secondary Ion Mass Spectrometry (SIMS II) Stanford University, Stanford, California, USA,
August 27–31, 1979
Editors: A. Benninghoven, C. A. Evans, Jr., R. A. Powell, R. Shimizu, H. A. Storms
1979. 234 figures, 21 tables. XIII, 298 pages
(Springer Series in Chemical Physics, Volume 9)
ISBN 3-540-09843-7

Contents: Fundamentals. – Quantitation. – Semiconductors. – Static SIMS. – Metallurgy. – Instrumentation. – Geology. – Panel Discussion. – Biology. – Combined Techniques. – Postdeadline Papers.

M. A. Van Hove, S. Y. Tong

Surface Crystallography by LEED

Theory, Computation and Structural Results

1979. 19 figures, 2 tables. IX, 286 pages
(Springer Series in Chemical Physics, Volume 2)
ISBN 3-540-09194-7

Contents: Introduction. – The Physics of LEED. – Basic Aspects of the Programs. – Symmetry and Its Use. – Calculation of Diffraction Matrices for Single Bravais-Lattice Layers. – The Combined Space Method for Composite Layers: by Matrix Inversion. – The Combined Space Method for Composite Layers: by Reverse Scattering Perturbation. Stacking Layers by Layer Doubling. – Stacking Layers by Renormalized Forward Scattering (RFS) Perturbation. – Assembling a Program: The Main Program and the Input. – Subroutine Listings. – Structural Results of LEED Crystallography. – Appendices. – References. – Subject Index.

Vibrational Spectroscopy of Adsorbates

Editor: R. F. Willis
With contributions by numerous experts
1980. 97 figures, 8 tables. XII, 184 pages
(Springer Series in Chemical Physics, Volume 15)
ISBN 3-540-10429-1

Contents: Introduction. – Theory of Dipole Electron Scattering from Adsorbates. – Angle and Energy Dependent Electron Impact Vibrational Excitation of Adsorbates. – Adsorbate-Induced Optical Phonons. – Inelastic Electron Tunnelling Spectroscopy. – Inelastic Molecular Beam Scattering from Surfaces. – Neutron Scattering Studies. – Reflection Absorption Infrared Spectroscopy: Application to Carbon Monoxide on Copper. – Raman Spectroscopy of Adsorbates at Metal Surfaces. – Vibrations of Monatomic and Diatomic Ligands in Metal Clusters and Complexes. – Analogies with Vibrations of Adsorbed Species on Metals. – Coupling Induced Vibrational Frequency Shifts and Island Size Determination: CO on Pt {001} and Pt {111}.

Springer-Verlag
Berlin
Heidelberg
New York

Electron Spectroscopy for Surface Analysis

Editor: H. Ibach
1977. 123 figures, 5 tables. XI, 255 pages
(Topics in Current Physics, Volume 4)
ISBN 3-540-08078-3

Contents:
H. Ibach: Introduction. – *D. Roy,
J. D. Carette:* Design of Electron Spectrometers for Surface Analysis. – *J. Kirschner:* Electron-Excited Core Level Spectroscopies. – *M. Henzler:* Electron Diffraction and Surface Defect Structure. – *B. Feuerbacher, B. Fitton:* Photoemission Spectroscopy. – *H. Froitzheim:* Electron Energy Loss Spectroscopy.

Interactions on Metal Surfaces

Editor: R. Gomer
1975. 112 figures. XI, 310 pages
(Topics in Applied Physics, Volume 4)
ISBN 3-540-07094-X

Contents:
J. R. Smith: Theory of Electronic Properties of Surfaces. – *S. K. Lyo, R. Gomer:* Theory of Chemisorption. – *L. D. Schmidt:* Chemisorption: Aspects of the Experimental Situation. – *D. Menzel:* Desorption Phenomena. – *E. W. Plummer:* Photoemission and Field Emission Spectroscopy. – *E. Bauer:* Low Energy Electron Diffraction (LEED) and Auger Methods. – *M. Boudart:* Concepts in Heterogeneous Catalysis.

Theory of Chemisorption

Editor: J. R. Smith
1980. 116 figures, 8 tables. XI, 240 pages
(Topics in Current Physics, Volume 19)
ISBN 3-540-09891-7

Contents:
J. R. Smith: Introduction. – *S. C. Ying:* Density Functional Theory of Chemisorption of Simple Metals. – *J. A. Appelbaum, D. R. Hamann:* Chemisorption on Semiconductor Surfaces. – *F. J. Arlinghaus, J. G. Gay, J. R. Smith:* Chemisorption on d-Band Metals. – *B. Kunz:* Cluster Chemisorption. – *T. Wolfram, S. Ellialtioğlu:* Concepts of Surface States and Chemisorption on d-Band Perovskites. – *T. L. Einstein, J. A. Hertz, J. R. Schrieffer:* Theoretical Issues in Chemisorption.

X-Ray Optics

Applications to Solids
Editor: H.-J. Queisser
1977. 133 figures, 17 tables. XII, 228 pages
(Topics in Applied Physics, Volume 22)
ISBN 3-540-08462-2

Contents:
H.-J. Queisser: Introduction: Structure and Structuring of Solids. –
M. Yoshimatsu, S. Kozaki: High Brilliance X-Ray Sources. – *E. Spiller, R. Feder:* X-Ray Lithography. – *U. Bonse, W. Graeff:* X-Ray and Neutron Interferometry. –
A. Authier: Section Topography. –
W. Hartmann: Live Topography.

Springer-Verlag   Berlin Heidelberg New York